Sentential Probability Logic

Sentential Probability Logic

*Origins, Development, Current Status,
and Technical Applications*

Theodore Hailperin

Lehigh
University
Press

Bethlehem: Lehigh University Press
London: Associated University Presses

Associated University Presses
440 Forsgate Drive
Cranbury, New Jersey 08512

Associated University Presses
16 Barter Street
London WC1A 2AH, England

Associated University Presses
P.O. Box 338, Port Credit
Mississauga, Ontario
Canada L5G 4L8

The paper used in this publication meets the requirements of the American National Standards for Permanence of Paper for Printed Library Materials Z39.48-1984.

Library of Congress Cataloging-in-Publication Data

Hailperin, Theodore.
    Sentential probability logic : origins, development, current status, and technical applications / Theodore Hailperin.
        p.   cm.
    Includes bibliographical references (p.   ) and index.
    ISBN 0-934223-45-9 (alk. paper)
    1. Algebraic logic. 2. Algebra, Boolean. 3. Probabilities.
    I. Title.
QA10.H344 1996
511.3—dc20   96-10867
                                                            CIP

PRINTED IN THE UNITED STATES OF AMERICA

CONTENTS

PREFACE

The present monograph is an outgrowth of my *Boole's Logic and Probability*.

During the course of working on Boole's probability ideas it gradually appeared that, despite his claim to have a new theory of probability, his work was best thought of as a probability logic. I included a tentative sketch of such a logic in the second edition of my Boole book. Full elaboration, the core of this book, took some time, in the end turning out to be significantly different from what others have called probability logic. The present volume, as the title indicates, is limited to the sentential aspects of this logic.

Since the topic of probability logic does not appear in standard histories of probability, I have here included a considerable amount of background history—perhaps with more detail than some readers may care for. Initial version of this historical material appeared in a series of two articles in the journal *History and Philosophy of Logic*. I am grateful to its editor and publisher for having had the opportunity of bringing the material out for scholarly vetting. In its current version, here chapters 1–3, in addition to some revision, much has been added.

The material in chapter 0 is not referred to until chapters 4 and 5, and if so desired need not be read until then. Most of the ingredients for chapter 4 should be fairly familiar to logicians, an exception perhaps being linear programming. Accordingly a summary of this algebraic subject is included in chapter 0. In chapter 5 the topic of suppositional logic, being new, will be unfamiliar. To help with the understanding a simple form of it used by

7

designers of switching circuits is described in chapter 0.

Chapter 6, on technical applications, will be seen to be on the thin side. It is my hope that other applications will soon be forthcoming, once general methods for logical reasoning under uncertainty become familiar and widespread.

It is a pleasure to acknowledge the help of my son Max. He urged the use of a computer on which to compose (in the typographic sense) the book, gave unfailingly expert advice on using TEX, and supplied programs for the more difficult typesetting situations. Additionally, using *Mathematica*, he computed the polytope vertices needed for the many examples in chapters 4 and 5. Continuing in this family vein my wife, Ruth, without any previous experience cheerfully volunteered to produce, using $\mathcal{AMS}$-TEX, the computer file from handwritten manuscript. By the time the book was finished she and the computer had reached an uneasy truce.

Sentential Probability Logic

INTRODUCTION

Since much of this book involves new and unusual items we believe a prospective reader would welcome a travel brochure type of description of what it contains. This Introduction is such a description.

There is a wide variety of studies in the literature which have referred to themselves as "probability logic". There are no standard names, it being left to the context to distinguish one from another. Our kind, restricted to the sentential level, may be characterized as follows.

We shall be presenting a *logic* in which probability values play a semantic role comparable to that of truth values in conventional verity logic. Just as truth values appear only in semantic statements of verity logic, so probability values appear only in semantic statements of probability logic. Verity and probability logic, though differing in their semantics, share a common formal syntax, i.e. the formal languages involved will be identical. When attention is restricted to sentences which can take on only the two extreme values 1 and 0, then the semantics as well as the syntax become identical. Thus probability logic includes verity logic. It is also a generalization of verity logic though how, and in what particular aspects, requires some explanation.

A key semantic notion serving to characterize a logic is that of logical consequence—a precise specification of the circumstances under which a sentence ψ is said to follow from sentences $\phi_1, \ldots, \phi_m$. Our definition of this notion for probability logic (given in §4.4) is quite different in character from that which is normally encountered in verity logic. But when verity logic is appropriately reformulated it then becomes apparent that its definition of logical consequence has as a natural generalization that for probability logic. This reformulation of verity sentential logic is carried out in §0.1.

11

As early as 1867 C. S. Peirce tried introducing a notion of conditional event, i.e. an event whose probability is a conditional probability. Since for us there is no difference on the syntactic level between verity and probability logic, it is clear that if one is to have conditional events in a logic its syntax has to be different, i.e. include more than the usual sentence connectives. Indeed, in chapter 5 we shall be presenting an extension of probability logic, a *conditional* probability logic which includes a new connective, the suppositional. This new connective is based on the semantic concept of an indefinite truth value, a notion we introduce and discuss in §0.3. We point out its relation to the 'don't-care' condition used by designers of combinational (switching) networks. chapter 0 concludes with a section describing some mathematical results from linear algebra and linear programming which we shall be referring to in later chapters.

Our historical account, comprising chapters 1–3, begins with a brief mention of Leibniz. Over the years, in letters and writings, he called attention to the desirability of the study of a "doctrine of degrees of probability". Leibniz envisioned that it would provide a method by which the likelihood or plausibility of a conclusion could be estimated, thus enabling rational adjudication between rival claims or opposing arguments. As with some others of his prescient ideas, he did little with it.

The first results of any substance relating probability and logic were obtained by Jakob Bernoulli. Arguments which produce "opinion or conjecture" were divided by Bernoulli into three types according as the premises, and the argumentation from premises to conclusion, are contingent or necessary (omitting the type when both are necessary). Using Huygens' notion of expectation (i.e., the value of a gamble) Bernoulli determines a numerical measure for the "force of proof" or "degree of certainty or probability which the argument generates". This is accomplished by assuming, as in games of chance, that there is a determined number of cases in some of which the premises "exist" (are true) and others not; similarly for the argumentations. He assumes that all cases are equally possible and, if not, then weighted appropriately. Although not named by Bernoulli it will be convenient to refer to the ratio of cases for which a proposition holds to the total number of cases as its *degree of contingency*, rather than 'probability', the term which Bernoulli is using for his force of proof, or degree of certainty. Bernoulli takes the degree of certainty or probability for an argument to be the value of a gamble which nets 1 (certainty) in those cases in which the conclusion is proved, and 0 otherwise. When both premise and argumentation are contingent this value turns out to be the product of their respective degrees of contingency; that is, representing an argumentation from A to C by '$A \to C$' and using 'P' for 'degree of contingency', Bernoulli shows, in effect, that the probability of premise A proving conclusion C is given by $P(A)P(A \to C)$. The question how this relates to $P(A(A \to C))$ is not considered. Bernoulli also discusses two kinds of arguments which he refers to as "pure" and "mixed". Pure arguments are those in which in certain cases a conclusion is proved but in other cases nothing positive is proved; in mixed arguments the conclusion is proved in some cases and the contrary in other cases. The results Bernoulli obtains for these types of arguments are described and discussed in our §§1.3, 1.4. Of special interest is his derivation of the force of proof when several arguments—pure, mixed, or both—for a conclusion are combined. The topic of combining arguments (also testimony, or evidence) for a conclusion was considered by a number of later writers. These considerations are referred to in our §1.5 (at the end) and in §§2.2, 2.3, 2.6.

As we have noted, the term 'probability' was not used by Bernoulli in reference to propositions but only to arguments. At that time expectation, or an equivalent, was the preferred way of talking about chance situations. By the time of J. H. Lambert probability, rather than expectation, had become the standard notion. Lambert applies it initially to propositions of the form 'A is B', with its probability taken to be the ratio of the number of A's which are B's to the number which are or are not "as they are encountered in experience without selection". The notion is then applied to propositions of other forms via rules of chance using lottery drawings as a paradigm.

The arguments Lambert considers are traditional syllogistic ones, though with an intensional, rather than the extensional interpretation now common. In his treatment of combining different arguments for one and the same conclusion, namely the conclusion 'A is B', he takes these arguments to be of the form

$$M_i \text{ is } B$$
$$A \text{ is } M_i$$
$$\text{therefore} \quad A \text{ is } B.$$

The minor premise 'A is M_i' is taken to be necessary but since M_i is only a "partial characteristic" of B, the major premise 'M_i is B' is taken to be contingent, and hence so also the conclusion 'A is B'. The result Lambert obtains for the probability of the conclusion when all such arguments are combined is similar to Bernoulli's, though Lambert is quite explicit about requiring independence of the 'M_i is B'. Lambert also introduces numerically quantified terms, e.g. '$\frac{3}{4}A$ are B', meaning that $\frac{3}{4}$ of the A's are B's, but with a tacit probabilistic meaning, namely that of any four B's selected there will be three A's among them. When the numerical factor is in the predicate position then the term is taken intensionally, e.g. 'C is $\frac{3}{4}A$' means that C has $\frac{3}{4}$ of A's attributes. Though Lambert's results involved material assumptions—e.g. that the A's are uniformly distributed among the B's—we have included them in our historical account of probability logic as they are readily converted to purely logical results.

Thomas Bayes' essay comes in for brief mention—not for its celebrated result but for the reason that it contains (implicitly) the first occurrence of a probability logic result involving conditional probability. Likewise, a contemporaneously appearing astronomical paper of John Michell comes in for discussion in connection with the relation between the two conditional probabilities, that of A, given B and that of not-B, given not-A.

Our chapter covering the nineteenth century opens with Bolzano. His *Wissenschaftslehre* (1837) contains a treatment of probability based on logical notions. Aspects of it are akin to what Keynes and Carnap did later, though neither of them was, apparently, aware of Bolzano's work. Bolzano bases his definition of probability on *relative validity*, a notion analogous to his *validity* (considered by some to be a forerunner of Tarski's model-theoretic concept of validity). These ideas of Bolzano are described and the results derived from them are examined. Some results are erroneous, the error generally stemming from neglect of possible logical interrelationships among component propositions. Noteworthy, on the other hand, is Bolzano's calling attention to a widespread error in probabilistic inference: that the probability of a conclusion in a necessary inference is *equal to* the product of the probabilities of the premises.

With De Morgan the association of probability with logical notions first appears in an 1837 encyclopedia article on the Theory of Probabilities. His few examples there are hardly more than a simple application of the product rule for the probability of a conjunction and in which, moreover, questions of dependence are almost always ignored. But by 1847 the preface of his book entitled—FORMAL LOGIC: or, *The Calculus of Inference*, Necessary and Probable—includes a vigorous defense for the numerical theory of probability as part of logic. Its chapter 10 is devoted to probable inference. There is, however, no systematic development but only illustration of inferences via six problems and their solution. The domain of application is restricted to argumentations in which the conclusion is a necessary consequence of the premises—a regression from Bernoulli's views. In the other direction, the domain is enlarged by De Morgan's inclusion of material notions such as *belief, authority* and *testimony.*

Boole's successful introduction of an algebra of logic (*The Laws of Thought*, 1854) enabled him to expeditiously handle probabilities of complex propositional compounds. His theory of probability, which he claimed had the ability to solve "any" problem in probability, is described. The theory is based on a special assumption, namely that any compound proposition can be analyzed into simple, stochastically independent components, or (a later view) that one could express such a problem in terms of "ideal" elements which were of this nature. Although Boole thought of his procedure as solving a problem in probability, it could also be viewed as establishing a probable inference: the events (logically expressed) whose probabilities are given can be considered as premises and the event whose probability is sought, as conclusion. Boole's idea of solving "any" problem in probability can be made viable (in a restricted sense) if one uses probability intervals in place of probability values. (See Theorem 4.63 below.)

We bring up for discussion (in §2.6) the question, at one time quite active, of how to combine two or more items of probabilistic evidence for a conclusion so as to have a single probability value. A number of proposed solutions are critically examined. When viewed as a problem in conditional probability logic, namely that of finding $P(C \mid AB)$ if given $P(C \mid A) = p$ and $P(C \mid B) = q$, it results that $P(C \mid AB)$ can take on any value in $[0, 1]$ if nothing further is assumed concerning the logical structure of A, B, and C (Theorem 5.45). The introduction of a symbol for conditional probability and a codification of its principal properties first occurs in Peirce's *1867a* and also, apparently independently, in MacColl's *1880*.

Chapter 3 opens with brief accounts of work associated respectively with Keynes, Carnap and Reichenbach, these being taken as representative of grand theories aiming at a combined system of probability and inductive inference. Critical discussion is limited to our topic and carried out to an extent sufficient to indicate what there is in common with, and what is different from, our conception of probability logic. Keynes takes his fundamental (undefined) probability relation $a/h = \alpha$ to mean that "knowledge of h justifies a rational belief in a of degree α". In his essay on induction Nicod takes Keynes' $a/h = \alpha$ to mean "if h is true, a is probable to a degree α" and cites some inference forms, to which we take exception finding them to be faulty. The notion, now called 'confirmation' ($c(a, b) =$ degree to which b confirms a) was axiomatized by Hosiasson. Its formal properties are the same as those of conditional probability as codified by Peirce and MacColl. Carnap found these properties too weak to base an inductive logic on and sought, rather, to define confirmation (called by him 'probability$_1$') in terms of the logico-syntactic structure of its components (relative to a given language). Carnap's conception, for the case of a very simple language, is compared with our semantically based probability logic. In Reichenbach's approach to establishing an inductive-probability logic we find numerous unclarities and were unable to accept his contention that what he has constructed is a multi-valued logic.

Probabilistic inference, despite the efforts of Bernoulli, Lambert and De Morgan, did not thrive. The primitive state of the formal logic at their disposal—not much more than the traditional syllogistic—may have been a hindrance to development. This, however, would not have applied to Boole. But in his case the probability theory he used was not comprehensible to others. Whatever the reason, not until 1966, in a paper by Suppes, do we find some notice of probabilistic inference, though subsidiary to other matters and hence not systematically developed. In some of Suppes' inference forms nearness to 1 of the premise probabilities was featured— that $P(A)$ is near 1 was written $P(A) \geq 1 - \epsilon$, that is, $1 - \epsilon \leq P(A) \leq 1$, which immediately suggests the more general use of intervals or upper and lower bounds, e.g. $a \leq P(A) \leq b$, in which the upper value need not be 1. Indeed, it will turn out that probability intervals, rather than exact values, are featured in a decision procedure for probability logic.

The problem of finding bounds on the probability of an event determined by given probabilities of other events, was first investigated by Boole in connection with his probability theory. We describe the three methods which he devised for obtaining such bounds, one of them not providing the best possible bounds; the other two, which do, are seen to be primal and dual forms of a linear program. A special case of the problem was treated by Fréchet in 1935. Our 1965 paper revived Boole's work on probability bounds, justified its results from a contemporary linear programming viewpoint, and extended them to more general situations. These results become a basis for a decision procedure for logical consequence in sentential probability logic (§4.6).

Independently of our 1965 paper two others appeared which also dealt with logico-probability problems using linear programming. In the first of these E. W. Adams and H. P. Levine address themselves to the question: What is the maximum uncertainty in the conclusion of an inference if the premises have uncertainties less than some specified amount? The uncertainty of a proposition is the probability of its denial; to say that the uncertainty of ϕ is less than ϵ is to say $P(\neg\phi) \leq \epsilon$, which is equivalent to $(1 - \epsilon) \leq P(\phi) \leq 1$. Moreover, the Adams-Levine notion of inference is a general one, not requiring that there be any formal connection between premises and conclusion. Hence their question is a special case of Boole's General Probability Problem (as interpreted by us in our *1965*) in which the given interval for a premise is of the form $[1 - \epsilon, 1]$, and the objective is to find the best possible lower bound for the probability of the proposition designated as the conclusion. The point of view, the methods, and the results obtained by Adams-Levine are described in some detail so as to compare them with our development. The author of the second paper in our pair, N. Nilsson, calls his contribution a "probabilistic logic". For the purposes at hand his semantic basis is unnecessarily complicated by the inclusion of a definition of the probability of a proposition in terms of 'real world' and 'possible worlds'. There is no need for a probability logician to define what is meant by the probability of a proposition, any more than for a verity logician to define what is meant by the verity (truth value) of a proposition. What matters is a specification of its (semantic) properties so that the relation of logical consequence can be properly defined for the logic.

Nilsson shows how to express, for a given set of sentences, the set of possible probability values they can take on for all possible probability assignments to the basic consistent (non-vanishing) conjunctions on these sentences. (The method is equivalent to what Boole describes on pp. 310–311 in his *1854*, though this would not be apparent to someone unfamiliar with Boole's peculiar way of doing logic.) To determine the range of probability values of a sentence S that is entailed by a set of sentences with given probability values Nilsson adjoins S and determines the region of consistent probability values for the enlarged set. The range of values for the probability of S in this region provides upper and lower bounds in terms of probability values of the others. As Nilsson notes, linear programming methods can be used to the same ends but doesn't go into the matter, declaring "the size of problems encountered in probabilistic reasoning is usually too large to permit a direct solution."

Our historical survey ends with a description of material from a 1966 paper of D. Scott and P. Krauss. It contains the earliest instance of a formal system which can justifiably be called a probability logic. The authors frame it in terms of an infinitary predicate language and hence full discussion falls outside the scope set for this monograph.[1] A highly reduced version of the Scott-Krauss formal system is described so as to bring out a difference in semantics with what is in our chapter 4.

Some of the writers whose views are examined in this survey considered their work to be an application of probability to logic; others that it was logic, i.e. probability logic. The formal developments in chapters 4 and 5 will show us joining this latter group, though the conceptual basis will be a new and different one. This portion of the monograph opens with a statement of requirements, i.e. sets standards, for something to be called a logic, one of these requirements being a suitable definition of 'logical consequence'.

The definition involves both syntactic and semantic notions. First the set of semantic values $\{0, 1\}$ ($1 = true$, $0 = false$) is enlarged to include all reals in the unit interval $[0, 1]$. The notion of a probability function is then introduced, namely as a function P from the set of sentences of a language to $[0, 1]$ having the well-known elementary probability properties—e.g. that if S_1 and S_2 are mutually exclusive (conjunction logically false) then $P(S_1 \vee S_2) = P(S_1) + P(S_2)$. Note that probability logic semantics includes two-valued notions such as 'logically false (in verity logic)'.

We first construct a probability logic for the simple language S_n, the set of sentences S_n consisting of all sentences obtainable from the atomic sentences $A_1, \ldots, A_n$ by recursive use of the sentence connectives $\neg$, $\vee$, $\wedge$. Let $K_1, \ldots, K_{2^n}$ be constituents (basic conjunctions) on $A_1, \ldots, A_n$. A model M for S_n is an assignment of values k_i to K_i ($i = 1, \ldots, 2^n$) where each $k_i \geq 0$ and the sum of all is 1. This assignment is uniquely extendible to a probability function P_M on S_n. The quantifier 'For all models M' is then replaceable by 'For all probability functions P_M'. The fundamental notion of logical consequence for probability logic can now be defined. Let $\phi_1, \ldots, \phi_m, \psi$ be sentences of S_n and $\alpha_1, \ldots, \alpha_m, \beta$ subsets of $[0, 1]$. Then ψ is a (probability) logical consequence of $\phi_1, \ldots, \phi_m$ (with respect to $\alpha_1, \ldots, \alpha_m, \beta$) if for all probability functions P_M,

$$P_M(\phi_1) \in \alpha_1, \ldots, P_M(\phi_m) \in \alpha_m \text{ implies } P_M(\psi) \in \beta.$$

---

[1] An exception to this restriction has been made in the case of term logic (e.g. as with Lambert), which is close enough to sentential logic not to require special treatment.

(Our main text provides motivation for this definition.) With an appropriate definition of model this definition is readily carried over so as to apply to the language S which is the union of all S_n, $n = 1, 2, \ldots$. This sentential probability logic is compared with the Scott-Krauss simplified-down-to-the-sentential-level version.

Interesting results ensue when the subsets $\alpha_1, \ldots, \alpha_m$ in the definition of logical consequence are subintervals of $[0, 1]$. Using linear programming theory we show that in this case there is an effective procedure for obtaining the set β providing the strongest conclusion, and that it is a subinterval. The method is illustrated with many examples of logical consequence.

Chapter 5 presents an approach to conditional probability based on a new logical notion. Our exposition begins with some historical material on conditional probability, using excerpts from De Moivre's *The Doctrine of chances*. The discussion of conditional probability leads to a recognition of the need for a notion of an event happening, supposing another event to have happened. This is to be a logical notion, not, *à la* Laplace, a causal one. Anticipating success, it is designated by '$\psi \mid \phi$' (read: ψ, supposing ϕ) and at the start it is assumed that ψ and ϕ are formulas which involve only the connectives $\neg$, $\vee$, $\wedge$. (We shall later see that this is no restriction of generality.) Introduction of $\psi \mid \phi$ into probability logic so as to extend it to a *conditional probability logic* requires specifying its semantic (probability) value in a model. If in a probability model (assignment of k_i to K_i, etc.) the P values of $\psi\phi$ and ϕ are, respectively, $\sum_{\psi\phi} k_i$ and $\sum_{\phi} k_i$, then the semantic value, $P(\psi \mid \phi)$, of $\psi \mid \phi$ is defined to be $\sum_{\psi\phi} k_i / \sum_{\phi} k_i$ if $\sum_{\phi} k_i \neq 0$, and otherwise c, an indefinite value in $[0, 1]$. This produces the usual general properties of $P(\psi \mid \phi)$. The notion of logical consequence generalizes readily to conditional probability logic, and here also intervals, rather than arbitrary sets of reals, are used. A number of logical consequences with conditional probabilities are cited, as well as a discussion of Boole's challenge Problem and associated matters.

At this stage in our exposition the expression '$P(\psi \mid \phi)$' could just as well have been written '$P(\psi, \phi)$', i.e. representing an operation on an ordered pair. Now we show (in §5.6) that '$\mid$' can be given meaning as a connective (we call it the suppositional) in a special kind of three-valued logic. In the case of $(0, 1)$-models, i.e. probability models in which all but one K_i is assigned the value 0, the connectives $\neg$, $\wedge$, $\vee$ turn out to be the same as the $\neg$, $\wedge$, $\vee$ of verity logic. This suggests looking at what happens to the conditional probability $P(\psi \mid \phi)$ under those circumstances. The reduction leads to a new kind of semantic table:

ϕ	ψ	$\psi \mid \phi$
1	1	1
1	0	0
0	1	u
0	0	u

where u is an unspecified (indefinite) member of the set $\{0, 1\}$. Adjoining u to $\{0, 1\}$ as a semantic value (with properties governed by its being an unspecified member of $\{0, 1\}$) a logic ('suppositional logic') is then constructed on this basis. The verity connectives $\neg$, $\wedge$, $\vee$ are readily generalized to suppositional logic. A key meta-theorem establishes the result that any formula constructed on $\neg$, $\wedge$, $\vee$, $\mid$ is effectively provable to be semantically equivalent to one of the form $\psi \mid \phi$, where neither ϕ nor ψ contain occurrences of ' $\mid$ '. In exactly the same manner in which probability logic extends verity logic, conditional probability logic extends suppositional logic.

In §5.8 we compare our logic of conditional events (i.e., $\psi \mid \phi$'s) with what others have written on the topic. Our semantically based sentential logic turns out to be equivalent to that of Goodman-Nguyen-Walker—allowing for the fact that their approach is an algebraically based one.

Our final chapter looks at some applications in which the treatment is most natural from the viewpoint of probability logic and linear programming. A theorem of Rényi's, for example, establishes a wide class of identities linear in the probabilities of given logical functions of n simple events (= atomic sentences). In terms of probability logic notions the theorem amounts to the following: if a linear relation holds for the 2^n special models obtained by assigning 1 to a constituent in the atomic sentences and 0 to all the others, then the relation is true in all models. When the special models are viewed as unit coordinate points in 2^n-dimensional space the set of all models corresponds to the points spanned by these coordinate points, i.e., to the set of points on or interior to the polytope they define. Then supposing that the linear relation is an inequality (defining a half-space) it is geometrically evident that if all the unit coordinate points are on or below the plane defining the half-space, then so are all the points of the polytope they span. This geometrical viewpoint suggests the possibility of replacing the plane by other surfaces to the same effect, so generalizing the theorem to other than linear relations.

Several papers written by statisticians come in for description as they employ linear programming methods for finding bounds on the probability of a logically expressed function of events. The interest is not in probabilistic inference but in obtaining estimates for a probability when there is insufficient data, especially on independence, for obtaining an exact value. On the other hand there are applications which, though seemingly unconnected with probabilistic inference, can be so viewed. Using the logical representation of combinational circuits having switches or gates, the problem of determining the probability of a gate being activated, given the probabilities of input gates being activated, is the same with whether a sentence is a probability logical consequence of certain other sentences. We describe examples of this in connection with circuit fault testing and fault trees.

Our final examples illustrate uses of probability logic in problems of network reliability. Its close affinity to the transmission of uncertainty in (verity) logical inference is noted.

Chapter 0

Groundwork

§0.1. Verity sentential logic

The topic of this section is ordinary sentential (or propositional) logic. We have added the modifier 'verity' to its usual name to distinguish it from the probability logic to be introduced later in chapter 4. Although this elementary part of standard logic is presumably well-known to readers of this book, we nevertheless wish to briefly sketch it in a modified form so that one can more readily see in what respects probability logic is its generalization. At the same time we take the opportunity to state some results of sentential logic to be used in later portions of the book. Assertions without accompanying justification are either well-known or easily established.

FORMAL SYNTAX

We assume a formal (i.e., uninterpreted) language S of sentences consisting of atomic sentences $A_1, \ldots, A_n, \ldots$ and compound sentences constructed from them in the usual fashion by means of the sentential connectives $\neg$ (not), $\wedge$ (and), $\vee$ (or). (Where convenient, $\wedge$ will be replaced by juxtaposition and $\neg$ by overline.)

FORMAL SEMANTICS

There are two *verity* (or *truth*) *values*, 0 and 1. We do not say what they mean (hence *formal* semantics) but we assume that they are endowed with the usual properties associated with these symbols in arithmetic. This is for convenience in stating their properties as verity values. In addition to subtraction from 1 we shall also use $\min(a, b)$ and $\max(a, b)$. These are sometimes written using infix notation, i.e., $a \wedge b$ for $\min(a, b)$ and $a \vee b$ for $\max(a, b)$ with $a, b \in \{0, 1\}$.[1] We also assume that we have *verity functions*, generically denoted by 'V'. These are functions from the set of sentences $\mathcal{S}$ into the set $\{0, 1\}$ which satisfy the following definition.

DEFINITION OF VERITY FUNCTION

For arbitrary sentences ϕ and ψ of $\mathcal{S}$:

$$V(\phi) \in \{0, 1\}$$
$$V(\neg\phi) = 1 - V(\phi)$$
$$V(\phi \wedge \psi) = \min(V(\phi), V(\psi)) \tag{1}$$
$$V(\phi \vee \psi) = \max(V(\phi), V(\psi)).$$

From (1) it follows immediately that with respect to any verity function the connectives have the properties commonly displayed in truth-tables; e.g., for $\neg$

$V(\phi)$	$V(\neg\phi)$
1	0
0	1

Moreover, it is easily seen that a value for each sentence of $\mathcal{S}$ is uniquely determined once V values are assigned to the atomic sentences. We refer to such an assignment as a (*verity*) *model*, and denote the uniquely determined verity function associated with a model M by 'V_M'.

The central notion of logic is defined in terms of models. A sentence ψ is a (*verity*) *logical consequence* of sentences $\phi_1, \ldots, \phi_m$ if the following holds:

For all models M,

$$\text{if } V_M(\phi_1) = 1, \ldots, V_M(\phi_m) = 1, \text{ then } V_M(\psi) = 1. \tag{2}$$

A standard notation for this consequence relation is

$$\phi_1, \ldots, \phi_m \vDash \psi. \tag{3}$$

However in order to free '$\vDash$' for a more general use in probability logic (where we shall also be using the notion expressed in (3)) we shall replace

---

1. One also finds in the literature ab, i.e., multiplication, for $a \wedge b$ and $a \oplus b \oplus ab$, $\oplus$ being addition modulo 2, for $a \vee b$.

it with '⊢' (the symbol normally used for deducibility in a formal system).
We shall from here on use

$$\phi_1, \ldots, \phi_m \vdash \psi \tag{4}$$

instead of (3) to represent (2). No harmful misunderstanding is likely to
occur since for any formal system of sentential logic the two relations are
equivalent (completeness theorem).

As we have noted, any model determines a unique verity function. Con-
versely, since the A_i have values specified by the function, each verity function determines a unique model. Hence in definition (2), in place of 'For
all models M', we could equivalently write 'For all verity functions V' and
drop the subscript M. Replacing in (2) the informal 'if, then' with '⊢' and
supressing the universal quantifier 'For all verity functions V' converts (2)
to

$$V(\phi_1) = 1, \ldots, V(\phi_m) = 1 \vdash V(\psi) = 1, \tag{5}$$

with free variable V representing an arbitrary verity function.

Property (5) can be trivially generalized by allowing '0' in place of any
or all of the '1's (trivially, since $V(\chi) = 0$ is equivalent to $V(\neg\chi) = 1$).
Adopting this generalization leads to

$$V(\phi_1) = v_1, \ldots, V(\phi_m) = v_m \vdash V(\psi) = v_0, \tag{6}$$

meaning that, where $v_1, \ldots, v_m, v_0$ are given verity values, for any verity
function V if $V(\phi_1) = v_1, \ldots, V(\phi_m) = v_m$, then $V(\psi) = v_0$. One can go
a step further in generalization. Since $x = a$ is equivalent to $x \in \{a\}$ the
equalities in (6) can be replaced by membership in subsets of $\{0, 1\}$. This
generalizes (6) to

$$V(\phi_1) \in \alpha_1, \ldots, V(\phi_m) \in \alpha_m \vdash V(\psi) \in \beta, \tag{7}$$

where $\alpha_1, \ldots, \alpha_m, \beta$ are any non-empty subsets of $\{0, 1\}$. The clauses sepa-
rated by commas occurring before the '⊢' will be referred to as the *premises*
of the logical consequence (7), and the clause coming after '⊢' as its *conclusion*. Any clause in the premise with its α_i being $\{0, 1\}$ may be deleted
as this clause would have no effect on the conclusion. But not so if β is
$\{0, 1\}$. For sometimes this is the *only* conclusion, and that is worthwhile
information. This would be the case, for example, with

$$V(A_1 \rightarrow A_2) \in \{1\} \vdash V(A_1) \in \{0, 1\}. \tag{8}$$

This information is obtainable from the usual truth-table

$V(A_1)$	$V(A_2)$	$V(A_1 \rightarrow A_2)$
1	1	1
1	0	0
0	1	1
0	0	1

by deleting the second row of truth values and reading the table in the reverse direction. However we have introduced (7) not simply for this reason (though we will make use of it in §0.4), but because the notion of probability consequence to be introduced in chapter 4 is a generalization of (7). In the generalization 'P' (probability function) takes the place of 'V', and subsets of $[0, 1]$ (the unit interval) take the place of subsets of $\{0, 1\}$. Or, putting it the other way around, when the probabilities of sentences can only be 0 or 1 then probability logical consequence reduces to (7).

In normal discourse the use of a semantic level language to express an objective proposition is rarely encountered. One says "John Doe has arrived" rather than "The truth value of 'John Doe has arrived' is 1". Accordingly, adopting the practice of writing simply 'χ' in place of '$V(\chi) = 1$' converts (5) to the standard form $\phi_1, \ldots, \phi_m \vdash \psi$. As is customary, we say ϕ is *valid*, or *logically true*, if $\vdash \phi$, i.e., if ϕ is a logical consequence of no premises; ϕ is *logically false* if $\vdash \neg\phi$, and is *logically contingent* if neither $\vdash \phi$ nor $\vdash \neg\phi$. We say ϕ *logically implies* ψ if $\vdash \phi \rightarrow \psi$, and ϕ is *logically equivalent to* ψ if $\vdash \phi \leftrightarrow \psi$.

The following theorems, to be referred to later, relate semantic and syntactic notions.

A *constituent on* sentences $A_1, \ldots, A_n$ is a conjunction of all of these in which some, none, or all are negated.[2]

Theorem 0.11. *A sentence ϕ which involves at most the letters $A_1, \ldots, A_n$, is either logically equivalent to $A_1 \overline{A}_1$ or else to a unique disjunction of constituents on $A_1, \ldots, A_n$.*

Let $\mathcal{S}_n$ be the set of sentences involving at most the letters $A_1, \ldots, A_n$. A *model for a $\phi \in \mathcal{S}_n$* is an equivalence class of models in which two models are equivalent if they agree on their assignments to $A_1, \ldots, A_n$. (Thus when

---

2. The term 'constituent' dates back to Boole. Also in use is 'basic conjunction' or, in computer engineering literature, 'fundamental product'. We shall often use Boole's overline on a letter (e.g., $\overline{A}$) to indicate its negation ($\neg A$).

considering models for a $\phi \in S_n$ one can effectively ignore assignments to A_i for $i > n$.) Then for elements of S_n there are only 2^n non-equivalent models since assignments to letters beyond A_n are irrelevant.

Theorem 0.12. *There is a one-to-one correspondence between the 2^n distinct models for an element of S_n and assignments of 1 to some one constituent on $A_1, \ldots, A_n$ and 0 to all the others.*

The result here is evident since assignment of a 1 to a constituent K_i (which entails 0 for all the others) corresponds to the model in which the unnegated letters of K_i have a 1 assigned, and 0 to the negated letters. Equivalently, then, a model for an element of S_n can be thought of as an assignment of truth values to $K_1, \ldots K_{2^n}$ in which one K_i is assigned the value 1 and all the others the value 0.

A sentence of S_n is a *positive sentence* (informally, a positive function of $A_1, \ldots, A_n$) if it is logically equivalent to a sentence containing at most just the two connectives $\wedge$ and $\vee$.

Let $X = (x_1, \ldots, x_n)$ and $Y = (y_1, \ldots, y_n)$ be n-tuples of truth values (i.e., $x_i, y_i \in \{0, 1\}$). We write $X \leq Y$ if $x_i \leq y_i$ for $i = 1, \ldots, n$. For $\phi \in S_n$ we write $\phi(X)$, or $\phi(x_1, \ldots, x_n)$, for the truth value of ϕ when A_i is assigned the value x_i $(i = 1, \ldots, n)$.

Theorem 0.13. *Let $\phi \in S_n$ have the properties*

(i) $\phi(0, 0, \ldots, 0) = 0$
(ii) $\phi(1, 1, \ldots, 1) = 1$
(iii) *If $X \leq Y$, then $\phi(X) \leq \phi(Y)$.*

Then ϕ is a positive function of $A_1, \ldots, A_n$.

Proof. Let A_i be an atomic sentence (letter) having a negated occurrence in ϕ. (By well-known properties it is sufficient to assume that negation signs occur at most on atomic sentence letters.) By development

$$\phi \leftrightarrow \phi[1/A_i]A_i \vee \phi[0/A_i]\overline{A}_i, \tag{9}$$

where $\phi[1/A_i]$ is the result of replacing all occurrences of A_i in ϕ by 1 (a logically true sentence), and similarly for $\phi[0/A_i]$, 0 a logically false sentence. We also have, by expanding ϕ into complete disjunctive normal form,

$$\phi \leftrightarrow \phi_1 A_i \vee \phi_0 \overline{A}_i, \tag{10}$$

where ϕ_1 is the disjunction of A_i coefficients, and similarly ϕ_0 the disjunction of $\overline{A}_i$ coefficients. By (ii) ϕ can't be logically false so that there is at least one disjunct on the right-hand side of (9). By (i) ϕ_0 can't be absent (i.e., $\phi_0\overline{A_i}$ can't be $\overline{A_i}$) nor can it have a disjunct of only negated letters. By considering all possible truth value assignments we see that

$$\phi_1 \leftrightarrow \phi[1/A_i], \text{ and}$$
$$\phi_0 \leftrightarrow \phi[0/A_i]. \tag{11}$$

Also, for any truth values $x_i \in \{0, 1\}$,

$$(x_1, \ldots, x_{i-1}, 0, x_{i+1}, \ldots, x_n) \leq (x_1, \ldots, x_{i-1}, 1, x_{i+1}, \ldots, x_n)$$

so that by (iii)

$$\phi(x_1, \ldots, x_{i-1}, 0, x_{i+1}, \ldots, x_n) \leq \phi(x_1, \ldots, x_{i-1}, 1, x_{i+1}, \ldots, x_n).$$

Hence, in view of (11), ϕ_0 (logically) implies ϕ_1 and

$$\phi_1 \leftrightarrow \phi_1 \vee \phi_0. \tag{12}$$

Substituting for ϕ_1 in (10) produces

$$\phi \leftrightarrow (\phi_1 \vee \phi_0)A_i \vee \phi_0\overline{A_i}$$
$$\leftrightarrow \phi_1 A_i \vee \phi_0.$$

Thus ϕ is equivalent to a sentence without negated A_i. A repetition of the process serves to eliminate all negated sentence letters.

§0.2. Indefinite descriptions

In this section we are adhering to the usual predicate-language quantifier logic. A dialect form of it which includes the notion of an indefinite description is presented. To make the exposition coherent we have had to include a bit more material than will be referred to in subsequent sections.

Bertrand Russell's 1905 paper 'On Denoting' is widely known as having introduced an analysis of the definite description ('*the x* such that ϕ') as an incomplete symbol explainable in context. At the beginning of his paper

Russell mentions four types of phrases, two of which are 'every [all, any] x such that ϕ' and 'some [a, an] x such that ϕ'. We refer to expressions of this type as, respectively, *general* and *particular indefinite descriptions*. For such phrases Russell suggested an interpretation by way of specific examples (modified inessentially by us):

$$C(\text{every human}) \quad \text{means} \quad \forall x(H(x) \to C(x))$$
$$C(\text{some human}) \quad \text{means} \quad \exists x(H(x) \wedge C(x)).$$

With this bare mention Russell goes on to his discussion of the definite description.[3] Our interest here is with the indefinite description. Since the very beginning of quantification theory a special, limited, form of the (general) indefinite description has been in use in mathematical logic. This is the free individual variable. I quote the opening paragraph of Frege's *Begriffsschrift* (*1879*, translation from *van Heijenoort 1967*, 10):

§1. The signs customarily employed in the general theory of magnitudes are of two kinds. The first consists of letters, of which each represents either a number left indeterminate or a function left indeterminate. This indeterminacy makes it possible to use letters to express the universal validity of propositions, as in

$$(a + b)c = ac + bc.$$

The other kind consists of signs such as $+$, $-$, $\sqrt{}$, 0, 1, and 2, of which each has its particular meaning.

Referring to Russell's example, if H were true of everything then the describing condition $H(x)$ could be dropped, and '$C(\text{everything})$' would be interpreted as '$\forall x C(x)$'. The natural way of formally representing '$C(\text{everything})$' would then be '$C(x)$' with 'x' in place of 'everything'. The deletion of a universal quantifier, to obtain a free variable standing for anything whatever, applies only to initially placed universal quantifiers. For example, '$\forall x C(x) \to \phi$' and '$C(x) \to \phi$' (no free x in ϕ) have different meanings. The former says 'ϕ, if C holds of everything' whereas the latter says 'ϕ, if C holds of anything whatever'. Or, as Quine puts it (*1972*, 120):

---

3. Modern logical treatments of the definite description are to be found in a number of books; for example, from the mathematical point of view in *Kleene 1967*, 167–71, and from the linguistic-philosophical point of view in *Quine 1972*, §41.

In general the difference in English usage between 'any' and 'every' may have struck many of us as unsystematic and even mysterious. Scope of the universal quantifier is the key to it. Where distinction is needed between broader and narrower scope, as between '$(x)(Fx \supset p)$' and '$(x)Fx \supset p$', the English speaker's unconscious understanding is that 'any' calls for the broader scope and 'every' for the narrower. This rule works not only in connection with the conditional but also elsewhere, notably in connection with negation. Thus take the universe of discourse as consisting of all poems. 'I do not know any poem', then, and 'I do not know every poem', call respectively for the broader and the narrower scope:

$$(x) - (\text{I know } x), \qquad -(x)(\text{I know } x).$$

The free individual variable is a degenerate indefinite description in that the (unmentioned, vacuous) description applies to anything. The deficiency that inhibits adoption in logic of the indefinite description with arbitrary description, namely, that for it to function properly as a term requires that there be something satisfying the description, does not affect the free individual variable. For first-order predicate logic is normally formulated with the (tacit) assumption that the domain of individuals is not empty. To achieve a full use of indefinite descriptions requires the introduction of the notion of a *restricted variable*: free restricted variables will then be identified with (general) indefinite descriptions. A formal theory of quantification over restricted variables was presented by us in our *1957* and, a simplified and generalized version, by Prullage in her *1976*. In the following brief sketch we shall be quite informal as we shall only have need of some of its simple features.

The expressions of the formal system—which we call $QR\nu$—consists of formulas, and variables of two kinds one of which involves formulas. In addition to the usual individual variables x, y, ... we also have restricted variables symbolized by νxR, νyS, ..., where R, S,... are formulas. Here 'νx' is a variable binding operator—all occurrences of x in νxR are bound occurrences. (We chose the letter nu as suggestive of a̲n̲y̲. We read 'νxR' as 'any x such that R'.) Formulas are built up from

(i) atomic formulas, i.e., predicates with argument places filled by variables of either kind,

(ii) sentential connectives, and

(iii) universal quantification, with quantifier variables being of either
kind.

Because nesting of restricted variables is allowed the definition of *formula*
has to take this into consideration. Thus it is legitimate to first apply the
quantifier $(\forall\nu x R(x, \nu y S))$ to

$$C(\nu x R(x, \nu y S)) \tag{1}$$

but not the quantifier $(\forall\nu y S)$—though it can be afterwards. The variable
$\nu y S$ is said to be adherent (or subordinate) in (1), but not in the formula
resulting after application of $(\forall\nu x R(x, \nu y S))$.

Details of the definition of *formula*, a list of axioms, and proofs of its
soundness and completeness for $\mathcal{QR}\nu$, may be found in the aforementioned
paper of Prullage. Since the range of a restricted variable can be empty—
e.g., if the R of $\nu x R$ asserts something contradictory of x—the axioms
are framed in accordance. Thus in place of the $\forall x A \rightarrow A[y/x]$ of usual
predicate logic, $\mathcal{QR}\nu$ has

$$(\forall\alpha)A \rightarrow (\neg(\forall\beta)f \rightarrow A[\beta/\alpha]), \tag{2}$$

where β is an alphabetic variant of α, and f is the 0-place always-false
predicate. The condition $\neg(\forall\beta)f$ insures, for the range of the restricted
variable β, that it be not empty before β can be inferred as an instance of
the universal quantification.

In this theory one can show that to each sentence of $\mathcal{QR}\nu$ there is a
provably equivalent ν-less sentence, its *ν-less transform*, i.e., a sentence
without restricted variables. The basis of this result is the following axiom
(and its converse, Theorem 2.12) in *Prullage 1976*, 594, 596:

$$(\forall\nu\alpha A)B \rightarrow (\forall\beta)(A[\beta/\alpha] \rightarrow B[\beta/\nu\alpha A]), \tag{3}$$

where β is an alphabetic variant of α. Note that the antecedent of the main
conditional in (3) is a universal quantification whose quantifier variable is
$\nu\alpha A$, whereas in the consequent the quantifier variable β, being a variety
of α, is of one 'level' lower, though with the compensating restriction that
it have the property A.

As already mentioned we are identifying free restricted variables and
(general) indefinite descriptions. In the case of usual predicate logic the
meaning of a formula with free variables is the same as that of its closure

(prefixing universal quantifiers for each free variable). This will also be the case with $\mathcal{QF}\nu$, *closure* being defined appropriately. Such a definition can be found in *Prullage*, p. 593. Here we shall content ourselves with a simple illustration, using the formula

$$C(\nu x R(x, \nu y S(y))). \tag{4}$$

Its closure is obtained by first prefixing the quantifier $(\forall \nu x R(x, \nu y S(y)))$ and then to this the quantifier $(\forall \nu y S(y))$. To see the meaning of this closed formula we look at its equivalent ν-less transform. This is obtained by first operating on the inner quantification to obtain

$$(\forall x)(R(x, \nu y S(y)) \rightarrow C(x))$$

and then the outer quantifier to obtain

$$(\forall y)(S(y) \rightarrow (\forall x)(R(x, y) \rightarrow C(x))). \tag{5}$$

We render all this in linguistic form by setting

$$S(y) = y \text{ is a sailor}$$
$$R(x, y) = x \text{ rescues } y$$
$$C(x) = x \text{ is to be commended.}$$

Then the indefinite description form (4) translates into

Anyone who rescues any sailor is to be commended

which is also the sense of (5).

Although attention will be confined in our work to the general indefinite description, we shall say a few words about the particular indefinite description.

One introduces a notation '$\sigma x R$' which is related to existential quantification the way $\nu x R$ is related to universal quantification. Then $C(\sigma x R)$, for example, will have the meaning of $(\exists x)(R \wedge C(x))$. Expressions having both νx and σy present may be ambiguous since the order in which universal and existential quantifiers are written is material. To be able to use both general and particular indefinite descriptions an appropriate definition of *formula* would have to be devised. It clearly would be a complicated one. Observe that asserting, for example, $\nu x R \neq a$ is not the same as denying the assertion of $\nu x R = a$. The former is $\forall x(R \rightarrow x \neq a)$, the latter is $\sigma x R \neq a$, i.e., $\exists x(R \wedge x \neq a)$.

§0.3. Unspecified truth value. Don't-care logic

In §0.1 we were led to consider on an informal basis a semantic notion, that of a sentence having a truth value which is a member of the set $\{0, 1\}$ but is otherwise unspecified. We now wish to hypostasize this kind of truth value, that is, have a value u of which it can be asserted that it is a member of $\{0, 1\}$ and yet neither $u = 0$ nor $u = 1$ is warranted. This is accomplished not by violating the laws of set-logic but by using indefinite descriptions and an idiomatic language. Rather than a strange new semantic concept we will then have one constructed from familiar material—truth and falsity.

The indefinite description $\nu x(x \in \{0, 1\})$ (see §0.2) has the properties described in the preceding paragraph. For, when asserted,

$$\nu x(x \in \{0, 1\}) \in \{0, 1\} \quad \text{means} \quad \forall x(x \in \{0, 1\} \rightarrow x \in \{0, 1\})$$
$$\nu x(x \in \{0, 1\}) = 0 \quad \text{means} \quad \forall x(x \in \{0, 1\} \rightarrow x = 0)$$
$$\nu x(x \in \{0, 1\}) = 1 \quad \text{means} \quad \forall x(x \in \{0, 1\} \rightarrow x = 1),$$

the first of these being true, and the second and third being false. (Note that the denial of $\nu x(x \in \{0, 1\}) = 0$ is not $\nu x(x \in \{0, 1\}) \neq 0$, i.e., $\neg(\nu x(x \in \{0, 1\})) = 0)$, but $\sigma x(x \in \{0, 1\}) \neq 0)$.)

In §0.1, for convenience, the semantic properties of 0 and 1 were specified by means of ordinary arithmetic operations. We wish to do the same for u. To this end it will be helpful to discuss a similar situation involving the constant of integration C encountered in calculus. The beginning student may be initially surprised by the peculiar arithmetic properties such as $C + 1 = C$, $C + C = C$, etc. However, he or she is quickly convinced of its meaningfulness by intuitive explanations. To put the matter more formally, C is actually the indefinite description $\nu x(x \in \mathcal{R})$, where $\mathcal{R}$ is the set of reals. Since C is not a real number the expression '$C + 1$' needs special interpretation. It stands for an unspecified member of the set whose elements are 1 added to each of the reals. Denoting this set by $\mathcal{R} + 1$, then

$$C + 1 \quad \text{is} \quad \nu \dot{x}(x \in \mathcal{R} + 1).$$

But as $\mathcal{R} + 1$ is the same set as $\mathcal{R}$, $C + 1$ is then the same as C. In a similar manner, $C + C$ is also seen to be C. For our unspecified truth value u we have the following 'arithmetic' properties.

Theorem 0.31.

$$0 \leq u \leq 1 \tag{1}$$

$$1 - u = u \tag{2}$$

$$\min(0,\, u) = (0 \wedge u) = 0, \quad \min(1,\, u) = (1 \wedge u) = u$$

$$\max(0,\, u) = (0 \vee u) = u, \quad \max(1,\, u) = (1 \vee u) = 1 \tag{3}$$

$$\min(u,\, u) = (u \wedge u) = u, \quad \max(u,\, u) = (u \vee u) = u.$$

Proof. By virtue of the theorem on ν-less transform alluded to in §0.2, asserting (1) is equivalent to asserting

$$\forall x (x \in \{0,\, 1\} \to 0 \leq x \leq 1),$$

which is an arithmetical truth. In (2) $1 - u$, just as $C + 1$ of our discussion, has an idiomatic meaning. It is an unspecified ('arbitrary') member of the set of complements with respect to 1 of elements of $\{0, 1\}$. But

$$\exists y (y \in \{0,\, 1\} \wedge x = 1 - y) \leftrightarrow x \in \{0,\, 1\}.$$

Hence $1 - u$ is u. As for (3) take $\min(0,\, u)$, for example. This is an unspecified member of the set of all $\min(0,\, x)$, for $x \in \{0,\, 1\}$. But this set is the singleton $\{0\}$. Hence $\min(0,\, u) = \nu x(x \in \{0\}) = 0$. Similar arguments apply to the other stated equalities.

Having described the semantic operations on specified and unspecified truth values, we can proceed to the formal semantics. But first the syntax.

FORMAL SYNTAX FOR DON'T-CARE LOGIC

This is the same as that for verity sentential logic but with the addition of a unary connective $\triangle$.[4] Let the set of such *formulas* be denoted by '$\mathcal{S}^u$' (reserving the word 'sentence' for elements of $\mathcal{S}$). Now for the semantics.

---

4. Read as 'don't-care'. The reason for the name 'don't-care logic' will be seen presently.

FORMAL SEMANTICS FOR DON'T-CARE LOGIC

As with verity logic we introduce a semantic function, which we shall call 'U', from formulas of S^u into the set[5] of unspecified and specified semantic values, with the following properties:

$$U(\Phi) \in \{0, 1, u\}$$
$$U(\neg\Phi) = 1 - U(\Phi)$$
$$U(\Phi \wedge \Psi) = \min(U(\Phi), U(\Psi))$$
$$U(\Phi \vee \Psi) = \max(U(\Phi), U(\Psi)) \tag{4}$$
$$U(\triangle\Phi) = \begin{cases} 0 & \text{if } U(\Phi) = 0 \\ u, & \text{otherwise.} \end{cases}$$

(As usual, where convenient $\wedge$ will be replace by juxtaposition and $\neg$ by overline.)

On the basis of (4), 3-valued semantic tables for $\neg$, $\wedge$, $\vee$ are readily constructed and seen to be the same as Kleene's *strong tables* for these connectives (*1952*, 334). Our interpretation for u stems from conceiving of it as $vx(x \in \{0, 1\})$. Kleene's interpretation, as given in (ii) of the following quotation, is substantially the same as ours (*Kleene 1952*, 335).

> We further conclude from the introductory discussion that, for definitions of partial recursive operations, t, f, u must be susceptible of another meaning besides (i) 'true', 'false', 'undefined', namely (ii) 'true', 'false', 'unknown (or value immaterial)'. Here 'unknown' is a category into which we can regard any proposition as falling, whose value we either do not know or choose for the moment to disregard; and it does not then exclude the other two possibilities 'true' and 'false'.

Properties listed in (4) determine a semantic value for each formula in S^u once an initial assignment has been made to $A_1, \ldots, A_n, \ldots$. Such assignments we call *models*. Two formulas of S^u are *semantically equivalent* if they have the same semantic value in each model. (If the formulas contain the same atomic sentence letters their tables would be identical.) Clearly, semantic equivalence is invariant under replacement of a formula-part by one which is semantically equivalent to it.

---

5. Since the unspecified truth value u is not a fixed entity like 0 or 1, the word 'set' here is being used idiomatically.

The following table shows that $\triangle\Phi$ and the conjunction $\Phi\triangle(\Psi\vee\overline{\Psi})$ are semantically equivalent:

Φ	$\triangle\Phi$	$\Psi\vee\overline{\Psi}$	$\triangle(\Psi\vee\overline{\Psi})$	$\Phi\triangle(\Psi\vee\overline{\Psi})$
1	u	1 or u	u	u
0	0	1 or u	u	0
u	u	1 or u	u	u

Thus, in place of the connective $\triangle$ operating on Φ one can equally well form the conjunction of the constant $\triangle(\Psi\vee\overline{\Psi})$ with Φ. It will be convenient to drop the argument $\Psi\vee\overline{\Psi}$ and use '$\triangle$' as a 'propositional' constant whose value is u in any model. One readily sees that ('$=$' here meaning 'is semantically equivalent to')

$$\neg\triangle = \triangle$$
$$\triangle\triangle = \triangle$$
$$\triangle(\Phi\Psi) = \Phi\triangle\Psi \tag{5}$$
$$\triangle(\Phi\vee\Psi) = \triangle\Phi\vee\triangle\Psi.$$

For don't-care logic we shall employ special definitions of model and validity. A $(0,1)$-*model* is a model in which each A_i is assigned either 0 or 1. We shall restrict our attention only to such models.[6] Thus a *semantic table for don't-care logic* will be a mapping from $\{0,1\}^n$ to $\{0,u,1\}$. For validity we shall require a formula to have the value 1 in do-care (non-u) cases: A formula $\Phi\in\mathcal{S}^u$ is u-*valid* if in no $(0,1)$-model is 0 the value of Φ, and in at least one model its value is 1. Since our interest will be exclusively with $(0,1)$-models it will simplify matters if, in addition to having the A_i take on only 0 or 1 as values, we also assume that the syntactic variables ϕ, ψ, $\chi\ldots$, range over elements of $\mathcal{S}$—hence, being constructed solely from the A_i and $\neg$, $\wedge$, $\vee$, their values will also be either 0 or 1. The unspecified truth value comes in with the occurrence of $\triangle$; for example $\triangle\phi$ has the semantic table

ϕ	$\triangle\phi$
1	u
0	0 .

Associated with any formula Φ of $\mathcal{S}^u$ there will be a unique (don't-care) semantic table entered from $A_1,\ldots,A_n$ (assuming these are the letters

---

6. Note this divergence from all other conceptions of many-valued logic. The semantics is not the same as that of a 3-valued logic as ordinarily conceived.

present in Φ) and with values assigned to each row that are either 1, 0, or u. As usual, with each row we associate a constituent K_i ($1 \leq i \leq 2^n$). Then the formula

$$K_{i_1} \vee \cdots \vee K_{i_r} \vee \triangle K_{j_1} \vee \cdots \vee \triangle K_{j_s}. \tag{6}$$

where $K_{i_1}, \ldots, K_{i_r}$ are the constituents for rows of Φ that are assigned 1, and $K_{j_1}, \ldots, K_{j_s}$ are those assigned u, is easily seen to have the same semantic table as Φ. An alternative form for (6) is, by (5),

$$K_{i_1} \vee \cdots \vee K_{i_r} \vee \triangle (\overline{\overline{K}_{j_1} \ldots \overline{K}_{j_s}}). \tag{7}$$

In addition to $\triangle$ we shall also be interested in a binary connective '$\dashv$' whose definition involves u. It is closely related to the verity conditional $\phi \to \psi$ (which may also be written as $\overline{\phi} \vee \phi\psi$) for we shall define

$$\Psi \dashv \Phi \quad \text{to be} \quad \triangle\overline{\Phi} \vee \Phi\Psi. \tag{8}$$

A comparison of the semantic tables for $\phi \to \psi$ and $\psi \dashv \phi$,

ϕ ψ	$\phi \to \psi$	$\psi \dashv \phi$
1 1	1	1
1 0	0	0
0 1	1	u
0 0	1	u

shows that when the antecedent (i.e., ϕ) is 0, then $\psi \dashv \phi$ has the value u rather than 1 as in the case of $\phi \to \psi$.

In terms of '$\dashv$' formula (7) can be written as

$$K_{i_1} \vee \cdots \vee K_{i_r} \dashv \overline{K}_{j_1}\overline{K}_{j_2} \ldots \overline{K}_{j_s}. \tag{9}$$

It corresponds to a semantic table in which rows $i_1, \ldots, i_r$ have been assigned the value 1 (these being the rows for which $\psi\phi$ has the value 1), and rows $j_1, \ldots, j_s$ the value u (these being the rows for which ϕ has the value 0).

The ideas introduced in this section will be used later in §5.6.

§0.4. Don't-care conditions in logic design

In this section we describe a use of don't-care conditions which strictly speaking digresses from our subject of probability logic (and hence can be omitted if the reader so chooses). The purpose of presenting it is to see the unspecified truth value being used in another context. We believe this will enhance understanding and appreciation of what is a relatively novel notion. What we shall look at occurs in a subject described as 'logic design of combinational networks (or, switching circuits)' and, in particular, when they involve so-called 'don't-care' conditions. First a quick explanation of these notions.

A two-valued (verity) truth-table can represent all possible actions of a combinational logic network whose components are binary gates (switches). The values 1 and 0 may, for instance, stand for transmission, and non-transmission, of a signal. Each line of the table describes the output (1 or 0) which results when inputs (1 or 0) are assigned to (or arrive at) the entering variables of the table. There are three simple types of gates (inverter, and-gate, or-gate) whose actions are represented by the tables for $\neg$, $\wedge$, and $\vee$. When such gates are connected so that their outputs become inputs to others (but acyclically) the effect is that of truth-functional composition. Accordingly, any such combinational logic network is representable mathematically by a logical function ('switching function') rendered symbolically by a sentence of verity logic. Such an 'analysis' of combinational networks into their representing logical functions, and then being able to work with these instead of actual physical apparatus, is of engineering importance. For example, recognizing that

$$A_1 A_2 \vee \overline{A_1} A_3 \vee A_2 A_3 \quad \text{and} \quad A_1 A_2 \vee \overline{A_1} A_3$$

have the same table enables a network described by the former to be replaced by the simpler one described by the latter, with a consequent saving of equipment.

The problem of 'synthesis'—designing a combinational network with prescribed characteristics—is theoretically simple. Supposing there are n input variables, a combinational table (truth-table) is constructed in which each of the 2^n rows (of 1's and 0's) is assigned a 1 or 0 in accordance with the desired operating characteristics. Each row of the table determines a fundamental product (constituent) and the disjunction of these to which a 1 has been asssigned is a logical function for the network. There are situations (this is where our application of unspecified truth values comes in) where the designer has no interest in, or doesn't care, whether a particular combination (row of the table) is assigned 1 or 0.

As an example, to code (represent) the 10 decimal digits by means of binary digits requires a minimum of four binary digits, since three produce only $2 \cdot 2 \cdot 2 = 8$ different (table) combinations. Now a combinational table with four inputs has 16 rows, of which only 10 are needed. The practice of a designer in such a situation is to place some symbol, e.g., 'd' ('don't care') alongside of the rows which won't be needed. Although not needed, the switching circuit logic designer can use them to simplify the logical function which descibes the action of the circuit. To see what properties are assumed for the d in carrying out such simplification, and to see that these are the same as for our unspecified truth value u we make use of an illustrative example discussed in *McCluskey 1986*, 203. The example is concerned with simplifying the logical function for a combinational table with some don't-care combinations. First an explanation of the terminology.

For tables with only 0 or 1 assignments to the rows a representing function can be obtained by writing down the logical sum of the fundamental products (constituents) to which a 1 has been assigned. Call these 1-terms. Sums of 1-terms can be simplified by combining any two which differ only in a letter being negated in one and unnegated in the other; that is, in general, replacing $\phi A_i \vee \phi \overline{A}_i$ by ϕ. Terms which can no longer be shortened in this way are called *prime implicants* ('implicants' since they still imply the original sum). A usual device for finding the prime implicants of a sum, and hence being able to find a minimal sum, is called a Karnaugh map. We explain it by using one that occurs in the example of McCluskey's to which we will be referring.

	wx			
yz	00	01	11	10
00			1	
01		d	1	
11	1	d	d	
10		d		

TABLE 0.41

In Table 0.41 are 16 cells corresponding to the sixteen rows of a truth-table with variables w, x, y, z. The more familiar way of writing the information in the table would be as in Table 0.42.

	w	x	y	z	f
K_0	0	0	0	0	0
K_1	0	0	0	1	0
K_2	0	0	1	0	0
K_3	0	0	1	1	1
K_4	0	1	0	0	0
K_5	0	1	0	1	d
K_6	0	1	1	0	d
K_7	0	1	1	1	d
K_8	1	0	0	0	0
K_9	1	0	0	1	0
K_{10}	1	0	1	0	0
K_{11}	1	0	1	1	0
K_{12}	1	1	0	0	1
K_{13}	1	1	0	1	1
K_{14}	1	1	1	0	0
K_{15}	1	1	1	1	d

TABLE 0.42

To see the connection between the tables consider the 1-term K_{12} in Table 0.42, that is, the term $wxy'z'$ (using primes to indicate negation). Writing the subscript 12 on K_{12} in binary form, 1100, locates the 1-term in the cell under '11' and across from '00'. The map is used to easily spot pairs of 1-terms that can be replaced by a simpler term since any two adjacent cells (including horizontal and vertical wrap-around) differ only in one place of the four binary digits asociated with each cell. Thus, consulting Table 0.41, we see that $K_{12} \vee K_{13} = K_{1100} \vee K_{1101} = wxy'z' \vee wxy'z$ is replaceable by wxy'. The sum of all such simplified 1-terms (after all possible simplifications), and which is such that every cell of the map with a 1 has been included, is a *minimal sum*. Now concerning maps that have d-terms (as in the table) McCluskey says (p. 203):

The addition of d terms does not introduce any extra complexity into the procedure for determining minimal sums. Any d terms that are present are treated as 1-terms in forming the prime implicants, *with the exception that no prime implicants containing only* d *terms are formed. The* d *terms are disregarded in choosing terms of the minimal sum.* No prime implicants are included in order to ensure that each d term is contained in at least one prime implicant of the minimal sum. The explanation of this procedure is that d terms are used to make the prime implicants as large as possible so as to include the maximum number of 1-cells and to contain as few literals as possible. No prime implicants need be included in the minimal sum because of the d terms, for it is not required that the function equal 1 for the d terms.

Although the example and discussion do not provide information on every possible kind of occurrence or use of d, there can be no doubt that it is compatible with its being an unspecified truth value:

(i) it is semantic-like in that it is attributable to logical terms, yet is neither 0 nor 1;

(ii) when a d-term is conjoined by $\vee$ with a 1-term it functions as a 1-term in that the result is taken to be a 1-term $(1 \vee d = 1)$;

(iii) yet not when conjoined with another d-term $(d \vee d = d)$ except if the result is then conjoined with a 1-term.

Unlike common truth-tables, to which one can associate a logical function (sentence of S) and operate algebraically with it, tables with d's have had so far no formal representation correlated with them. We now show, by identifying d with the unspecified truth value u and resorting to our don't-care logic, that this correlation can be accomplished.

With the identification of the don't-care d with the unspecified truth value u, the table in the example (Table 0.41) is expressible as a formula of S^u:

$$K_3 \vee K_{12} \vee K_{13} \dashv \overline{K}_5\overline{K}_6\overline{K}_7\overline{K}_{15}, \tag{1}$$

where the K's preceding '$\dashv$' are the 1-terms and the K's succeding '$\dashv$' are the d-terms. It is evident that replacing any S sentence-part of such a formula by a logically equivalent one results in one that is u-logically equivalent to the original formula. Thus any disjunctions of 1-terms of (1) can be replaced by a (simpler) equivalent sentence and the resulting formula will depict the same 'activity' as the original one. Also, adjoining to the 1-terms a d-term, e.g., K_7,

$$K_7 \vee K_3 \vee K_{12} \vee K_{13} \dashv \overline{K}_5\overline{K}_6\overline{K}_7\overline{K}_{15} \tag{2}$$

again results in a formula u-logically equivalent to the original (1). For when K_7 takes on the value 1 then $\overline{K}_7$ takes the value 0 and hence, by the definition of $\dashv$, the value of (2) is $u(=d)$.

Simplifying the part of the formula succeding '$\dashv$' (i.e., the part with the d-terms) has not been considered in the literature, nor, of course, interactions of such formulas with others. If this were to come about logico-algebraic operations on formulas of this type could become of interest.

§0.5. Elimination. Linear programming

Every sentential combination of elementary events is logically equivalent to a disjunction of constituents based on these events. The probability of a disjunction is equal to the sum of the probabilities of the (mutually exclusive) disjuncts. It is not surprising, then, that linear algebra turns out to be a significant feature of probability logic. This section is devoted to presenting some results which we shall be needing in later chapters. In the main they are basic matters which can be found in any number of treatises or textbooks, though not all of it in any one place. They are stated here for convenience of reference.

Our first item is a necessary and sufficient condition that a system S of m inequations in n unknowns,

$$
\begin{aligned}
a_{11}x_1 + a_{12}x_2 + \cdots + a_{1n}x_n &\geq d_1 \\
a_{21}x_1 + a_{22}x_2 + \cdots + a_{2n}x_n &\geq d_2 \\
&\vdots \\
a_{m1}x_1 + a_{m2}x_2 + \cdots + a_{mn}x_n &\geq d_m
\end{aligned}
\tag{1}
$$

being solvable, i.e., having a solution. The result is derived by a repetition of the process of elimination: replacing the system by one having one fewer variables, and which is such that it has a solution if and only if the original system does. Since systems of equations are more familiar than inequations, and the process of elimination being also simpler, we shall first describe it for this case.

Let E be a system of m equations in n unknowns. Suppose that of its m equations there are $m - m_0 \ (\geq 1)$ in which the coefficient of x_n is not

0. Select one of these, say it is the j-th, and, solving for x_n, obtain

$$x_n = -\frac{1}{a_{jn}}(a_{j1}x_1 + \cdots + a_{jn-1}x_{n-1} - d_j). \tag{j}$$

After doing the same with each of the other of the $m - m_0$ equations with non-zero x_n coefficient, equate their right hand sides with that of (j), so obtaining a set of $m - m_0 - 1$ equations in the variables $x_1, \ldots, x_{n-1}$. (If $m - m_0 = 1$ then this set is empty.) Let E' be the system of equations which is the union of this set with the set of m_0 (≥ 0) equations whose x_n coefficient is 0. It readily follows that E has a solution if and only if E' does. Recursive repetition of this procedure produces a sequence of sets $E, E', \ldots, E^{(n)}$, with $E^{(n)}$ having no variables. Then E has a solution if and only if each equation in $E^{(n)}$ is a true statement or else $E^{(n)}$ is the empty set.[7]

Turning now to systems of linear inequations of type S as depicted in (1), we describe for it an elimination procedure called *Fourier elimination*. To each $a_{in} \neq 0$ there are now two possible types of solution for x_n according to whether $a_{in} > 0$ or $a_{in} < 0$:

$$x_n \leq -\frac{1}{a_{in}}(a_{i1}x_1 + \cdots + a_{in-1}x_{n-1} - d_i) \tag{i}$$

$$-\frac{1}{a_{in}}(a_{i1}x_1 + \cdots + a_{in-1}x_{n-1} - d_i) \leq x_n. \tag{i'}$$

If m_0 of the inequations in (1) have $a_{in} = 0$, there will be $m - m_0$ inequations of type (i) or (i'). Let there be r of type (i) and l of type (i').

Case 1. Both r and l are ≥ 1.

Suppose the right hand sides of type (i) are $R_1, \ldots, R_r$ and the left hand sides of type (i') are $L_1, \ldots, L_l$. Consider the set of rl inequations obtained by asserting that each of the L_i ($1 \leq i \leq l$) is less or equal to every R_j ($1 \leq j \leq r$). We obtain the set S' by adjoining these rl inequations to the m_0 inequations having $a_{in} = 0$. It is readily seen that if an n-tuple $(x_1^0, \ldots, x_n^0)$ satisfies S, then the $(n-1)$-tuple $(x_1^0, \ldots, x_{n-1}^0)$ satisfies S'; and for any $(x_1^0, \ldots, x_{n-1}^0)$ satisfying S' there is an x_n^0 such that the n-tuple $(x_1^0, \ldots, x_{n-1}^0, x_n^0)$ satisfies S—for we have

$$\max(L_1, \ldots, L_l) \leq \min(R_1, \ldots, R_r),$$

---

7. As an example of $E^{(n)} = \emptyset$, let E be the system of one equation $\{3x_1 - x_2 = 7\}$. Then $x_2 = 3x_1 - 7$ and $E' = \emptyset$. Clearly $(x_1^0, 3x_1^0 - 7)$ is a solution for any value x_1^0.

and hence any x_n^0 such that

$$\max(L_1, \ldots, L_l) \leq x_n^0 \leq \min(R_1, \ldots, R_r)$$

provides a value for which $(x_1^0, \ldots, x_{n-1}^0, x_n^0)$ satisfies S.

Case 2. Either r or l is 0.

Then we take S' to be the set of m_0 inequations with $a_{in} = 0$, that is, we adjoin nothing to them. To see why let it be, for example, the (i') which are absent. Then there is an x_n satisfying the (i) type inequations if and only if there is an $(n-1)$-tuple $(x_1^0, \ldots, x_{n-1}^0)$ satisfying S'. If there is no such $(n-1)$-tuple, then clearly S can't be satisfied, and if there is then, letting $R_1^0, \ldots, R_r^0$ be the values of the right hand sides of the (i) type inequations, we see that any x_n^0 such that

$$x_n^0 \leq \min(R_1^0, \ldots, R_r^0)$$

will provide a value such that $(x_1^0, \ldots, x_{n-1}^0, x_n^0)$ satisfies S. We sum up this discussion in an official statement:

Theorem 0.51. *A linear inequational system S of the form*

$$a_{11}x_1 + a_{12}x_2 + \cdots + a_{1n}x_n \geq d_1$$
$$a_{21}x_1 + a_{22}x_2 + \cdots + a_{2n}x_n \geq d_2$$

$$\vdots$$

$$a_{m1}x_1 + a_{m2}x_2 + \cdots + a_{mn}x_n \geq d_m$$

is solvable (has a solution, is consistent) if and only if the procedure of Fourier elimination applied successively to the variables $x_1, \ldots, x_n$ results in a sequence of linear inequation systems $S, S', \ldots, S^{(n)}$ such that either $S^{(n)}$ is the empty set, or else each relation in $S^{(n)}$ is true.

Relations of the form $L \leq d$ and $L = d$ can be included in a system of form S; for the first of these is equivalent to $-L \geq -d$, and the second to the pair of relations $-L \geq -d$, $L \geq d$.

Linear programming.[8] The problem of finding a solution of a linear system (of equations and/or inequations) which minimizes (or maximizes)

---

8. This is a summary of the elements of the subject, with particular reference to its use in probability logic. It is reproduced from our *1986*, §0.8. The statement of the theorem on linear fractional programming has been revised in accordance with a result in *Schechter 1989*.

a given linear function is called a *linear programming problem*. A particular type of linear program, of the kind we shall be interested in, is the following:

$$\text{minimize} \quad \sum_{j=1}^{m} c_j x_j,$$

$$\text{subject to} \quad \sum_{j=1}^{m} a_{ij} x_j = d_i, \quad i = 1, \ldots, n,$$

$$x_j \geq 0, \qquad j = 1, \ldots, m$$

where the c_j, a_{ij}, d_i are given constants and the x_j are variables. Use of matrix notation enables us to express this succinctly as

$$\text{minimize} \quad \mathbf{cx},$$

$$\text{subject to} \quad \mathbf{Ax} = \mathbf{d},$$

$$\mathbf{x} \geq \mathbf{0},$$

where $\mathbf{c} = \begin{bmatrix} c_1 & \cdots & c_m \end{bmatrix}$ is a row vector ($1 \times m$ matrix), where

$$\mathbf{x} = \begin{bmatrix} x_1 \\ \vdots \\ x_m \end{bmatrix} = \begin{bmatrix} x_1 & \cdots & x_m \end{bmatrix}^{\mathrm{T}}$$

is a column vector, where $\mathbf{d} = \begin{bmatrix} d_1 \ldots d_n \end{bmatrix}^{\mathrm{T}}$, and where $\mathbf{A}$ is the $n \times m$ matrix

$$\mathbf{A} = \begin{bmatrix} a_{11} & a_{12} & \cdots & a_{im} \\ a_{21} & a_{22} & \cdots & a_{2m} \\ \vdots & \vdots & & \vdots \\ a_{n1} & a_{n2} & \cdots & a_{nm} \end{bmatrix}.$$

The superscript T indicates matrix transpose. The function $\mathbf{cx}$ is called the *objective function*. Vectors $\mathbf{x}$ satisfying the constraints, i.e., in this case

$$\mathbf{Ax} = \mathbf{d} \quad \text{and} \quad \mathbf{x} \geq \mathbf{0},$$

are called *feasible solutions*. Feasible solutions which minimize (or maximize) the objective function are called *optimal solutions*. In linear programs which are of interest it is generally the case that the number of unknowns exceeds the number of equations ($m > n$), and we assume that there are no redundant equations in the system $\mathbf{Ax} = \mathbf{d}$ (so that rank $\mathbf{A} = n$). A set of n independent column vectors of the matrix $\mathbf{A}$ is called a *basis* for the system (and the linear program); the variables x_j associated with these columns are called *basic variables* (the remaining variables are *non-basic*). A *basic feasible solution* is a feasible solution in which the values of the non-basic variables are 0, and an *optimal basic solution* is a basic feasible solution which optimizes $\mathbf{cx}$.

The understanding of linear programming is enhanced if relations and properties are visualized geometrically in m-dimensional vector space. Since common 3-dimensional geometrical notions (line, plane, convex set) generalize readily, we shall not trouble to state definitions of these. In-equations $\mathbf{ax} \leq d$ or $\mathbf{ax} \geq d$ specify *half-spaces*. The intersection of finitely many half-spaces is a *convex polyhedron* (*polytope*, if bounded). A *supporting half-plane* to a convex set S is a hyperplane containing a point of S and having all of S on one side of the hyperplane.

In terms of this geometric language some of the fundamental facts of linear programming can now be stated.

The set of points corresponding to the feasible solutions of a linear program is a convex polyhedron P. A basic feasible solution corresponds to an extreme point (vertex or 'corner point') of P. If there are any optimal solutions then there are optimal extremal solutions. If non-empty, the set of optimal solutions (also a convex polyhedron) is the intersection of the polyhedron P with a supporting hyperplane whose equation is of the form $\mathbf{cx} = q_0$, where $\mathbf{cx}$ is the objective function (so that the supporting hyperplane is a member of the family of parallel planes $\mathbf{cx} = q$, q a parameter) and q_0 is the optimal value. If $\mathbf{cx}$ has a maximum and a minimum in the polyhedron then these values are attained at vertices of the polyhedron (at least—it could also be at all points of an edge, or of a face, etc.). The maximum and minimum are *global*, i.e., all values of $\mathbf{cx}$ for x in P lie between these two values.

For future reference we point out that the vertices of the polyhedron are obtainable from the linear conditions by rational operations (addition, multiplication, division). Hence the coordinates of a vertex will be rational functions of the coefficients and constant terms entering into the linear conditions.

There is an important result in the theory which relates the so-called primal and dual forms of a linear program. For the types which will be of particular interest to us, namely

(i)   minimize $\mathbf{cx}$,   (ii)   maximize $\mathbf{cx}$,
      subject to $\mathbf{Ax = d}$,   subject to $\mathbf{Ax = d}$,
                   $\mathbf{x \geq 0}$, $\mathbf{x \geq 0}$,

we define their respective duals to be

(i)   maximize $\mathbf{d}^T\mathbf{u}$,   (ii)   minimize $\mathbf{d}^T\mathbf{v}$,
      subject to $\mathbf{A}^T\mathbf{u} \leq \mathbf{c}^T$,   subject to $\mathbf{A}^T\mathbf{v} \geq \mathbf{c}^T$,
                   $\mathbf{u}$ arbitrary, $\mathbf{v}$ arbitrary.

An alternative version is the pair

   Primal: minimize $z = \mathbf{cx}$, subject to

(iii) $\mathbf{Ax \geq b}$. $(x_i \geq 0)$

   Dual: maximize $v = \mathbf{b}^T\mathbf{y}$, subject to

(iv) $\mathbf{A}^T\mathbf{y} \leq \mathbf{c}^T$. $(y_i \geq 0)$

We now state the result (*Dantzig 1963*, 125, *Stoer-Witzgall 1970*, 28):

Theorem 0.52 (Duality Theorem of Linear Programming).
For dual pairs of a linear program the following hold:

(a) *The value of the objective function at a feasible solution in a minimization program is greater or equal to the value of the objective function for any of the feasible solutions of the maximization program.*

(b) *If both primal and dual forms have feasible solutions then both have optimal solutions, and the respective objective functions are equal for these optimal solutions; and*

(c) *If one of the programs has an optimal solution, then so does the other (and by (b) their objective functions are equal at these optimal solutions).*

Fourier (also called *Fourier-Motzkin*) *elimination* can be made the basis for a direct solution of a linear programming problem. But since at each stage the number of inequations added to the system is of the form rl (see preceding discussion), it is generally not a practical method for solving problems, as uncontrollable large numbers of inequations could arise. However we shall find it useful when parameters are involved in the coefficients of the constraint equations. The method is quite simple to describe: add the equation $z = \mathbf{cx}$, where $\mathbf{cx}$ is the objective function, to the system of constraints and eliminate all variables but z. If the resulting sets of inequations are

$$L_1 \leq z, \; L_2 \leq z, \; \ldots, \; L_l \leq z$$
$$z \leq R_1, \; z \leq R_2, \; \ldots, \; z \leq R_r \tag{2}$$

then

$$\max(L_1, \ldots, L_l) \leq z \leq \min(R_1, \ldots, R_r) \tag{3}$$

provides the least and greatest value for z while subject to the constraints (*Dantzig 1963* §4-4). Elimination of z produces

$$\max(L_1, \ldots, L_l) \leq \min(R_1, \ldots, R_r), \tag{4}$$

which then provides the consistency conditions, i.e., the necessary conditions for the existence of a solution. If the L_i and R_j involve parameters, and hence have no fixed numerical values, then condition (4) converts to a set of conditions on these parameters, namely that each L_i $(i = 1, \ldots, l)$ shall be less than or equal to every R_j $(j = 1, \ldots, r)$.

FRACTIONAL LINEAR PROGRAMMING

This is an extension of linear programming in which the objective function, now a linear fractional form

$$\frac{\mathbf{cx} + d}{\mathbf{ax} + b}, \tag{5}$$

is to be optimized subject to linear constraints. A fundamental result of Charnes and Cooper in their *1962* shows how to reduce such a problem to a related ordinary linear programming problem with an additional variable. The Charnes-Cooper result includes the hypothesis that $\mathbf{ax} > \mathbf{0}$. *Schechter 1989* shows that this assumption can be dropped in favor of its inclusion as part of the specification of the feasible region. In our application of this material (5) will reduce to $\mathbf{cx}/\mathbf{ax}$, with $\mathbf{ax}$ never negative.

For this situation the Charnes-Cooper-Schechter result may be stated as follows:

Theorem 0.53. *The linear fractional programming problem:*

$$\text{optimize } \frac{\mathbf{cx}}{\mathbf{ax}}$$

subject to

$$\mathbf{Ax} = \mathbf{b}$$
$$\mathbf{x} \geq \mathbf{0}, \ \mathbf{ax} > \mathbf{0}$$

is equivalent to the linear programming problem:

$$\text{optimize } \ \mathbf{cy}$$

subject to

$$\mathbf{Ay} = t\mathbf{b}$$
$$\mathbf{ay} = 1$$
$$t, \mathbf{y} \geq \mathbf{0}.$$

Chapter 1

Early History

§1.1. Leibniz's vision

The following is an excerpt from Leibniz's *Nouveau essais* (*1962*, 466):[1]

> I have said more than once that there is need for a *new kind of logic*
> which would treat of degrees of probability. For what Aristotle has in
> his *Topics* is quite different. He is satisfied with arranging a few famil-
> iar rules according to common patterns; these could serve on the oc-
> casion when one is concerned with amplifying a discourse so as to give
> it some likelihood. No effort is made to provide a balance necessary
> to weigh the likelihoods in order to obtain a firm judgement. Anyone
> wishing to treat these matters would do well to examine games of
> chance; in general, I wish that some skillful mathematician would be
> willing to produce a detailed, systematic and extensive work on all
> the varieties of games, . . .

The *Topics* of Aristotle, to which Leibniz is here referring, has been described as a handbook for the guidance of contestants in public debating
contests. In contrast Leibniz, whose early training was in jurisprudence,
is primarily interested in the problem of rationally adjudicating between
opposing views or conflicting claims. In this connection, as our quotation
indicates, Leibniz believed that the recently developed mathematical methods for calculating chances well worth looking at. He gives the example of

---

1. This is a modern edition. Although the *Nouveau essais* were first published in
1765, long after Leibniz's death, he had ideas for a *doctrina de gradibus probabilitatis* at
least as early as 1670 (*Schneider 1981*, 204). These ideas are extensively described and
discussed in *Couturat 1901*, 238–55, 552–55. *Schneider 1981* also covers this ground in
less detail, but from the advantage of an additional 80 years of Leibniz scholarship.

two players, one winning if a 7, the other if a 9, is the outcome of a throw of
a pair of dice. The likelihoods of winning are in proportion to the number
of ('equally possible') outcomes favorable to the respective values. Aside
from such simple examples Leibniz seems not to have pursued the matter
any further.

The *Nouveaux essais* were written in Leibniz's mature period, the first
decade of the eighteenth century. His much earlier baccalaureate thesis of
1665, *De conditionibus* (an improved version in *Specimen juris* of 1669),
written before he became aware of the currently developing theory of probability, is a theoretical legal treatise on conditional rights.[2] It contains the
suggestion of ordering conditional rights on the scale of 0 to 1 (*Couturat
1901*, 552. *Hacking 1975*, chapter 10, 'Probability and the law'). A right,
for example to an estate or a throne, may not be absolute but dependent
upon a condition. A statement of the right would be a conditional sentence
whose antecedent embodies the condition. With regard to such statements
Leibniz (*1971*, 420) presents the following schema:

Antecedent:	impossible	contingent	necessary
	0	$\frac{1}{2}$	1
Right:	non-existent	conditional	absolute

This indicates that a right is non-existent, conditional, or absolute depending on whether the antecedent is impossible, contingent or necessary. The
fraction $\frac{1}{2}$ is used by Leibniz as standing for some unspecified value between 0 and 1. In other words Leibniz is allowing a right to have a range of
values between 0 and 1, but he gives no indication of how such values are
to be determined. There is perhaps some basis for Couturat's enthusiastic
assertion (*1901*, 553):

> En effect, ces valeurs 0, θ [arbitrary fraction between 0 and 1], et
> 1 mesurent précisément la probabilité de ce droit dans les trois cas
> ... Ainsi Leibniz a entrevu ici, d'une part, le Calcul des probabilités,
> ....

Indeed, involved here is the idea of a numerical measure of the likelihood of the consequent of a (necessary) conditional on the basis of that
of the antecedent—but only inchoately. So far as we know Leibniz never
developed the idea. Jakob Bernoulli did.

---

2. Apart from the matter of interest to us, it contains a nice exposition of the conditional proposition and its principal logical features. Except for Couturat (*1901*, 553–54),
this seems not to have been noticed by historians of logic.

§1.2. Jakob Bernoulli—probability logic via the fates of gamblers

Bernoulli's *Ars conjectandi 1713*, left unfinished at his death and published posthumously, contains four parts. Part I is an annotated presentation of Huygens' *De ratiociniis in ludo aleae*, part II is a development of mathematical properties of permutations and combinations, and part III states and solves a variety of problems concerned with players expectations in games of chance. Part IV, entitled 'Use and application of the preceding theory in civil, moral and economic affairs', is famous for containing the law of large numbers. However there are no realistic applications of the sort mention in its title—presumably it had been Bernoulli's intention to include some. The material which is the subject of this and the next section is contained in chapter 3 of part IV. Yet, despite its innovative character, it has been largely neglected. For example, Todhunter (*1865*, 70–71) devotes three short paragraphs to it; van der Waerden, in the introduction to Volume 3 of Bernoulli's collected works, doesn't mention it in his description of *Ars conjectandi*; Maistrov in *1974* quotes one sentence from it. Not until we come to *Hacking 1974* and *Shafer 1978* do we have substantial discussion; however the conclusions we come to will be distinctively different from theirs.

In this chapter 3, 'On various kinds of arguments, and how their weights are estimated in order to compute the probability of things', Bernoulli investigates the 'force of proof' of an argument and the degree of certainty of an opinion or conjecture on the basis of arguments for it. As with Leibniz's conditional rights, these investigations involve conditional sentences and the likelihood of their consequents. With Bernoulli, however, we find a substantial advance beyond Leibniz's strongly expressed hope. Conceivably Leibniz may have had some influence on Bernoulli's ideas, for they did correspond with each other. But as far as our topic is concerned, there seems to be little evidence for it. For example, Bernoulli's repeated requests to Leibniz for a copy of his *De conditionibus* finally ended when Leibniz simply acknowledged that he had none. (See *Schneider 1981b*, 212.)

Bernoulli's treatment distinguishes three kinds of arguments which produce opinion or conjecture: those which exist necessarily and imply [the conjecture] contingently (*necessariò existunt & contingenter indicant*), those which exist contingently and imply necessarily, and those which both exist and imply contingently. These notions he illustrates by the following examples (*1713*, 217–18).[3]

---

3. We avail ourselves here, and elsewhere in this section, of the excellent translations in *Shafer 1978*. However, Shafer uses 'to prove' as a translation for both Bernoulli's *indicare* and *probare*. We believe 'to imply' more suitable for most instances of *indicare* and accordingly have made the substitution in our transcriptions. (This is not trivial

My brother has not written me for a long time; I am not sure whether his indolence or his business is to blame; also I fear he might in fact have died. Here there are three arguments concerning the interrupted writing: indolence, death, and business. The first of these exists necessarily (by a hypothetical necessity, since I know and assume my brother to be lazy), but implies contingently, for it might have happened that this indolence did not keep him from writing. The second exists contingently (for my brother may still be among the living), but implies necessarily, since a dead man cannot write. The third both exists contingently and implies contingently, for he might or might not have business, and if he has any it may not be so great as to keep him from writing. Another example: I consider a gambler who, by the rules of a game, would win a prize if he threw a seven with two dice, and I wish to conjecture what hope he has of so winning. Here the argument for his winning is a throw of the seven, which implies it necessarily (by a necessity from the agreement entered into by the players) but exists only contingently, since other numbers of points can occur besides seven.

The term 'argument' has a number of meanings, two of which are closely related: the word is used (i) as a substantive representing a statement (a premise) which, if true, serves to establish or justify another statement (the conclusion), and also (ii) a notion involving two statements when there is a deduction (argumentation) from one to the other. The items which in the preceding quotation Bernoulli refers as arguments ('indolence, death, and business', and 'throw of a seven') show that he is there using the term in the first sense since he refers to arguments as implying [the conclusion] necessarily, or contingently; and when he says an argument 'exists' (note, separate from 'and implies') we take it to mean that the argument-as-premise is true. But in his description of the three types of argument both components, argument-as-premise and argumentation, are involved. (However Bernoulli doesn't concern himself with deductions but with the implication relation between premise and conclusion.) Using A for the premise and C for the conclusion we may characterize Bernoulli's three

---

as it makes for a major difference between Shafer's and our interpretation of Bernoulli's ideas.) For want of a better word we have kept Shafer's 'to exist' for Bernoulli's *existere* although the Latin has an active sense (to arise, come forth, appear). Likewise we use Shafer's 'thing' for *res* but with the meaning not of 'an object' but in the sense of 'the matter, affair or circumstance'. The page numbers of Bernoulli's *1713* are keyed in *Bernoulli 1975*.

types of arguments as:

> Type 1. A necessary, A-implies-C contingent
> Type 2. A contingent, A-implies-C necessary,
> Type 3. A contingent, A-implies-C contingent.

Note that Bernoulli is allowing contingent argumentations as well as contingent argument-premises. We shall later on encounter writers who, unaware of Bernoulli's analysis, restricted themselves to necessary argumentations, i.e., to Bernoulli's Type 2. This would exclude, for example, argumentations with insufficient premises.

Probability, Bernoulli's synonym for degree of certainty, is brought into the discussion (1713, 218–19):

> It is clear from what has been said thus far that the force of proof by which any given argument avails depends on the large number of cases whereby it can exist or not exist, imply or not imply, or even imply the contrary of the thing. Indeed, the degree of certainty or probability which the argument generates can be computed from these cases by the doctrine of the first part [of this book], just as the fates of gamblers in games of chance are usually investigated. In order to show this, we assume b is the number of cases where a given argument exists, c is the number where it does not exist, and $a = b+c$ is the number of both together. Similarly, we assume β is the number of cases where it implies, γ is the number where it does not imply or else implies the contrary of the thing and $\alpha = \beta + \gamma$ is the number of both together. Moreover, I suppose that all the cases are equally possible, or can happen with equal ease. Otherwise, discretion must be applied and in the place of any case that happens more easily than the others one must count as many cases as it happens more easily. For example, in place of a case that happens three times more easily than the others, I count three cases which can happen equally as easily as the others.

We see that Bernoulli is going to provide a numerical measure for what he refers to as the 'force of proof by which any given argument avails' or as the 'degree of certainty or probability which the arguments generates', and that he will do this by use of the doctrine of the first part. This doctrine is Huygens' theory, which has as its basic concept the value of a gamble in games of chance. In particular he will be using the following result (1713, 7):

> PROPOSITION III. If the number of cases which result in my getting g is p and, moreover if the number which results in my getting

l is q; (then) taking all cases to be of equal proclivity, my expectation is worth

$$\frac{gp + lq}{p + q}.$$ (1)

When rewritten in the form

$$g \cdot \frac{p}{p + q} + l \cdot \frac{q}{p + q}$$

it would appear—from the contemporary point of view—that (1) is the expected value of a random variable whose two values are g and l with respective associated probabilities $p/(p + q)$ and $q/(p + q)$. Bernoulli will be using (1) for the case of $g = 1$ and $l = 0$. Calling those cases for which $g = 1$ 'favorable' and those for which $l = 0$ 'unfavorable', then $p/(p + q)$ is the ratio of the number of favorable cases to the sum of the number of favorable and unfavorable cases. Thus the expectation which Bernoulli computes would be a probability in the 'classical' definition sense—but only if the cases are not only of 'equal proclivity' but also are mutually exclusive and exhaustive.

Bernoulli's '[argument] implies [conclusion]' we shall abbreviate to '$A \rightarrow C$', not thereby necessarily attributing to him the use of 'implies' in a truth-functional sense. When Bernoulli refers to an argument as being necessary we shall take it to mean that $A \rightarrow C$ is true in all cases. Although it will not be clear until we see how he computes it, what Bernoulli uses as his measure of the 'degree of certainty or probability generated by an argument' evidently depends on the cases of A and of $A \rightarrow C$.

In addition to the tripartite classification of types of argument mentioned earlier, Bernoulli also distinguishes arguments as to being pure or mixed (*1713*, 218):

> I call those arguments *pure* which prove [*probant*] a thing in certain cases in such a way that they prove nothing positively in other cases; I call those *mixed* which prove the thing in some cases in such a way that they prove the contrary in the remaining cases.

Here he is talking about an argument which proves rather than, as has been the situation up to now, of implying a thing in a case. We take this to mean that both A and $A \rightarrow C$ are true for the case. He elucidates the notions of pure and mixed by an example (*1713*, 218):

> A certain man has been stabbed with a sword in the midst of a rowdy mob, and it is established by the testimony of trustworthy men who were standing at a distance that the crime was committed by a

man in a black cloak. If it is found that Gracchus and three others in the crowd were wearing tunics of that color, this tunic is something of an argument that the murder was committed by Gracchus, but is mixed: for in one case it proves his guilt, in three cases his innocence, according to whether the murder was perpetrated by himself or by one of the remaining three; for it is not possible that one of these perpetrated it without Gracchus being thereby supposed innocent. But if indeed in a subsequent hearing Gracchus paled, this pallor of face is a pure argument; for it proves Gracchus' guilt if it arises from a guilty conscience, but does not, on the other hand, prove his innocence if it arises otherwise; for it could be that Gracchus pales from a different cause yet is still the murderer.

The distinction may be illustrated by the following diagrams in which inclusion of regions corresponds to implication (Figures 1.21 and 1.22). In this situation 'case' refers only to A since in Bernoulli's example the $A \to C$ part is necessary, i.e., true in any case. In Figure 1.21 A is resolvable into four mutually exclusive cases, the argument $(A, A \to C)$ proving C in case 1 and proving $\neg C$ in cases 2, 3, 4. In Figure 1.22 for the argument $(B, B \to C)$ there are two cases; when the first case holds the argument $(B, B \to C)$ proves C, but in the second case the argument proves 'nothing'. This latter case corresponds, in Bernoulli's example, to Gracchus's pallor, which is compatible with his being guilty or not guilty.

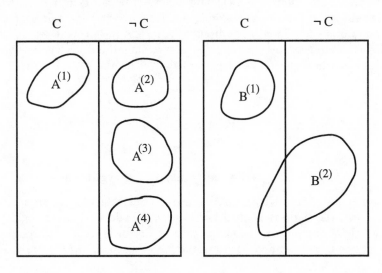

FIGURE 1.21: MIXED   FIGURE 1.22: PURE

Exactly what are Bernoulli's cases? In parts I and III of *Ars conjectandi*, which deals with expectations in connection with games of chance, a case corresponds to a possible (chance) outcome. These can all be explicitly listed and there is one, and only one, outcome in a given case. But when Bernoulli carries the notion over from games of chance to propositions we are no longer sure of what this notion means. There are three aspects in which this new usage differs from that in games of chance:

1. The element of chance (as an outcome in a dice roll, or in a blind drawing from an urn) is absent. But this is not a material difference since Bernoulli speaks of cases which are "equally possible or can happen with equal ease".

2. It isn't clear that there is only one set of possible cases associated with a given argument $(A, A \rightarrow C)$ as there would be for a given game. Bernoulli allows for a set of cases for A (a in number) and another set of cases for $A \rightarrow C$ (α in number). That the cases in each set are mutually exclusive is tacitly assumed, but nothing is indicated concerning their being exhaustive of all possibilities. Moreover, as we shall see, he tacitly assumes that the cases of the two sets are independent of each other.

3. For a pure argument a case for $A \rightarrow C$ need not determine a truth value for $A \rightarrow C$ (e.g., $B^{(2)}$ in Figure 2). In a game of chance, however, if one outcome is the case no other is, i.e., all outcomes are uniquely determined as to happening or not.

Despite these differences Bernoulli believes the theory of part I (which "determines the fates of gamblers") applies to computing the degrees of certainly of an argument. We discuss his results in our next section and, unlike *Hacking 1974* and *Shafer 1978*, offer explanations of what Bernoulli's results mean in terms of standard probability theory.

§1.3. Degree of certainty of an argument

Bernoulli's computations for the degree of certainty of an argument is apparently the earliest application of probabilistic notions to logic. He carries these computations out for each of the three types of arguments, and for the pure and mixed kind in the following manner (*1713*, 218-19):

1. So first let the argument *exist contingently and imply necessarily*. By what has just been said, there will be b cases where the argument exists and thus proves the thing (or 1), and c cases where it

does not exist and thus proves nothing. By Corollary I of Proposition III of part I, this is worth

$$\frac{b \cdot 1 + c \cdot 0}{a} = \frac{b}{a},$$

so that such an argument establishes b/a of the thing, or of the certainty of the thing.

2. Next let the argument *exist necessarily and imply contingently*. By hypothesis, there will be β cases where it implies the thing, and γ cases where it does not imply or implies the contrary; this now gives a force of argument for proving the thing of

$$\frac{\beta \cdot 1 + \gamma \cdot 0}{\alpha} = \frac{\beta}{\alpha}.$$

Therefore an argument of this kind establishes β/α of the thing; and moreover, if it is mixed it establishes (as is clear in the same way)

$$\frac{\gamma \cdot 1 + \beta \cdot 0}{\alpha} = \frac{\gamma}{\alpha}$$

of the certainty of the contrary.

3. If some argument *exists contingently and implies contingently*, I suppose first that it exists, in which case it is judged in the manner just shown to prove β/α of the thing and moreover, if it is mixed, γ/α of the contrary. Hence, since there are b cases where it exists and c cases where it does not exist and hence cannot prove anything, this argument is worth

$$\frac{b \cdot \frac{\beta}{\alpha} + c \cdot 0}{a} = \frac{b\beta}{a\alpha}$$

for proving the thing, and if it is mixed, is worth

$$\frac{b \cdot \frac{\gamma}{\alpha} + c \cdot 0}{a} = \frac{b\gamma}{a\alpha}$$

for proving the contrary.

In items 1. and 2. Bernoulli is computing the value of a gamble where the pay-off is 1 (certainty) if the "thing is proved" and 0 otherwise. He uses the corollary to Huygens' rule, cited above, in which $g = 1$ and $l = 0$. In item 3. the b cases for which the argument exists are valued at β/α rather that 1, and the remainder at 0. Despite his numerous references to the "degree of certainty of the thing" it is important to note that what Bernoulli computes is not the probability of the conclusion but the "force of proof of the argument", i.e., the probability that the conclusion is proved by the argument.

To see this we compare his results with a modern treatment of the problem, namely computing the expected value of an appropriate random variable with a given probability distribution. As such a random variable we choose X, where

$$X = \begin{cases} 1, & \text{if } A \text{ proves } C \\ 0, & \text{otherwise.} \end{cases}$$

Its expected value is

$$\begin{aligned} E(X) &= 1 \cdot P(X = 1) + 0 \cdot P(X = 0) \\ &= P(X = 1) \\ &= P(A \text{ proves } C). \end{aligned}$$

Assume, first of all, that the argument under consideration is of the mixed kind (Fig. 1.21). Then 'A proves C' is equivalent to 'A holds and A implies C', i.e., to the conjunction $A(A \rightarrow C)$, supposing '$\rightarrow$' to be the usual truth-functional conditional. Thus

$$E(X) = P(A(A \rightarrow C)). \tag{1}$$

Now for the probability distribution. As Bernoulli is assuming a set of equi-possible cases which determine when the argument exists (A holds), and also a set of equi-possible cases which determine when it implies the thing, one can have probabilities defined for each of these events using the 'classical' definition. Bernoulli's assumptions then allow us to write

$$P_H(A) = b/a, \qquad P_I(A \rightarrow C) = \beta/\alpha, \tag{2}$$

where P_H and P_I are the respective probability functions connected with the cases for A and for $A \rightarrow C$. But for $P(A(A \rightarrow C))$ to have a meaning we need to have a conjoint probability distribution for A and $A \rightarrow C$. This is something not considered by Bernoulli and, in effect, he assumes that it is the product distribution, i.e., that A and $A \rightarrow C$ are stochastically independent, with P_H and P_I the marginal probabilities.[4] Granted this independence, so that

$$P(A(A \rightarrow C)) = P_H(A)P_I(A \rightarrow C),$$

<hr/>

4. In §5.3 below we show, in the comment following Theorem 5.34, that A and $A \rightarrow C$ (with $\rightarrow$ taken as the 2-valued conditional) are stochastically independent if and only if either $P(A) = 1$ or $P(A \rightarrow C) = 1$, i.e., if and only if one or the other is necessary.

and noting that $b = a$ if A is necessary, and that $\beta = \alpha$ if $A \to C$ is necessary, the results Bernoulli obtains for the degree of certainty [of proving] the thing for the three types of arguments 1.–3. then follow (for the mixed kind). As for proving the contrary (end of item 2.), since it is being assumed that the argument is of the mixed kind, $A \to \neg C$ is equivalent to $\neg(A \to C)$. Hence (assuming independence)

$$
\begin{aligned}
P(A \text{ proves } \neg C) &= P(A(A \to \neg C)) \\
&= P(A\neg(A \to C) \\
&= P(A)P(\neg(A \to C)) \\
&= \frac{b}{a}(1 - \frac{\beta}{\alpha}) = \frac{b\gamma}{a\alpha}.
\end{aligned}
$$

We next consider the argument to be of the pure kind (Fig. 1.22). Here 'A proves C' has a new meaning occasioned by the occurrence of indecisive cases (overlapping both C and $\neg C$ region in Fig. 1.22). These cases "prove nothing", hence the β count in β/α treats such cases as though they implied the contrary. Let '$A \overset{*}{\to} C$' designate the implication resulting when the indecisive cases are made into ones implying the contrary. Then, assuming independence of A and $A \overset{*}{\to} C$,

$$
\begin{aligned}
P(A \text{ proves } C) &= P(A(A \overset{*}{\to} C)) \\
&= P(A)P(A \overset{*}{\to} C).
\end{aligned}
$$

And since Bernoulli uses β/α for $P(A \overset{*}{\to} C)$ as well as for $P(A \to C)$ the same expression, namely, $b\beta/a\alpha$, results for the pure as for the mixed argument kind. Nevertheless there is a difference. For

$$
P(A \text{ proves } C) = \begin{cases} P(A(A \to C)) = P(AC), & \text{if mixed} \\ P(A(A \overset{*}{\to} C)) \leq P(AC), & \text{if pure.} \end{cases}
$$

Shafer (1978) argues for a different interpretation of Bernoulli's probabilistic analysis of contingent arguments. His contention is that Bernoulli has an epistemic notion of probability which is non-additive (i.e., probabilities for and against need not add up to 1). Admittedly there are difficulties in understanding Bernoulli's ideas, but in view of his evident reliance on games of chance and gambling as his paradigm and the absence of any indication of a conceptual change, we find Shafer's position not convincing. Additionally, as Shafer acknowledges, his explanation encounters difficulties in accounting for some aspects of Bernoulli's treatment (e.g., the matters discussed in *Shafer 1978*, 332 and 334). As we have just seen, Bernoulli gives the value $b\gamma/a\alpha$ for an argument proving the contrary of the thing in the mixed type case. Concerning this Shafer says (*1978*, 332):

> So in the case of a mixed type argument which exists only contingently, we do indeed have a positive probability p for a thing and a positive probability q for its contrary such that $p + q < 1$.
> For
>
> $$\frac{b\beta}{a\alpha} + \frac{b\gamma}{a\alpha} = \frac{b}{a},$$
>
> which is less than 1 if the argument really exists only contingently.

Shafer is considering Bernoulli's worth of an argument for, and against, a conclusion as being complementary probabilities which in standard probability theory should add up to 1. Since they don't he concludes that nonadditive probability must be involved. But from our point of view Bernoulli is computing expected values of proving the thing, and its contrary—in the one case this is

$$P(A \text{ proves } C) = P(A(A \rightarrow C)) = P(AC)$$

and in the other

$$P(A \text{ proves } \neg C) = P(A(A \rightarrow \neg C)) = P(A\neg C).$$

The sum of these is $P(A)$, the b/a which Shafer computes, and there is no special need to account for its value being less than 1.

§1.4. On combining two or more arguments

After obtaining the results just discussed (i.e., items 1., 2., and 3.) Bernoulli then goes on to the question of computing the force of proof [for a conclusion] when there are two or more arguments for the same thing. In this connection he has the table:

arguments	1st	2nd	3rd	4th	5th	etc.
total number of cases	a	d	g	p	s	etc.
proving......................	b	e	h	q	t	etc.
non-proving or else proving the contrary.................	c	f	i	r	u	etc.

Considering first pure arguments, for which the last line in the table refers to the number of non-proving cases, Bernoulli obtains for the force of proof in combining the first and second arguments,

$$\frac{e \cdot 1 + f \cdot \frac{b}{a}}{d} \quad (= \frac{e}{d} + \frac{f}{d} \cdot \frac{b}{a}). \tag{1}$$

His reason is that for the second argument there are e of the d cases that prove the thing (with weight value 1) and of the remaining f cases there is a weight of b/a of proving the thing by virtue of the first argument.

Introducing 'A proves C' and 'B proves C' for the two arguments, and P_I and P_{II} as the (classical) probability notions pertinent to the two sets of cases, we can write in place of (1)

$$P_{II}(B \text{ proves } C) + P_{II}(B \text{ does not prove } C) \cdot (P_I(A \text{ proves } C). \tag{2}$$

This can also be written as

$$P((B \text{ proves } C) \text{ or } (A \text{ proves } C)), \tag{3}$$

where P is the product probability function of which P_I and P_{II} are its *independent* marginals. In obtaining (1) Bernoulli implicitly assumes independence of the cases, that is that there are no inter-relations. Moreover, the arguments being pure, 'proves' has a special meaning, namely 'A proves C' means '$A(A \overset{*}{\to} C)$'. Note that elements of the product probability space, which are ordered pairs of cases one from each of the two sets of cases, are to be considered equally possible. Moreover, to get Bernoulli's result one needs to consider a pair having at least one proving case to be a proving case for the combination. Without a proper clarification of 'cases' Bernoulli's result, i.e., (1), has dubious reliability.

A different approach is needed for the combining of mixed type arguments, in which case the last line in Bernoulli's table refers to the number of cases proving the contrary. Here he says (*1713*, 220–21):

> 5. Next let all the arguments be mixed. Since the number of proving cases in the first argument is b, in the second e, in the third h, etc., and the number proving the contrary, c, f, i, etc., the probability of the thing to the probability of the contrary is as b is to c on the strength of the first argument alone, as e is to f on the strength of the second alone, and as h is to i on the strength of the third alone, etc. Hence it is evident enough that the total force of proof resulting from the assemblage of all the arguments should be composed of the forces of all the arguments taken singly, i.e.,, that the probability of the thing to the probability of its contrary should be in the ratio of

$beh\dots$ to $cfi\dots$.Hence the absolute [i.e., not relative] probability of
the thing is $\frac{beh}{beh+cfi}$, and the absolute probability of the contrary is
$\frac{cfi}{beh+cfi}$.

A present day argument for this result would run somewhat as follows.
For a mixed type argument if in any given case the argument does not
prove then the contrary is proved. Hence for multiple mixed type arguments for the same thing all arguments (in any given case) must prove,
or all arguments must prove the contrary—otherwise the case leads to an
impossibility. The number of cases in which all prove is (for three arguments) beh, and in which all prove the contrary is cfi. (Note the tacit use
of independence.) Hence Bernoulli's result

$$\frac{beh}{beh+cfi}, \quad \frac{cfi}{beh+cfi}$$

for the probabilities for, and against, the thing being proved. These are,
of course, conditional probabilities, the conditions being that all three arguments prove the thing or all three prove the contrary. In terms of the
product probability space for the three sets of cases, only those ordered
triples are counted in which all three cases prove, or all three cases prove
the contrary, since the other triples represent impossible situations. Next
Bernoulli treats the situation when there are both pure and mixed arguments for a conclusion. He supposes, as an example, that there are five
arguments of which the first three are pure and the last two are mixed.
From his result on combining pure arguments he had that the first three
(pure) arguments provide $(adg - cfi)/adg$ of a certainty of the thing, so
that only cfi/adg of the certainty remains, there being $adg-cfi$ cases that
prove and cfi cases that do not prove. The two mixed arguments (taken by
themselves) provide a weight of $qt/(qt+ru)$, hence resulting in an expected
value of

$$\frac{(adg - cfi) \cdot 1 + cfi \cdot \frac{qt}{qt+ru}}{adg}$$

that is

$$\frac{adg - cfi}{adg} + \frac{cfi}{adg} \cdot \frac{qt}{qt + ru}, \tag{4}$$

which is the probability of the three pure arguments proving or, if not, then
the two mixed proving. In addition to Bernoulli's usual tacit assumption of
(stochastic) independence of the individual arguments of each kind proving,
there is also the tacit use of independence of the pure arguments from the
mixed ones.

A half century later Lambert (*1764*, 402) declared that this result of
Bernoulli's (i.e., (4)) must be erroneous since if $q = 0$, or $t = 0$, (no

case of a mixed argument proving, i.e., all cases proving the contrary) the formula doesn't reduce to 0, as it ought to, but to $1 - cfi/adg$. R. Haussner, translator of *Ars conjectandi* into German, comes to Bernoulli's defense. He remarks, in an end-note to this passage (*Bernoulli 1899*, 153) that Lambert's objection is not admissible for the following reason. If $q = 0$, or $t = 0$, i.e., if the contrary is absolutely certain by virtue of a mixed argument, then the pure arguments can't change anything—are made powerless (*entkräftet*) by the mixed arguments—and therefore should not be taken into consideration. Accordingly it is the formula for combining mixed arguments alone which should be used, and this gives the value 0; when using (4) the values $q = 0$ and $t = 0$ must, according to Haussner, be excluded.

To discuss the issue we point out that when combining pure arguments (for the same thing) no impossible combinations of cases can arise since all cases (of any of the arguments) are either proving or non-proving, i.e., there are no cases proving the contrary. But impossible combinations do arise when combining mixed arguments. As we have seen, Bernoulli takes care of this by using, in effect, conditional probabilities. However, his (4) is defective in not taking into account the possibility of impossible combinations between cases for pure and mixed. To illustrate the matter we use a simple form of Lambert's example in which there is one argument of each kind, as shown in Figure 1.41.

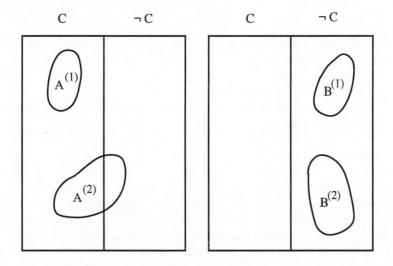

FIGURE 1.41

For this situation, with P_I and P_{II} the probability functions defined for the two (separate) arguments, Bernoulli's result gives

$$P_I(A \text{ proves } C) + P_I(A \text{ does not prove } C) \cdot P_{II}(B \text{ proves } C)$$

$$= \frac{1}{2} + \frac{1}{2} \cdot 0 = \frac{1}{2}. \tag{5}$$

But in order to correctly assess the chances of both proving we have to use a probability space allowing for all possible combinations of the two sets of cases. Here there are four such combinations, namely

	$B^{(1)}$	$B^{(2)}$
$A^{(1)}$	$(1,0)$	$(1,0)$
$A^{(2)}$	$(?,0)$	$(?,0)$

where in place of having an ordered pair $(A^{(i)}, B^{(j)})$ we use the character '1' to indicate a proving case (based on the one argument), '0' to indicate a proving-the-contrary case, and '?' to indicate a non-proving case. For '(1, 0)' we associate no probability since it represents an impossibility; for (?, 0) we associate, in keeping with Haussner's view, the probability 0. Thus we have the probability distribution

	$B^{(1)}$	$B^{(2)}$
$A^{(1)}$	$0/0$	$0/0$
$A^{(2)}$	0	0

where '0/0' means not a possible occurrence. Letting P be its probability function we compute

$$P((A \text{ proves } C) \text{ or } (B \text{ proves } C))$$

$$= P(A \text{ proves } C) + P(A \text{ does not prove } C) \cdot P(B \text{ proves } C)$$

$$= \frac{0}{2} + \frac{2}{2} \cdot \frac{0}{2}$$

$$= 0,$$

in agreement with Lambert and Haussner's interpretation. Bernoulli's result would be correct—as Haussner intimates—if there were no impossible combinations and if, also, the marginal probabilities were independent. This latter condition is not mentioned by either Bernoulli or Haussner.

We would be remiss in our account of Bernoulli's ideas on combining arguments if we left the impression that he was naive about the need for considering the inter-relationships of arguments. At the end of his chapter 3 he cautions against carelessly applying the rules for combining arguments without heed to the nature of the arguments. Some arguments, he says, which appear to be different may really not be so; vice versa, arguments appearing to be different may be the same; or arguments may be such as to make the contrary impossible, and so forth. He illustrates these matters with story-like examples. Such informal cautions are needed since Bernoulli's formal theory for combining arguments does not include a representation of the logical structure of the sentences involved in premises and conclusions, and hence no way of indicating inter-relationships.

Bernoulli also discusses having pure arguments on both sides of a question, i.e., for C and for $\neg C$, remarking that the resulting probabilities on combination could for each side considerably exceed $\frac{1}{2}$. This astonishes Shafer who says (*1978*, 336): "Not only does Bernoulli allow the probability of a thing and its contrary to add to less than one, he also allows it to add to more than one!". Bernoulli's actual words are (*1713*, 210):

> Here it should be noted that if the arguments adduced on each side are strong enough, it may happen that the absolute probability of each side significantly exceeds half of certainty, i.e., that both of the contraries are rendered probable, though relatively speaking one is less probable than the other. So it is possible that one thing should have $\frac{2}{3}$ of certainty while its contrary will have $\frac{3}{4}$; in this way both contraries will be probable, yet the first less probable than its contrary, in the ratio $\frac{2}{3}$ to $\frac{3}{4}$, or 8 to 9.

It would appear from this quotation that Bernoulli is content with a situation in which 'super-additive' probabilities arise. However at the end of his chapter he presents some examples (artificially constructed) to illustrate the need for care in applying the rules. One of these contains super-additive probabilities. It is instructive to see what he says about it. We quote in full the 'story' and his discussion (*1713*, 222–23):

> With regard to a written contract doubts are raised as to whether the appended date is fraudulent (predated). An argument against this could be that the document bares the signature of a notary, i.e., an official who takes an oath of office, who would be unlikely to commit fraud since he couldn't do this without greatly endangering his honor and position; so that of 50 notaries scarcely one is to be found who would venture to be so base. And yet arguments for the affirmative [that it was fraudulent] could be that the notary has a bad reputation, that he could expect to profit greatly by fraud and, especially, that he had attested to what had no probability—as for example that someone had loaned another 10,000 gold coins at a time when by every estimation he could scarcely have had a 100 to his name. If we here consider, by itself, the argument from the office and position of

the attestor, we can estimate the probability of the authenticity of the document at $\frac{49}{50}$ of certainty. But when we evaluate the arguments for the contrary, we must grant that it could hardly be unfalsified and therefore that fraud was committed is morally certain, i.e., has $\frac{999}{1000}$ of certainty. But there is no need to conclude from this that the probability of authenticity to the probability of fraud is (by §7) in the ratio of $\frac{49}{50}$ to $\frac{999}{1000}$, i.e., that they are about equal [and both near 1]. Certainly when we suppose that the notary has a bad reputation we suppose also that he is not among the 49 righteous notaries who abhor fraud, but that he is the 50th whose conscience doesn't bother him when he is faithless in office. But then that argument which could otherwise prove the authenticity of the document loses all force and is valueless.

So Bernoulli doesn't consider super-additivity as acceptable: it arises when there is a failure to analyze data sufficiently. One shouldn't just count cases but also look at the contents of the arguments for possible inter-relationships. When they exist a different set of cases could be appropriate. Expressed in present-day terms the argument for the conclusion, and that for its contrary, were based on different conditions, i.e., what is involved are the conditional probabilities

$$P(C \mid H_1) \quad \text{and} \quad P(\neg C \mid H_2);$$

and if H_1 and H_2 can differ there is no reason why the two probabilities couldn't both be near 1. But with a common condition, e.g., $H_1 H_2$ (assuming $H_1 H_2$ consistent) then $P(C \mid H_1 H_2) + P(\neg C \mid H_1 H_2) = 1$. Moreover, it is hard to reconcile acceptance of non-additivity with Bernoulli's belief, based on the law of large numbers which he was so proud of, that probabilities could be obtained *a posteriori* from frequencies. For relative frequencies for and against an event necessarily add to 1. We think a conscious use of conditional probability in non-games-of-chance situations clarifies the matter.

As a side observation it is interesting to note the two different ways in which Bernoulli estimates his probabilities. For the honesty of the notary it is statistical: "... of 50 notaries scarcely one is to be found ..."; while for the dishonesty it is moral certainty, to which (in his chapter 1, part IV) he has assigned the ratio $\frac{999}{1000}$.

§1.5. J. H. Lambert—probabilistic syllogisms

The material to be discussed in this section comes from Lambert's *Neues Organon oder Gedanken über die Erforschung und Bezeichnung des Wahren und dessen Unterscheidung von Irrtum und Schein (1764)*. It is an extensive two-volume work of which only the fifth chapter in volume 2, entitled 'Von dem Wahrscheinlichen', will be of interest to us. And even then we shall be selective, extracting from it those portions deemed to be relevant to probability logic.

It is clear that Lambert was thoroughly familiar with chapter 3, part IV of Bernoulli's *Ars conjectandi*. For example in §239 of his *Neues Organon* we find him discussing Bernoulli's result on combining pure and mixed arguments and, as we mentioned in our preceding section, disputing its correctness. The connection between logic and probability begun by Bernoulli becomes more visible and strengthened in Lambert. Where Bernoulli refers to the probability of a thing (*res*), Lambert refers to the probability of a proposition (*Satz*). Moreover, probability is a notion in its own right, not derivative from expectation. The logical structure of simple propositions is symbolically rendered, the basic form considered being 'A is B' (e.g., Man is mortal). To such propositions Lambert assigns a 'natural' degree of probability, this being the number of A's which are B's, to the total number which are or are not, as they are encountered in experience without selection. In addition to the traditional syllogistic forms Lambert also introduces—long before De Morgan did—numerically quantified forms such as '$\frac{3}{4}A$ are B' meaning that three quarters of the A's are B's. But, in addition, such sentences are endowed with a probabilistic significance which we shall be discussing presently.

Preliminary to the first of Lambert's probabilistic results there is the following description of a necessary inference form (*1764*, vol. 2, 336):

The premises

> B is C, D, E, F, etc. [i.e., B is C, and B is D, and etc.]
>
> A is C, D, E, F, etc.

do not yield 'A is B' unless the predicates C, D, E, F, etc. are, singly or in combination [*zusammengenommen*], a characterizing attribute [*eigenes Merkmal*] of B. If there is one such then, calling it M, one has an identity 'B is M' or, by simple conversion, 'M is B'. The second premise gives 'A is M', and hence the conclusion 'A is B'. For each such M there is a valid argument establishing 'A is B'.

From this necessary type of inference a transition is made to a probabilistic one (*1764*, vol. 2, 338–39):

§169. If the predicates C, D, E, F, etc. are not characterizing attributes of B, then they apply to additional subjects [beyond B]. Accordingly if one takes into account, as considered in §154, *et seq.*, the enumeration of cases in which they apply to B and in which they do not apply to B, or finds on other grounds the ratio between the two, then the degree of probability that accrues from them [towards a proof] is calculable. We content ourselves with reducing this calculation to the theory of games of chance. One imagines as many piles of tickets as there are arguments. In each pile let this number of valid or marked tickets to the number of unmarked tickets be in exactly the same ratio as that of the cases in which the argument [associated with that pile] is valid to that in which it is invalid. One supposes then that Caius blindly takes a ticket from each pile; the question is, how probable is it that among these selected tickets there is no valid one? It would be as probable or as improbable that all the arguments found on behalf of the proposition do not prove it. The theory of games of chance specifies the following rule for this calculation. *One multiplies together the number of tickets in each pile, and likewise, one multiplies together the number of non-valid or unmarked tickets in each pile; then dividing the latter product by the former yields the degree of probability that the arguments do not prove. And if this degree, which is necessarily a fraction, is subtracted from 1, then the remainder is the degree of probability that the arguments prove.*

Lambert does not explicitly say what the arguments that he is working with are, but from the context we infer that they are arguments whose conclusion is 'A is B' and whose two premises are of the form 'M is B' and 'A is M', where M is a predicate formed by combination (i.e., conjunction) from one or more of the predicates C, D, E, F, etc. and both of 'B is C, D, E, F, etc.' and 'A is C, D, E, F, etc' are necessary. Then for any M as described, 'A is M' is necessary but 'M is B' is only contingent (has a "degree of probability"). If there are n of the predicates C, D, E, F, etc. then there are $2^n - 1$ ways of forming a combination of one or more of the predicates. Not all of these lead to independent arguments—for example, if 'M_a is M_b' then 'M_b is B' implies 'M_a is B'. Lambert is aware of this and assumes that he has only independent arguments. Designating these combinations which lead to independent arguments as $M_1, \ldots, M_m$, we formulate his question as follows:

Given the independent arguments α_i, where α_i is the conjunction

$$(M_i \text{ is } B) \text{ and } (A \text{ is } M_i) \qquad (i = 1, \ldots, m) \qquad (1)$$

with the second component 'A is M_i' being necessary and the first component 'M_i is B' contingent with probability

$$p_i = P(\alpha_i) = P(M_i \text{ is } B),$$

what is the probability that 'A is B' is proved by the arguments?

As each α_i logically implies 'A is B' the statement 'α_i proves $(A$ is $B)$', i.e., '$\alpha_i(\alpha_i \rightarrow (A \text{ is } B))$' is equivalent to α_i. Thus to ask if an α_i proves 'A is B' is to ask if $\alpha_1 \vee \cdots \vee \alpha_m$ holds, and the probability of this is $P(\alpha_1 \vee \cdots \vee \alpha_m)$. The answer Lambert gives,

$$1 - (1 - p_1)(1 - p_2)\ldots(1 - p_m), \tag{2}$$

is indeed, by virtue of the assumed independence, equal to $P(\alpha_1 \vee \cdots \vee \alpha_m)$. (This is not the probability of 'A is B', for although $\alpha_1 \vee \cdots \vee \alpha_m$ implies 'A is B' it need not be equivalent to it—its probability is lower bound for that of 'A is B'.) Note that Lambert uses the metaphor of urns with lottery tickets to enable him to speak of the probability of compound propositions, since initially only the simple (universal) categorical proposition has a 'natural' probability assigned. To a present-day probabilist he is constructing an m-fold product space of the m independent arguments.

The inference form just discussed is contrasted with another type of probabilistic inference (*1764*, vol. 2, 355–56):

§185. On the other hand it is quite otherwise when the probability of a conclusion [*Schlußsatz*] is to be determined from the probability of the premises [*Vordersätze*]. For the premises cannot be viewed as arguments which are separate and independent of each other, since the conclusion depends on both conjointly; the conclusion then holds when all premises do. This being presupposed, calculation of the probability of the conclusion for an entire sorites [*Schlußkette*] can also equally well be reduced to the theory of games of chance. To this end we will again take piles of tickets and, indeed, as many as there are premises in the sorites. In each pile let the number of valid to nonvalid be in the same ratio as the cases in which the premises is true to those in which it is not. One then supposes that Caius blindly takes a ticket from each pile. The question is: How probable is that among the drawn tickets no nonvalid [one] occurs, or that all are valid? This is the degree of probability which the conclusion of the given sorites would have. To calculate this the theory of games gives the following rule: *One multiplies together the number of tickets in each pile; and likewise one multiplies together the number of valid tickets in each pile; the division of the latter by the former specifies the degree of probability of the conclusion.*

We note, first of all, that Lambert is considering only necessary inferences (and, of course, syllogistic ones since no others were recognized). When he refers to the (collection of) premises of the sorites we assume that he means with deletion of those which are consequences of others. The answer he gives for the probability of the conclusion, namely

$$P(\pi_1)(P(\pi_2)\dots P(\pi_m),$$

where $\pi_1,\dots,\pi_m$ are the premises in the sorites, is the probability of the conjunction of these premises only if one assumes their stochastic independence. Moreover $P(\pi_1\pi_2\dots\pi_m)$, or $P(\pi_1)P(\pi_2)\dots P(\pi_m)$ if one assumes independence, is not necessarily the probability of the conclusion since the conclusion is only implied by this conjunction, the value $P(\pi_1\pi_2\dots\pi_m)$ is just a lower bound.

It is interesting to note that Lambert believed that a contingent (probable) proposition could follow from non-contingent premises. He gives the example, for 'C ein Individuum' (*1764*, vol. 2, 359):

$$\tfrac{3}{4}\,A \text{ sind } B$$
$$C \text{ ist } A \qquad\qquad (3)$$
$$folglich \quad C\,\tfrac{3}{4}\text{ ist } B.$$

That is to say, translating his special symbolism, three quarters of the A's are B's, C is an A, therefore with probability $\tfrac{3}{4}$, C is a B. His justification for the inference form is the following (pp. 358–59):

§189. One has, then, the two propositions

$$\tfrac{3}{4}A \text{ are } B$$
$$C \text{ is } A,$$

and the question is: what kind of conclusion, since they have the common middle term A, can be drawn? We suppose that both propositions are true and definite; namely, that with regard to the upper one, one is assured that neither more nor less that $\tfrac{3}{4}$ of the A's have predicate B; and that with regard to the lower one C is an individual and that it is an A. If one knows no more than this it remains absolutely undetermined whether C is among the $\tfrac{3}{4}$ of the A's which are B's or among the $\tfrac{1}{4}A$'s which are not B's. Were this to be determined one could forthwith conclude whether C is a B or not, and there would be complete certainty. But as we are supposing that we know of C only that it is an A, we can determine no conclusion other than: it is more likely [*vermuthlicher*] that B applies to C than it doesn't. Inasmuch as among four A's there are always three which have predicate B, and since with regard to C no selection is allowed, it is then three times more likely that C is among the A's which are B's than among those which are not. Accordingly the conclusion that C is a B is not fully certain but deviates by $\frac{1}{4}$ from certainty, that is to say its probability is $\frac{3}{4}$.

To us this demonstration of Lambert involves a (tacit) assumption of a uniform probability distribution, namely the assumption that any $\frac{3}{4}$ of the A's is equally likely of being in B as any other. Then the 'classical' definition of probability applies and we may compute the probability that any preselected A (e.g., the individual C) is in B. A simple combinatorial calculation shows this to be $\frac{3}{4}$.

We contrast Lambert's 'syllogistic' inference form (3) with the following probabilistic one which we believe reproduces the essence of it:

$$\text{For all } x, \quad P(x \in B \mid x \in A) = \tfrac{3}{4}$$
$$P(C \in A) = 1$$
$$\text{therefore} \quad P(C \in B) = \tfrac{3}{4}.$$

This inference is valid since, on instantiating in the first premise the x to C, we have

$$\frac{P(C \in B \wedge C \in A)}{P(C \in A)} = \tfrac{3}{4},$$

from which the conclusion follows since $P(C \in A) = 1$ and hence $P(C \in B \wedge C \in A) = P(C \in B)$.

Lambert goes on to consider probabilistic inference for a syllogism when there is a limitation on the middle term in the minor premise (*1764*, vol. 2, 360–61):

§191. In all these cases [so far considered] the degree which probability [theory?] determines [for the conclusion] comes to the conclusion from the major premise. We will now invert the situation and show also how it stems from the minor premise. Let $MNPQ$ be the attributes of the concept B which fill up its extension [*die seinen Umfang ausfüllen*], it being left undetermined whether there is present a characterizing attribute of B. One then has the two propositions

$$MNPQ \text{ is } B$$
$$C \text{ is } MNP,$$

which again allows only a probable conclusion

$$C \text{ is } B,$$

since we are supposing it left undecided as to whether C also has the attribute Q. Here the degree of probability is assigned in proportion as the magnitude [*grösse*] and the number of attributes MPQ which one already has found in C is to the magnitude and number of those which are yet to be found. One sets, e.g.,

$$MNPQ = A$$
$$MNP = \tfrac{2}{3}A$$

and then has the inference

all A are B

C is $\tfrac{2}{3}A$

therefore $C\,\tfrac{2}{3}$ is B.

The inference still holds [not only for singular C but also] for C general, or particular, or with a definite degree of particularity.

   In our discussion of this passage we take the case of C an individual. In the first place we interpret his requirement, that the attributes M, N, P, Q fill up B's extension, to mean that the intersection of the extensions of M, N, P, Q include that of B. Hence if, as here, Lambert has '$A = MNPQ$' and 'All A are B', then A and B have the same extension. As for the minor premise 'C is $\tfrac{2}{3}A$', there are fewer attributes for $\tfrac{2}{3}A$ than for A (MNP versus $MNPQ$) so that $\tfrac{2}{3}A$ has a *larger extension* than A; let us designate this extension by '$(\tfrac{2}{3}A)_p$'. In addition to Lambert's second premise meaning '$C \in (\tfrac{2}{3}A)_p$' there is also the tacit assumption that an artitrarily chosen member of $(\tfrac{2}{3}A)_p$ has a $\tfrac{2}{3}$ chance of being an A, i.e., that

For all x, $P(x \in A \mid x \in (\tfrac{2}{3}A)_p) = \tfrac{2}{3}$.

Consequently, the premise 'C is $\tfrac{2}{3}A$' implies that $P(C \in A) = \tfrac{2}{3}$. This, together with the premise that B has the same extension as A, yields the conclusion $P(C \in B) = \tfrac{2}{3}$.

   From this inference form Lambert goes on to the more general form in which there is restriction on the middle term in both premises. He cites as valid

$\tfrac{3}{4}A$ are B

C is $\tfrac{2}{3}A$ (4)

therefore $C\,\tfrac{1}{2}$ is B.

He also cites as valid a form with major premise complementary to that of (4) (*1764*, vol. 2, 362):

§193. Replacing the major premise in this inference [i.e., (4)] by its negative, one obtains

$$\tfrac{1}{4}A \text{ are not } B$$
$$C \text{ is } \tfrac{2}{3}\, A \tag{5}$$
$$\text{therefore} \quad C\,\tfrac{1}{6} \text{ is not } B.$$

Hence the probability of the denial of the conclusion [C is B] is $\tfrac{1}{6}$, and on the other hand that it is affirmed is $\tfrac{1}{2}$. Both probabilities together give $\tfrac{1}{6} + \tfrac{1}{2} = \tfrac{2}{3}$, which is the probability of the minor premise.

Shafer (*1978*, 355 *et seq.*), finds this (and similar results) to be evidence for Lambert's adherence to a non-additive concept of probability. Lambert does in fact assert that the sum of the probabilities for a proposition and its denial add up, in his example, to less than 1. We contend, however, that the basis on which he makes this assertion is faulty. Specifically, his inference forms (4) and (5) are invalid, requiring in place of the predicate B in the conclusion the predicate $A \cap B$, respectively, $A \cap \overline{B}$. For if we take Lambert's major premise to mean

$$\text{For all } x, \; P(x \in B \mid x \in A) = \tfrac{3}{4}, \tag{6}$$

so that, instantiating x to C,

$$\frac{P(C \in B \wedge C \in A)}{P(C \in A)} = \tfrac{3}{4}, \tag{7}$$

and the minor premise to mean $P(C \in A) = \tfrac{2}{3}$, then

$$P(C \in B \wedge C \in A) = \tfrac{2}{3} \cdot \tfrac{3}{4} = \tfrac{1}{2}. \tag{8}$$

Similarly, for the complementary inference form (5)

$$P(c \notin B \wedge C \in A) = \tfrac{1}{4} \cdot \tfrac{1}{2} = \tfrac{1}{6}. \tag{9}$$

Adding (8) and (9) gives $P(C \in A)$ with the value $\tfrac{1}{2} + \tfrac{1}{6} = \tfrac{2}{3}$. This is the probability of the minor premise, and our explanation accounts for the fact that the two probabilities involved do add up to this value and not to 1.

Continuing his investigations of the probabilistic syllogism Lambert introduces a generalization in which individuals can be one of three kinds: those to which a given term applies, those to which it does not, and those to which it is undetermined as to whether it does or does not apply. He illustrates the kind of inference he has in mind with the following example:

$$(\tfrac{2}{3}a + \tfrac{1}{4}e + \tfrac{1}{12}u)A \text{ sind } B$$

$$C \text{ ist } (\tfrac{3}{5}a + \tfrac{2}{5}u)\, A \qquad\qquad (10)$$

$$\text{folglich,} \quad C\,(\tfrac{2}{5}a + \tfrac{3}{20}e + \tfrac{9}{20}u) \text{ ist } B.$$

The major premise here has the meaning that of the totality of individuals which are A's, $\tfrac{2}{3}$ are certainly B's, $\tfrac{1}{4}$ certainly are not, and for the remaining $\tfrac{1}{12}$ it is indefinite, or not determined, as to whether they are, or are not, B's. In the minor premise the coefficient $(\tfrac{3}{5}a + \tfrac{2}{5}u)$ represents the attributes of B divided into those which apply to C, weighted at $\tfrac{3}{5}$, and a remaining part, weighted at $\tfrac{2}{5}$, for which it is undetermined whether they do or do not. The conclusion of (10) says "of 20 unselected cases that appear with conclusion of this kind and degree, 8 affirm, 3 deny, and 9 remain undetermined; or rather, in a single case, there are 8 grounds to affirm the conclusion, 3 to deny it, and 9 to conclude nothing, or leave it uncertain."

As with the preceding inference forms, we present a reformulation in contemporary terms. Lambert's new feature, the inclusion of indeterminacies, will be replaced by inequalities. Analogous to our expressing '$\tfrac{3}{4}A$ are B' by 'For all x, $P(x \in B \mid x \in A) = \tfrac{3}{4}$' we shall express the major premise of (10) by

$$\text{For all } x, \quad \tfrac{2}{3} \leq P(x \in B \mid x \in A) \leq \tfrac{9}{12}, \qquad (11)$$

from which, using

$$P(x \in B \mid x \in A) = 1 - P(x \notin B \mid x \in A)$$

and simple algebra, we have

$$\text{For all } x, \quad \tfrac{1}{4} \leq P(x \notin B \mid x \in A) \leq \tfrac{1}{3}. \qquad (12)$$

Then for arbitrary x the value of $P(x \in B \mid x \in A)$ is at least $\tfrac{2}{3}$, and might be up to $\tfrac{1}{12}$ more, while $P(x \notin B \mid x \in A)$ is at least $\tfrac{1}{4}$, and might be up to $\tfrac{1}{12}$ more. Thus (11) encapsulates all the information in the major premise. As for the minor premise, just as we rendered 'C is $\tfrac{2}{3}A$' by '$P(C \in A) = \tfrac{2}{3}$' so for '$C$ is $(\tfrac{3}{5}a + \tfrac{2}{5}u)\,A$' we shall write

$$\tfrac{3}{5} \leq P(C \in A). \qquad (13)$$

To obtain the coefficients of a, e, and u in the conclusion of (10) Lambert 'multiplies' the two expressions $(\tfrac{2}{3}a + \tfrac{1}{4}e + \tfrac{1}{12}u)$ and $(\tfrac{3}{5}a + \tfrac{2}{5}u)$, and arranges the result in this manner:

$$\frac{2}{5}aa + \frac{3}{20}ae + \frac{3}{60}au$$
$$+ \frac{4}{15}au$$
$$+ \frac{2}{20}eu$$
$$+ \frac{2}{60}uu$$
$$\overline{\frac{2}{5}a \quad + \frac{3}{20}e \quad + \frac{9}{20}u}$$

We compare Lambert's conclusion in (10) with what we can obtain from our reformulated version. From (11) and (12) we have, for arbitrary C,

$$\frac{2}{3} \cdot P(C \in A) \leq P(C \in B \wedge C \in A)$$
$$\frac{1}{4} \cdot P(C \in A) \leq P(C \notin B \wedge C \in A)$$

so that, by use of (13), we obtain

$$\frac{2}{5} \leq P(C \in B \wedge C \in A)$$
$$\text{and} \quad \frac{3}{20} \leq P(C \notin B \wedge C \in A), \tag{14}$$

whereas Lambert's conclusion would, in our interpretation, be

$$\frac{2}{5} \leq P(C \in B)$$
$$\text{and} \quad \frac{3}{20} \leq P(C \notin B). \tag{15}$$

Since (14) implies (15), we see that Lambert's conclusion, though correct, is weaker than ours. On the other hand might not Lambert's (15), or its equivalent

$$\frac{2}{5} \leq P(C \in B) \leq \frac{17}{20},$$

be the strongest conclusion about $P(C \in B)$ which the premises warrant? Clearly Lambert hasn't addressed himself to this type of question. We shall in our chapter 5; in particular, the result of Theorem 5.48 provides the strongest conclusion one can draw about $P(A_1)$ if given only that

$$p_1 \leq P(A_1 \mid A_2) \leq p_2 \quad \text{and} \quad q_1 \leq P(A_2) \leq q_2.$$

In our preceding section we have mentioned Lambert's objection to Bernoulli's result on combining pure and mixed arguments. Here is a further comment.

Lambert says, first of all, that all of Bernoulli's types of arguments can be encompassed in the one form

$$\text{All } A\,(Ma + Nu + Pe) \text{ are } B$$

(a = affirming, u = indefinite, e = denying). For example, if $P = 0$ then we have the pure type and if $N = 0$ then the mixed type. Combining a pure and a mixed argument by Lambert's scheme requires forming the product

$$(Ma + Nu)(ma + pe)$$
$$= (Mm)aa + (Nm)au + (Np)ue + (Mp)ae,$$

concerning which Lambert says "We have, however (§237), entirely omitted the ae cases, as they are impossible, and this makes the product here given different from that of Bernoulli."

In concluding this section we mention that there is a discussion, from the viewpoint of inductive logic, of this Bernoulli-Lambert disagreement in *Hacking 1974* but with the epistemological notion of evidence in place of argument(-premise).

§1.6. Thomas Bayes—and a problem
"... no less important than curious"

The phrase quoted in our title comes from a letter of Richard Price to the Secretary of the Royal Society in which he transmits an essay "found among the papers of our deceased friend Mr. Bayes". The letter and essay were read to the Society December 23, 1763. Some eleven years later Laplace's memoir of 1774 treated essentially the same problem. It is believed that Laplace was unaware of Bayes' essay (*Stigler 1978*). Over the past two centuries a vast literature has been generated by the novel idea in Bayes' essay and Laplace's memoir. We shall restrict our attention to a small aspect of it relevant to our subject.

The problem in Bayes' essay which excited Price is the following (*Bayes 1763*, 376):

PROBLEM
Given the number of times in which an unknown event has happened and failed: *Required* the chance that the probability of its happening in a single trial lies somewhere between any two degrees of probability that can be named.

There are two striking features about this problem aside from its evident relevance to statistical inference:

   (i) it is concerned with the chance (= probability) *of a probability* lying between two given values, and

(ii) this chance to be determined supposing that the probability is that of an event of which *it is only known* that in $p+q$ trials it happened p times and failed q times.

Feature (i) involves the notion of a probability distribution of a probability value, a notion not within the scope of this monograph. The second feature concerns the probability of an event on given information, i.e., a conditional probability. Nothing is said about a (prior) probability distribution though, in effect, in the course of his solution Bayes attributes one to the $\frac{p}{p+q}$ values, namely that they are uniformly distributed over $[0,1]$ (see *Stigler 1986b*, 128). Before undertaking the solution of his PROBLEM Bayes develops from basic principles some results on conditional probability. One of these is an item for our historical account of probability logic.

Bayes seems to have been the first to make significant use of the multiplication rule (for the probability of the conjunction of two events) to gain information about the probability of one of the two happening when it is known that the other has happened, provided one also knows the probability of their conjunction. Here is his statement of the rule (*1763*, 378):

PROP. 3

The probability that two subsequent events will both happen is a ratio compounded of the probability of the 1st, and the probability of the 2d on supposition that the first happens.

Bayes doesn't say what he means by a 'subsequent' event. However at the beginning of the essay he explains "An event is said to be determined when it has either happened or failed." We surmise that Bayes is using 'subsequent' to imply that the event is not yet determined, i.e., that it is a contingent or chance event.[5]

Solving in his PROP. 3 for the quantity of interest, i.e., the conditional probability, Bayes then has the corollary

Hence if of two subsequent events the probability of the 1st be $\frac{a}{N}$, and the probability of both together $\frac{P}{N}$, then the probability the 2nd on supposition of the first happens is $\frac{P}{a}$.

We contrast this corollary with his

---

5. *Shafer 1982* has a different view of what Bayes means here by 'subsequent'. On the basis of this view he constructs a mathematical framework for probability which takes the timing of events into account. Such a framework, he believes, helps to explain Bayes' ideas. Shafer's paper includes a useful Appendix on the history of conditional probability (though it omits mention of C. S. Peirce's contribution. See our *1988*, 184–85).

PROP. 5

If there be two subsequent events, the probability of the 2nd being $\frac{b}{N}$ and the probability of both together $\frac{P}{N}$, and it being discovered that the second has happened, from hence I guess that the first has also happened, the probability that I am right is $\frac{P}{b}$.

This PROP. 5 appears to be a restatement of his corollary to PROP. 3, with the roles of 1st and 2nd interchanged and, as such, has puzzled some readers. (See *Dale 1991*, 33, for an account). What exactly Bayes has in mind isn't clear, but his restatement (if that is what it is) involves an epistemological element ("...it being discovered that...") which is absent from the corollary to PROP. 3. On the basis of this PROP. one is obtaining new (probability) knowledge about the happening of an event. In an explanatory footnote Price, a minute reader of the essay, remarks: "what is proved by Mr. Bayes in this and the preceding proposition [PROP. 4] is the same with the answer to the following question. What is the probability that a certain event, when it happens, will be accompanied with another to be determined at the same time?" In other words, in terms of the notion of conditional probability, not then yet codified, the question is What is $P(E_1 E_2 \mid E_2)$? When Bayes' proposition gives as the answer $P(E_1 E_2)/P(E_2)$ it implies that

$$P(E_1 E_2 \mid E_2) = P(E_1 \mid E_2). \tag{1}$$

Historically, (1) is the earliest result in probability logic involving conditional probability. (See Theorem 5.30(e) below.)

The solution of Bayes' PROBLEM—which we shall not go into—consists in finding $P(A \mid B)$ where B is 'an event E has, in $p + q$ independent trials, happened p times and failed q times' and A is '$a \leq P(E) \leq b$' for some $[a, b]$ in $[0, 1]$. In his demonstration Bayes assumes that, to put it in modern language, the probability distribution of $P(E)$ is a uniform one. For a detailed discussion see *Dale 1991*, chapter 2.

The first problem considered in Laplace's memoir *1774* is equivalent to Bayes' PROBLEM. Here is how Laplace puts it:[6]

> If an urn contains an infinity of white and black tickets in an un-known ratio, and we draw $p + q$ tickets from it, of which p are white and q are black, then we require the probability that when we draw a new ticket from the urn, it will be white.

(Having an "infinity" of tickets serves to insure that the ratio of white to black is unchanged as the tickets are drawn.)

---

6. Our English translation comes from *Stigler 1986*.

Whereas Bayes uses (in effect)

$$P(A \mid B) = \frac{P(AB)}{P(B)}$$

in his demonstration and in the course of it elaborates the right-hand side, Laplace does the elaboration initially, formulating it as a general principle:

> If an event can be produced by a number n of different causes, the probabilities of these causes given the event are to each other as the probabilities of the event given the causes, and the probability of the existence of each of these is equal to the probability of the event given that cause, divided by the sum of all the probabilities of the event given each of these causes.

(Laplace's use of the term 'cause' is no longer current.) In modern notation the principle is:

$$If \qquad E \to C_1 \vee C_2 \vee \cdots \vee C_n$$

then [if the $P(C_i)$ are all equal,] for $i, j = 1, \ldots, n$,

$$\frac{P(C_i \mid E)}{P(C_j \mid E)} = \frac{P(E \mid C_i)}{P(E \mid C_j)};$$

and [if the C_i are mutually exclusive] then for $j = 1, \ldots, n$,

$$P(C_j \mid E) \left[= \frac{P(C_j E)}{P(E)} \right] = \frac{P(E \mid C_j)}{\sum_{i=1}^{n} P(E \mid C_i)},$$

a special case of the Bayes' Rule found in just about every text on probability or statistics. Its derivation from basic probability principles requires only very simple logical properties, e.g.,

$$E \leftrightarrow EC_1 \vee \cdots \vee EC_n \quad \text{if} \quad E \to C_1 \vee \cdots \vee C_n.$$

Conditional probabilities will be the focus of our attention in chapter 5.

§1.7. John Michell on the distribution of the fixed stars

Our topic for this section is a probabilistic inference inconspicuously buried in an astronomical paper, *An Inquiry into the Probable Parallax and Magnitude of the Fixed Stars, from the Quantity of Light which they afford us, and the Particular Circumstance of their Situation* (*Michell 1767*). This paper, appearing in the Transactions of the Royal Society three years after Bayes' essay, is celebrated in the history of astronomy as having provided the first realistic estimate of the distance of the fixed stars, and for establishing the existence of physical (as opposed to optical) double stars by means of a theoretical argument involving probability. The latter argument (*not* its conclusion) was criticized at great length in *Forbes 1850*. Forbes' paper, though prominently appearing in the *Philosophical Magazine* and occasioning considerable discussion, was apparently unknown to the author of the biography of Michell in the *Dictionary of Scientific Biography*, who writes:

> ... The directness of Michell's language [describing his results] leaves something to be desired; but the unimpeachable logic of his arguments gave a convincing theoretical proof of the existence of physical binary stars in the sky long before Herschel (1803) provided a compelling observational proof.

Since Michell's conclusion about binary stars turned out to be true, it could be that the author of this biography was lulled into uncritical acceptance of the argument, for Forbes' criticisms are quite telling. To examine this argument we open with a quote from Michell (*1767*, 428):

> It has always been usual with astronomers to dispose the fixed stars into constellations: this has been done for the sake of remembering and distinguishing them, and therefore it has in general been done merely arbitrarily, and with this view only; nature herself however seems to have distinguished them into groups. What he [Michell] means is, that from the apparent situation, of the stars in the heavens, there is the highest probability, that either by the original act of the Creator, or in consequence of some general law, such perhaps as gravity, they are collected together in great numbers in some parts of space, while in others there are either few or none. The argument he [Michell] intends to make use of, in order to prove this, is of the kind which infers either design, or some general law, from a general analogy, and the greatness of the odds against things having been in the present situation, if it was not owing to some such cause.

On the assumption of a random [actually, uniform] distribution of the stars over the celestial sphere, each star being in any of the 13,131 sub-regions of 2° diameter with equal probability, Michell computes the odds that, of the 230 stars comparable in brightness to the double star β Capricorni, no two should fall within that angular distance. He finds the odds to be 80 to 1. When more stars are taken into account, e.g., the six brightest stars of the Pleiades, the odds against such a close grouping amount to 500,000 to 1. Since there are a large number of such groupings Michell's conclusion is that it is "next to certainty" that there is a cause for these groupings and that it is not a matter of chance. To outline the probabilistic element of Michell's argument in contemporary language, let S = present situation of the fixed stars, L = existence of some general law or original act of the Creator. Then since $P(S \mid \neg L) = 1 - P(\neg S \mid \neg L)$,

$$P(S \mid \neg L) \approx 0 \to P(\neg S \mid \neg L) \approx 1$$
$$\to P(L \mid S) \approx 1,$$

and since $P(S \mid \neg L) \approx 0$ (β Capricorni, the Pleiades, etc.), one has then the conclusion $P(L \mid S) \approx 1$. Note, in particular, the inference

$$P(\neg S \mid \neg L) \approx 1$$
$$\text{therefore} \quad P(L \mid S) \approx 1. \tag{1}$$

We shall only state Forbes' two principal objections:

(i) Michell takes the high improbability of an event's happening, when it is one of a great many possibilities, as that of the event when it is already the case. ("The improbability, for instance, of a given deal producing a given hand at whist is so immense, that were we to assume Mitchell's [sic] principle, we would be compelled to assign to it as the result of an active Cause with far more probability then even found by him for the physical connection of the six stars of the Pleiades.")

(ii) Michell's assumption of a uniform probability distribution for any star of a given magnitude and any subregion of the celestial sphere "leads to conclusions obviously at variance with the idea of random or lawless distribution, and is therefore not the expression of that Idea." Forbes likens it to assuming that any face of a die has an equal chance of coming up before one knows whether the die is loaded or not.

Boole's *1851a*, his first published paper on probability, was occasioned by the appearance of Forbes' paper. It presents a quite different objection to the Michell argument (Boole *1851a = 1952*, 249–50):

> The proper statement of Mr. Mitchell's problem, as relates to β Capricorni, would therefore, be the following: (1.) Upon the hypothesis that a given number of stars have been distributed over the

heavens according to a law or manner whose consequences we should
be altogether unable to foretell, what is the probability that such a
star as β Capricorni would nowhere be found?   (2.) Such a star as
β Capricorni having been found, what is the probability that the law
or manner of distribution was not one whose consequences we should
be altogether unable to foretell?   The first of the above questions
certainly admits of a perfectly definite numerical answer. [Forbes
denied that this was possible unless one had a specific probability
distribution. Michell, in effect, does assume one.] Let the value of
the probability in question be p. It has then generally been maintained that the answer to the second question is also p, and against
this view Prof. Forbes justly contends. [Forbes, who is not explicit
on this point, is being given more credit than is warranted.]

Boole then goes over to an abstract formulation:

Let us state Mr. Mitchell's problem, as we may now do, in the
following manner: There is a calculated probability p in favour of the
truth in a particular instance of the proposition. If a condition A has
prevailed, a consequence B has not occurred. Required the similar
probability for the proposition, if a consequence B has occurred, the
condition A has not prevailed.   Now, the two propositions are logically connected. The one is the "negative conversion" of the other;
and hence, it either is *true* universally, the other is so. It seems hence
to have been inferred, that if there is a probability p in a special instance in favour of the former, there is the same probability p in a
special instance in favour of the latter. But this inference would be
quite erroneous. It would be an error of the same kind as to assert
that whatever probability there is that a stone arbitrarily selected is
a mineral, there is the same probability that a non-mineral arbitrarily
selected is a non-stone. But that these probabilities are different will
be evident from their fractional expressions, which are—

1. $\dfrac{\text{Number of stones which are minerals}}{\text{Number of stones}}$

2. $\dfrac{\text{Number of non-minerals which are not stones}}{\text{Number of non-minerals}}$

It is true that if either of these fractions rises to 1, the other does
also; but otherwise, they will, in general, differ in value.

Note that Boole seems to think that conditional probabilities involve
conditional (i.e., if—, then—) sentences since he says "the two propositions

are logically connected", one being the "negative conversion" of the other. Nevertheless, as his example shows, he does render the probabilities as conditional probabilities, not as probabilities of conditional sentences. Of special interest is his pointing out that the inference

$$P(\neg B \mid A) = p$$
$$\text{therefore} \quad P(\neg A \mid B) = p, \tag{2}$$

is not valid, except when $p = 1$. Boole either didn't realize, or didn't make clear that it is only this special (valid) case which Michell's argument needs.[7] To see this note that Michell claims, in effect, that for a large number of S_i,

$$P(S_i \mid \neg L) = p_i,$$

but doesn't use 'negative conversion' on these separately but first obtains for a conjunction $S = S_1 S_2 \ldots$ of many of the S_i that, independence of the S_i being (tacitly) assumed,

$$P(S \mid \neg L) = p_1 p_2 \ldots, \tag{3}$$

Since $p_1 p_2 \cdots \approx 0$,

$$P(\neg S \mid \neg L) \approx 1,$$

and now, since $p \approx 1$, 'negative conversion' (i.e., (2)) can be used to obtain

$$P(L \mid S) \approx 1.$$

Clearly it would be worthwhile having a general method such than one could obtain relationships between probabilities such as $P(\neg B \mid A)$ and $P(\neg A \mid B)$, i.e., between probabilities whose arguments are logically related. Boole asserts (in *1851a*) that he has had such a method "for a considerable time." We shall be discussing this method in §§2.4, 2.5 below.

---

7. We had not appreciated this fact when writing §6.1 of our *1986*. Michell's argument lacks cogency on other grounds. See *Dale 1991*, §4.3.

Chapter 2

The Nineteenth Century

§2.1. Probability in Bolzano's *Wissenschaftslehre (1837)*

During the late eighteenth and early nineteenth century there was considerable activity in probability theory and its applications. Yet there seems tò have been no contribution to probability logic as such. Not until Bolzano is there something for inclusion in our history. Before considering the specific items of interest in Bolzano's work we describe his conception of probability, as it involves some unusual features. Our account is based on §161 of his *1837*, which was written in the decade 1820–1830.

For Bolzano probability was a part of logic, but for him logic was *Wissenschaftslehre*, i.e., theory of science, in which he made no sharp distinction between material and formal components.[1] Almost a century before Keynes (whose *1921* makes no mention of him), Bolzano takes probability to be a relation between propositions but, unlike Keynes for whom it was an undefined, informally described, primitive notion, Bolzano provides a definition. To state his definition requires the introduction of some of Bolzano's special terminology. We first define his notion of 'validity' (of a proposition) since he will be identifying probability with 'relative validity', and a comparison of the two is relevant.

Bolzano speaks of taking an *idea*[2] in a proposition to be *variable* and considers how the truth [value] of the proposition behaves as the selected idea is varied, i.e., replaced by other ideas [according to a specified rule]. Only those propositions are counted which "have a reference, i.e., whose

---

1. There is an exposition and critical evaluation of the *Wissenschaftslehre* from a mid-twentieth century logician's point of view in *Kneale and Kneale 1962* (part 5, chapter 5).

2. A special notion. Roughly, Bolzano's 'idea' is to 'term' as 'proposition' is to 'sentence'.

subject-idea have a referent". Then the *validity* of a proposition is the relation between the number of [elements in the set of] all [referring] propositions which can be generated by varying (one or more) selected ideas, and the number of true propositions in that set. The *degree of validity* [with respect to the selected ideas] is the ratio of the two numbers. By way of illustration take the proposition:

In a drawing of five balls from an urn containing 90 consecutively numbered balls, one of the five is no. 8.

If the idea '8' is taken as variable, this proposition has degree of validity $\frac{5}{90}$, since when '8' is replaced by each of the ideas '1', '2', ..., '90', exactly 5 of the 90 propositions result in a true proposition. To put this illustration into contemporary language let U be the set of balls in the urn, $\phi(8)$ the proposition, and let $|A|$ denote the number of elements in a set A. Then

$$\text{degree of validity of } \phi(8) = \frac{|\{\, x \mid x \in U \wedge \phi(x) \,\}|}{|\{\, x \mid x \in U \,\}|}$$

Thus Bolzano's degree of validity is actually not a property of the proposition $\phi(8)$ but of the predicate ϕ (or sentential form $\phi(x)$). Also, in contemporary terms, we note that the degree of validity of $\phi(a)$ for any $a \in U$ would be the probability of $\phi(a)$ if all $\binom{90}{5}$ possible drawings of 5 balls from the urn were equally likely.

Bolzano's definition of *relative validity* is contained in the following passage (*1972*, 238–39):[3]

... One of the relations among several propositions is very reminiscent of this concept of validity: given that in an individual proposition A or several propositions A, B, C, D, ... certain ideas $i, j, \ldots$ are variable, and given that A, B, C, D, ... are compatible with respect to these ideas, it will often be important to find the proportion between cases in which A, B, C, D, ... all become true and in which a certain other proposition M becomes true. For, if we accept A, B, C, D, ... then this proposition tells us whether we should also accept M. If M becomes true in more than half of the cases in which A, B, C, D, ... are true, the truth of A, B, C, D, ... entitles us to accept

---

3. We are taking our English quotations in this section from *Bolzano 1972*, edited by Rolf George. This one-volume edition of the four-volume *1837*, in part translated and in part summarized, does contain most of the §161 of interest, though certain parts of significance to us were omitted. Throughout this translation Bolzano's *Gültigkeit* was rendered as 'satisfiability'. We believe that 'validity' is a more appropriate translation and, accordingly, have taken the liberty of making the change. Also, we have chosen to render Bolzano's *Voraussetzung* by 'supposition' rather than 'premise' or 'assumption'.

M as well, otherwise not. I wish to call this relation between the indicated classes the relative validity of proposition M with respect to
propositions A, B, C, D, ... or the probability which accrues to M
from the supposition of A, B, C, D, ... I call this relation relative
validity because of the similarity it has to the relation which I have
called validity of a proposition in §147.

After a sentence explaining validity, Bolzano continues:

> ... In the case of relative validity of a proposition M with respect
> to certain others A, B, C, D, ..., we want to know the relation be-
> tween the cases where A, B, C, D, ... together with M becomes true.
> I call this relation *probability* [*Wahrscheinlichkeit*] since it seems to me
> that there is an increasingly wide-spread usage where by 'probability'
> we mean nothing but such a relation between given propositions, ...
> 2. Being a relation between two sets, the relative validity or prob-
> ability of a proposition has a certain magnitude which, if it is deter-
> minable at all, can be represented by a fraction whose denominator
> and numerator are related as these two sets.

Aside from the obscurity of what Bolzano's 'cases' means his defnition of
relative validity is the same as the usual conditional probability of M given
the conjunction of A, B, C, D, ... As we shall presently see, Bolzano
simply counts the number of cases (i.e., treats them as equal) when the
'suppositions' A, B, C, D, ... (as he refers to them) do not favor one case
over any other. Since it has the same formal properties as conditional probability it will be convenient for us to symbolize it by '$P(M \mid A,B,C,D,...)$'
suppressing, as does Bolzano in much of his discussion, mention of the ideas
$i, j, ...$.

We turn to the clause "or the probability which accrues to M from the
supposition A, B, C, D, ..." which Bolzano considers to be descriptive of
his notion of relative validity. We wish to note the implied connection to
probabilistic inference which his use of the word 'supposition' suggests. An
explanation for this usage can be inferred from Bolzano's remark:

> For if we accept propositions A, B, C, D, ... then this proposition
> tells us whether we should also accept M. If M becomes true in more
> than half the cases in which A, B, C, D, ... are true, the truth of A,
> B, C, D, ... entitles us to accept M as well, otherwise not.

We assume 'more than half' is to be taken liberally as 'practically all'.

This suggests the inference rule:

$$P(M \mid ABCD \ldots) \approx 1$$
$$ABCD \ldots$$

therefore accept M.

Inference rules of this nature will be discussed below in §3.3.

Bolzano is well aware that it is meaningless to speak of ratios (or quotients) when infinite numbers are involved and limits his considerations to 'determined' probabilities. Such probabilities are obtained when the propositions generated by varying the ideas all have equal probability and, additionally, the set of all such is divisible into finitely many subsets which are mutually exclusive, [exhaustive,] and of equal size. He also points out that it is possible to determine whether propositions have the same degree of probability without knowing what this degree is (*1972*, 240-41):

> ... For example, let the probability of a proposition M be relative to a premise A, namely that Caius has drawn a ball from an urn in which, among others, there is a ball numbered 1 and another numbered 2. I claim that, *if no other premises are given* [emphasis added], the following two propositions: 'Caius has drawn ball no. 1' and 'Caius has drawn ball no. 2' both have exactly the same degree of probability provided that the ideas 'no. 1' and 'no. 2' are among those which are envisaged as variable in the context of this enquiry. For if we compare these propositions with the given premise A, we notice that both of them have exactly the same relation to A, since the only difference between them is that one of them has the idea 'no. 1', where the other has the idea 'no. 2'. Both of these ideas are contained in proposition A in one and the same way.

One clearly sees here the adherence to an epistemic rather than an aleatoric conception of probability: Bolzano equates the respective degrees of probability of 'Caius has drawn ball no. 1' and 'Caius has drawn ball no. 2' merely on the basis of the premise A containing the ideas 'no. 1' and 'no. 2' in one and the same way. How to determine the degree of probability is then described (*1972*, 241-42):

> If we are asked to determine the probability of a proposition M from premises A, B, C, D, ..., with variables $i, j, \ldots$, we should first try to find a number of propositions K, K', K'', ..., all of which have the same degree of probability under the given assumptions A, B, C, D, ..., and which are such that every class of ideas whose substitution for $i, j, \ldots$ makes all of A, B, C, ... true will make

one and only one of the propositions K, K', K'', ... true. Let it furthermore be true that proposition M, whose probability we wish to determine, stands to the just-indicated propositions K, K', K'', ... in the following relations: none of these propositions leaves the truth of M undetermined; rather, from every one of them we can deduce either M or Neg. M. In this case we merely have to count the number of propositions K, K', K'', ..., and determine how many of them make M true in order to find the relation between the set of cases in which the assumptions A, B, C, ... are true, and the set of cases in which both, these assumptions and M, becomes true. If the total number of propositions K, K', K'', ... $= k$; and if the number of these propositions from which M is deducible $= m$, then it is clear that the total infinite set of cases in which the assumptions A, B, C, ... become true can be divided into k equal parts, and that m of these parts represent the infinite set of cases in which both A, B, C, ... as well as M become true. Hence the desired degree of probability of proposition M is m/k.

From his definition of relative validity and the method just described Bolzano derives a number of elementary theorems regarding the probability of propositions. Using our introduced symbol for his notion we may restate Bolzano's verbally expressed results succinctly as follows:

a) $P(M \mid A, B, \ldots) \leq 1$

$$P(M \mid A, B, \ldots) = \begin{cases} 1 & \text{if } M \text{ is a consequence of } A, B, \ldots \\ 0 & \text{if } M \text{ is incompatible with } A, B, \ldots \end{cases}$$

b) If supposing A, B, ... is equivalent to supposing A', B', ... then
   $$P(M \mid A, B, \ldots) = P(M \mid A', B', \ldots).$$

c) If M is equivalent to M', then
   $$P(M \mid A, B, \ldots) = P(M' \mid A, B, \ldots).$$

d) $P(\text{Neg}.M \mid A, B, \ldots) = 1 - P(M \mid A, B, \ldots)$

e) If R is deducible from M, then
   $$P(R \mid A, B, \ldots) \geq P(M \mid A, B, \ldots).$$

Bolzano shows that the inequality in e) could be a strict one by the example of an urn containing 40 blue-yellow striped balls, 40 red-green striped balls and 20 monochromatic. Letting R be 'a ball chosen is striped' and M be 'a ball chosen is blue-yellow striped', then $P(R) = \frac{8}{10} > \frac{4}{10} = P(M)$.

Bolzano's version of the product rule is significantly different from the usual formulations. We quote his statement of it but, for clarity, reducing the number of factors to two (*1972*, 243):

11. Let proposition M have a degree of probability μ under the supposition A, B, C, ... and with respect to the ideas $i, j, \ldots$. Let proposition N have degree of probability ν under the supposition D, E, F, ... with respect to the same ideas $i, j, \ldots$, and let D, E, F, ... be compatible with A, B, C, ... Let K, K', ... be propositions of equal probability generated from A, B, C, ... and let L, L', ... be similarly generated from D, E, F, ... If we can now generate propositions of the following kind: 'K and L is true', [K and L' is true,] K' and L is true', etc., and *if these propositions are again all propositions of equal probability,* [our emphasis] then we can assert that the degree of probability that propositions M, N, are both true under the combined supposition A, B, C, ... D, E, F, ... and with respect to the same ideas $i, j, \ldots$ is equal to the product $\mu \times \nu$.

Bolzano's conclusion here is,

f) $P(M \text{ and } N \mid A, B, C, \ldots D, E, F, \ldots)$
$$= P(M \mid A, B, C, \ldots) \cdot P(N \mid D, E, F, \ldots).$$

But this cannot be true in general. For taking M and N to be the same proposition, and likewise A, B, C, ... and D, E, F, ... to be the same, f) then says that $\mu = \mu^2$ for $P(M \mid A, B, C, \ldots) = \mu$. What Bolzano may have had in mind is that the mutually exclusive K, K', K'', ... generated from the A, B, C, ... and the mutually exclusive L, L', L'', ... generated from D; E, F, ... are *logically* independent, for in his proof (omitted in the condensed *1972*) he counts the number of KL (logical) products as kl in number, where k is the number of K's and l the number of L's. Not only is Bolzano's theorem not in general correct but the example with which he illustrates it is faulty. The example is of an urn of six balls of which four are black and five are fragrant (and no other information). The probability of selecting one which is both black and fragrant is sought. His answer is that it is the product $(\frac{4}{6})(\frac{5}{6}) = \frac{5}{9}$, as if 'black' and 'fragrant' were independent. But the assumptions of the example are compatible with either of two possible constitutions of the urn, namely:

(I) two non-black balls fragrant (i.e., 3 black and fragrant)
(II) one non-black ball fragrant (i.e., 4 black and fragrant).

Thus the compound property 'black and fragrant', having two possible values, is not well-defined for the contents of the urn.

Bolzano's §161 continues with a number of theorems involving logico-probability relations. We shall not detail these as they suffer from defects of the kind exposed in item 11. However there is an interesting remark in connection with the following:

12. Under the condition in item 11, if R is deducible from 'M and N', then

$$P(R \mid A, B, \ldots D, E, \ldots) \geq P(M \mid A, B, \ldots) \cdot P(N \mid D, E, \ldots).$$

In Note 3 at the end of his §161 Bolzano remarks that in many textbooks this result is stated: "The probability of a conclusion is equal to the product of the probabilities of its premises". He points out that on the basis of item 12 the product is a value below which the probability could not fall, and gives a specific example where it is indeed larger. Unfortunately his result in item 12 is invalid and, moreover, his counterexample to the textbook rule is defective for the same reason as the previously described one. The defect, however, is easily removed and one can indeed give a correct counterexample. But that this counterexample is in accord with item 12—as Bolzano would have us think—would not be so, since *that* result is invalid. We should like to disentangle this bit of confusion.

To get to the nub of the matter let us treat all probabilities as being unconditional ones. Then Bolzano's item 12 result is that

$$\text{If } (M \text{ and } N) \text{ implies } R, \text{ then } P(R) \geq P(M)P(N). \tag{1}$$

That this is invalid can be seen by taking R to be the conjunction MN. Then we get from (1)

$$P(MN) \geq P(M)P(N), \tag{2}$$

to which counterexamples are easily obtained—take M and N to be mutually exclusive, for example. As we shall see (Theorem 4.43 (a_1) below) in place of (1) one should have

$$\text{If } MN \text{ implies } R, \text{ then } P(R) \geq P(MN), \tag{3}$$

i.e., the probability of the conclusion is not less than then probability of the *conjunction* of the premises—only if M and N are stochastically independent can one replace this conjunction with the product of the premise probabilities. Thus, giving a counterexample to

$$P(R) = P(M)P(N) \tag{4}$$

is weaker than one for $P(R) = P(MN)$—though sufficient to render invalid an argument using (4).

This error to which Bolzano called attention was widespread, even among logically sophisticated writers. We quote from *De Morgan 1864 = 1966*, 218:

And I hold the supreme *form* of syllogism of one middle term to be as follows;—There is the probability α that X is in relation L to Y; there is the probability β that Y is in relation M to Z; whence there is the probability $\alpha\beta$ that X has been proved in these premises to be in relation L of M to Z.

And even at the end of the century (*Jevons 1892*, 209):

If the probability is $\frac{1}{2}$ that A is B, and also $\frac{1}{2}$ that B is C, the conclusion that A is C, on the ground of these premises, is $\frac{1}{2} \times \frac{1}{2} = \frac{1}{4}$.

Bolzano's idea of probability as relative (or degree of) validity was revived in the twentieth century by Carnap (see §3.1 below). Since in his *1950* Carnap makes no mention of Bolzano we presume that Bolzano's ideas were unknown to him.

§2.2. De Morgan—testimony and argument

In De Morgan's writings we find no mention of earlier work on our topic. It is hard to believe that he, a bibliophile especially interested in probability and its applications,[4] did not see Bernoulli's *Ars conjectandi*, Part IV. Yet nothing seems to indicate that he was aware of its chapter 3 (discussed above in our §§1.2–1.4). And, as for Lambert's work, apparently the *Neues organon* was unknown to De Morgan until some nine years after his first writings on the subject.[5] The presumption is then that De Morgan's work was independent of Lambert's (and Bernoulli's ?). His ideas on the topic first make their appearance in 1837—the same year that Bolzano's *Wissenschaftslehre* was published—in an article on Probability in the *Encyclopedia metropolitana.*[6]

---

4. Todhunter (*1865*, 49) mentions having been sent by De Morgan a (presumably rare) copy of Arbuthnot's 1692 English translation of Huygens' *De ratiociniis*.

5. De Morgan's *1847b*, 323–24, mentions, in connection with diagramming syllogistic inferences, having just recently seen the *Neues organon*.

6. Issued in 50 parts, spanning the period 1817–45. The 1837 date for De Morgan's article we take from the bibliographical list in his wife's *Memoir of Augustus De Morgan* (1872). The article was reprinted as part of the *Encyclopedia of Pure Mathematics* (London and Glasgow, 1847). Not having available a copy of the 1837 printing (an offprint version from which Mrs. De Morgan apparently took her date), I shall, when quoting material from it, use the page numbers of the *Encyclopedia of Pure Mathematics* reprinting, i.e., of *1847a*.

In this article of De Morgan's items relating to probability logic do not occur as part of a systematic development but as examples illustrating various probability principles. Thus to illustrate the product rule for the probability of a conjunction of events (stated incompletely, the requirement of independence being omitted) De Morgan gives the following example (*1847a*, 399):

> For instance, suppose the following syllogism: A is B; B is C; therefore A is C. Suppose the first assertion is considered as having ten to one in its favour, and the second three to one. Now the probability of the conclusion is entirely that of both premises being true, from which it necessarily follows. The probabilities for the single premises are $\frac{10}{11}$ and $\frac{3}{4}$, the product of which $\frac{15}{22}$ is that of the conclusion, or 15 to 7, a little more than two to one. This, ...

In addition to his neglect of the independence condition, De Morgan here joins Lambert in making the error of taking the probability of [independent] premises which necessarily imply a conclusion, as being the probability of the conclusion instead if just a lower bound for it. Our quotation continues with "This, it must be observed, is independent of any force the conclusion may derive from other sources, a case we shall presently consider." Foreshadowed here is De Morgan's interest in the general problem of combining evidence (e.g., from testimony or authority) with argument. The case he considers is posed as a question which, in addition to a syllogistic inference with contingent premises, there is a further datum (*1847a*, 400):

> (15.) Let us now consider the following case. It is an even chance that A is B, and the same that B is C; and therefore 1 to 3 on these grounds only, that A is C. But other considerations of themselves give an even chance that A is C. What is the resulting degree of evidence that A is C?

By 'the resulting degree of evidence that A is C' De Morgan means the probability of 'A is C' since that is what he goes on to compute. The problem is solved by enumeration of cases (*1847a*, 400):

> ... Let us take all the possible cases.
>
> 1. An even chance that A is B, from the premises
> 2. An even chance that B is C, from the premises
> 3. An even chance that A is C, from other evidence

Let (1) denote the truth of 1, and [1] its falsehood. The following are possible cases:

1. $(1)(2)(3)$ 2. $(1)(2)[3]$ 3. $[1](2)(3)$ 4. $[1](2)[3]$
5. $[1][2](3)$ 6. $[1][2][3]$ 7. $(1)[2](3)$ 8. $(1)[2][3]$

Here are eight cases, equally probable, each consisting of a result compounded of three results, and therefore having probability of $\frac{1}{8}$. Of these cases 1, 2, 3, 5 and 7 make the conclusion 'A is C' necessary, either by the establishment of the argument, or of the independent evidence. That is, five out of eight equally possible combinations are in favour of A being C, or the odds for it are 5 to 3.

There are a number of things which have to be straightened out here. Although De Morgan says that he is letting '(1)' denote the truth of 1, i.e., of 'An even chance that A is B', his solution shows that it is the proposition 'A is B' which it designates. Similarly '[1]' is actually 'not-(A is B)'. Furthermore, it is not 'An even chance that A is C from other evidence' which '(3)' designates but the evidence itself (call it E) for the proposition—he says, e.g., that case 5 (namely $[1][2](3)$) makes the conclusion 'A is C' necessary, which means then that (3) implies 'A is C'. Thus De Morgan's actual premises, in modern notation, are:

1. $P(A \text{ is } B) = \frac{1}{2}$
2. $P(B \text{ is } C) = \frac{1}{2}$
3. $P(E) = \frac{1}{2}$,

together with the tacit assumptions

(i) $((A \text{ is } B) \text{ and } (B \text{ is } C))$ implies $(A \text{ is } B)$,
(ii) E implies $(A \text{ is } C)$,
(iii) $(A \text{ is } B)$, $(B \text{ is } C)$, and E, are stochastically independent.

With this understanding the conclusion that De Morgan draws, that the odds in favor of 'A is C' are 5 to 3, is incorrect—all that one can conclude is that there are 5 cases, out of 8 equally likely cases, in favor of 'A is C', but we do not know if the other 3 are unfavorable, i.e., that 'A is C' is not the case. Just because the evidence for 'A is C' is false doesn't make 'A is C' false. All that one can conclude is that $\frac{5}{8}$ is a lower bound for the probability of 'A is C'. The problem of combining argument and evidence is then treated in general with arbitrary probability values assigned to the argument-premises and to the evidence, tacitly assuming them to be independent. The answer De Morgan gives is the probability of the alternation (non-exclusive or) of the two (*1847a*, 400):

(16) The above, generalized, is as follows. Let a and $(1 - a)$ be the probabilities for and against the argument [actually, argument-premises], and let e and $1 - e$ be the probabilities from any other source. Then the chances that they are both wrong is $(1 - a)(1 - e)$, and of the contradictory, namely, that the result 'A is C' follows from one or the other is

$$1 - (1 - a)(1 - e) \quad \text{or} \quad a + e - ae.$$

De Morgan here misspeaks: what he seeks is the probability of the result, not the probability that it *follows from* the alternatives—we know that it does follow.) But that the argument (call it G) or the evidence (call it E) implies the result is only warrant for $P(G$ or $E)$ being a lower bound for the result. He goes on to express dissatisfaction (*1847a*, 400):

The preceding result evidently fails to represent the character of usual impressions by probable reasoning. Let there be $\frac{3}{4}$ of probability for 'A is C' from argument, but suppose other grounds of evidence give only $\frac{1}{1000}$ of probability for it, or 999 to 1 against it. Can it be believed that the latter adds probability to the former, or that the total probability is $\frac{3}{4} + (\frac{1}{1000})(1 - \frac{3}{4})$, which is greater that $\frac{3}{4}$?

This result is admissible, provided the methods by which $\frac{3}{4}$ is established have a probability which is not shaken by the smallness of the probability for the result independently of them. But this can hardly be the case, and the preceding theorem therefore omits a necessary consideration which applies with more or less force whenever e is less than a.

De Morgan does not say how to take this omitted consideration into account, but goes on to present an instance where he says it does not matter. At this point we break off the discussion as we shall be resuming it in our next section after presenting De Morgan's treatment of his *Problem 3.*

Lamentably some of the probability principles which De Morgan states in this encyclopedia article are either unclearly or carelessly worded. Take for example (*1847a*, 401):

Principle IV. Knowing the probability of a compound event, and that of one of its components, we find the probability of the other by dividing the first by the second. This is a mathematical result of the last [i.e., his *Principle III*] too obvious to require further proof.

Taken literally the principle ignores the possibility of dependency of the events and, indeed, Boole so read it (see our §2.4). However from the context, and De Morgan's subsequent use of the principle, we gather that one of the components is to be a conditional probability. Clearly, it would have been helpful to have a term and notation for the notion of conditional probability, but this was late in coming (see our §2.7).

De Morgan's ideas on probable inference are taken up again as part of a memoir in the Cambridge Philosophical Society Transactions (*1849*, read 9 November 1846); the part on probability grew into chapter 10, On Probable Inference, of his *Formal Logic (1847b)*, to which we now turn.

The Preface of *Formal Logic*, wherein its novel features are enumerated, contains the following arresting statement (*1847b*, v):

> The old doctrine of modals is made to give place to the numerical theory of probability. Many will object to this theory as extralogical. But I cannot see on what definition, founded on real distinction, the exclusion of it can be maintained. When I am told that logic considers the validity of the inference, independently of the truth or falsehood of the matter, or supplies the conditions under which the hypothetical truth of the matter of the premises gives hypothetical truth to the matter of the conclusion, I see a real definition, which propounds for consideration the forms and laws of inferential thought. But when it is further added that the only hypothetical truth shall be absolute truth, certain knowledge, I begin to see arbitrary distinction, wanting the reality of that which preceded. Without pretending that logic can take cognizance of the probability of any given matter, I cannot understand why the study of the effect which partial belief of the premises produces with respect to the conclusion, should be separated from that of the consequences of supposing the former to be absolutely true. Not however to dispute upon names, I mean that I should maintain, against those who would exclude the theory of probability from logic, that, call it by what name they like, it should accompany logic as a study.

Hopes aroused by these remarks for a substantial developed theory to appear are not fulfilled. Even making allowance for the relatively undeveloped state of the formal logic he uses, De Morgan's actual accomplishments are somewhat disappointing. He restricted himself to inference forms in which the conclusion necessarily follows from the premises, a restriction which even Bernoulli had dispensed with; independence requirements needed to justify replacing the probability of a conjunction by the product of the probabilities of the conjuncts are persistently ignored; and there are other probability principles stated (and used) without appropriate justifying hypotheses. Much of the chapter on probable inference involves notions we consider to be non-logical such as authority, testimony, and combination of testimony of witnesses with given credibilities. Nevertheless we shall have to take these into consideration since they are part and parcel of his presentation.

De Morgan refers to two sources of conviction: argument and testimony. Of the two meanings for 'argument' which we have earlier recognized, his usage is generally, though not always, confined to the active sense which involves reasoning. Unlike Bernoulli, who allowed for contingent arguments, De Morgan requires them to be necessary (*1847b*, 191-192):

> I shall suppose all the arguments I speak of to be logically valid; that is, having conclusions which certainly follow from the premises. If then the premises be all true, the conclusion is certainly true. If a, b, c, &c. be the probabilities of the independent premises, or the independent propositions from which premises are deduced, then the product $abc...$ is the probability that the argument is every way good.

With reference to this remark, let A, B, C, ... be the 'independent' premises whose respective probabilities are a, b, c, ..., and let K be a conclusion following from A, B, C, ... It is not clear what De Morgan means by an argument being 'every way good'. If it merely means that the conclusion holds then, as we have noted earlier, the product $abc...$ is only a lower bound for this probability (assuming his independence means stochastic independence). On the other hand if we suppose 'every way good' to mean that the premises *prove* the conclusion, then we can agree with De Morgan's assertion since if $ABC \cdots \rightarrow K$ is necessary,

$$P(ABC \ldots \text{ proves } K) = P(ABC \cdots \wedge (ABC \cdots \rightarrow K))$$
$$= P(ABC \ldots)$$
$$= abc \ldots.$$

In preparation for the next section we need to discuss a weak form of the inverse probability principle which De Morgan states, though without a complete set of hypotheses needed to justify its validity. This is his formulation of the principle (*1847b*, 190):

> If the probability of the observed event, supposed still future, from the several possible precedents, severally supposed actually to exist, be a, b, c, &c: then, when the event is known to have happened, the probabilities that it happened from the several precedents are
>
> $\dfrac{a}{a+b+c+\ldots}$ for the first, $\dfrac{b}{a+b+c+\ldots}$ for the second, &c.

Denoting the event referred to by E and the 'precedents' by $A_1, \ldots, A_n$, then, as is well known, a correct form of the inverse principle ('Bayes' rule') is:

$$P(A_k \mid E) = \frac{P(E \mid A_k)P(A_k)}{\sum_{i=1}^n P(E \mid A_i)P(A_i)}, \qquad (k = 1, \ldots, n),$$

where the A_i are mutually exclusive and E can happen only if some one of the A_i does (i.e., if E implies $A_1 \vee A_2 \vee \cdots \vee A_n$). If we take De Morgan's 'probability of the observed event, supposed still future, from the several precedents' to be the $P(E \mid A_i)$ then one can obtain his formula by assuming, in addition to the above italicized assumptions, that the 'prior' probabilities $P(A_i)$ are all equal.[7]

He cites a second situation which results in the same rule (1847b, 190):

> Again, if there be several events, which are not all that could have happened; and if, by a new arrangement (or by additional knowledge of old ones) we find that these several events are now made all that can happen, without alteration of their relative credibilities: their probabilities are found by the same rule. If a, b, c, &c. be the probabilities of the several events, when not restricted to be the only ones: then, after the restriction, the probability of the first is $a \div (a + b + \ldots)$ and so on.

To have the matter before us in symbols, let the events be $A_1, \ldots, A_n$. Then De Morgan's desiderata are the conditional probabilities of these events on condition that some one of them happens. Using the definition of conditional probability, one has (assuming that $P(A_1 \vee \cdots \vee A_n) \neq 0$),

$$P(A_k \mid A_1 \vee \cdots \vee A_n) = \frac{P(A_k(A_1 \vee \cdots \vee A_n))}{P(A_1 \vee \cdots \vee A_n)}$$

$$= \frac{P(A_k)}{\sum_{i=1}^n P(A_i)}, \qquad (k = 1, \ldots, n),$$

the second line following from the first if the A_i are mutually exclusive. This is the formula De Morgan will be using. Although his statement of the rule does not include the italicized hypothesis, it is the case that, in the applications where he uses it, the components A_i are mutually exclusive. This formula is also derivable from Bayes' rule, cited above, by replacing

---

7. De Morgan's version of the weak inverse probability principle closely follows that of *Laplace 1820*, 182, which also fails to mention the hypothesis of equal prior probabilities, though it does get mentioned in Laplace's discussion which follows the statement. Further on, p. 184, Laplace does consider the general case of non-equal prior probabilities. De Morgan reviewed *Laplace 1820* for the *Dublin Review*, vol. 3, April 1837, 338–54, July 1837, 237–48.

E by $A_1 \vee \cdots \vee A_n$ and deleting all occurrences of $P(E \mid A_i)$ which, for this E, are equal to 1.

§2.3. Probable inference via six problems

De Morgan's ideas on the nature of probable inference are not presented in systematic form but are embodied in the statement of six *Problems* and their solutions. The first of these concerns the combining of testimonies, a topic with a long history which has been treated by many writers.[8] Although we do not consider this topic to be in the purview of probability logic we shall nevertheless discuss it so as to make its material assumptions explicit and, moreover, to contrast it with a similar problem which can be formulated without such assumptions. Here is De Morgan's statement of it (*1847b*, 195):

> *Problem* 1. There are independent testimonies to the truth of an assertion, of the value μ, ν, ρ, &c. (one of them being the initial testimony of the mind itself which is to form the judgment): required the value of the united testimony.
>
> Let μ' be $(1 - \mu)$, &c. as in page 187. Here is a problem of the same class as in page 190 [our last quote in the preceding section]; the restrictions are, that all the testimonies are right, or all wrong, the independent chances of which are $\mu\nu\rho \ldots$ and $\mu'\nu'\rho' \ldots$ [i.e., μ, ν, ρ, $\ldots$ and μ', ν', ρ', $\ldots$]. Hence the probabilities are
>
> $$\frac{\mu\nu\rho \ldots}{\mu\nu\rho \cdots + \mu'\nu'\rho' \ldots} \quad \text{for;} \qquad \frac{\mu'\nu'\rho' \ldots}{\mu\nu\rho \cdots + \mu'\nu'\rho'} \quad \text{against.}$$

In examining De Morgan's solution we take the values μ, ν, ρ, etc. to be credibilities of the witnesses, credibility of a witness being the probability that what is asserted is correct. We set[9]

$$T(W, A) = \text{Witness } W \text{ testifies to the truth of '}A\text{'}$$

---

8. For a brief account of the history, and a critical discussion of this and similar problems, see *Keynes 1921*, 180-85.

9. Strictly speaking we ought to write '$T(W, 'A')$' (note, 'A' in single quotes) since W may not be willing to testify to B even if B is logically equivalent to A. As is well-known, some humans can be illogical.

and then have

$$P(A \mid T(W, A)) = W\text{'s credibility with regard to } `A\text{'}$$

For two witnesses, W_1 and W_2, we have, abbreviating $T(W_1, A)$ to T_1 and $T(W_2, A)$ to T_2,

$$\mu = P(A \mid T_1), \quad \nu = P(A \mid T_2).$$

This interprets $\mu\nu/(\mu\nu + \mu'\nu')$, i.e., De Morgan's 'the value of the united testimony,' as

$$\frac{P(A \mid T_1)P(A \mid T_2)}{P(A \mid T_1)P(A \mid T_2) + (1 - P(A \mid T_1))(1 - P(A \mid T_2))}. \tag{1}$$

On making use of

$$P(A \mid T_i) = \frac{P(AT_i)}{P(T_i)}, \quad 1 - P(A \mid T_i) = P(\neg A \mid T_i),$$

this becomes

$$\frac{P(AT_1)P(AT_2)}{P(AT_1)P(AT_2) + P(\neg AT_1)P(\neg AT_2)}. \tag{2}$$

But if we take 'the value of the united testimony' as $P(A \mid T_1 T_2)$, then by the definition of conditional probability,

$$P(A \mid T_1 T_2) = \frac{P(AT_1 T_2)}{P(AT_1 T_2) + P(\neg AT_1 T_2)}.$$

This does not agree with De Morgan's result (1). We can make it agree, via (2), by adjoining two additional assumptions:

$$P(AT_1 T_2) = P(AT_1)P(AT_2)$$
$$P(\neg AT_1 T_2) = P(\neg AT_1)P(\neg AT_2).$$

But these are *ad hoc* and not consequences of any reasonable meaning for two witnesses to be independent. (For a detailed discussion of such matters see *Keynes 1921*, 181.)

Although, as we have just seen, the solution of De Morgan's *Problem* 1 hinges on non-logical matters and requires material assumptions, there is a purely logico-probability problem which it suggests, namely: For arbitrary propositions A, T_1, T_2, with given conditional probabilities

$$\mu = P(A \mid T_1), \quad \nu = P(A \mid T_2),$$

to what extent, and how, is the value of $P(A \mid T_1 T_2)$ expressible in terms of μ and ν? Boole's method, to be discussed in the next two sections, claims to be able to answer such questions. The problem of 'combining two probabilities for an event, so as to have one definite probability' will be the subject of §2.6 below.

De Morgan's next *Problem* considers a set of assertions whose respective probabilities are determined from testimony, and asks for the conditional probability of each under a variety of logically expressed conditions involving these assertions. But that the probabilities of the assertions that are involved arise by way of testimony is irrelevant to the solution. Hence in our discussion we will ignore this fact and consider them to be assertions each with a given probability. His statement and solution are as follows (*1847b*, 200):

> *Problem* 2. Let there be any number of different assertions, of which one must be true, and only one: or of which one may be true, and not more than one: or of which any given number may be true, but not more: required the probability of any one possible case.
>
> The solution of all these varieties depends on one principle, explained in page 190 [quoted in our preceding section]; requiring the previous probabilities of all the consistent cases to be compared. As an instance, suppose four assertions, A, B, C, D, and suppose μ, ν, ρ, σ, to be the probabilities from testimony, for each of them. If either of them have several testimonies, their united force must be ascertained by the last problem. First, let it be that one of them must be true, and one only. The probabilities in favour A, B, C, D, are in the proportion of $\mu\nu'\rho'\sigma'$, $\nu\mu'\rho'\sigma'$, $\rho\mu'\nu'\sigma'$, and $\sigma\mu'\nu'\rho'$. Either of these, divided by the sum of all, represents the probability of its case.

For the first of De Morgan's listed varieties the definition of conditional probability gives (using A' in place of $\neg A$, etc.)

$$P(A \mid AB'C'D' \lor A'BC'D' \lor A'B'CD' \lor A'B'C'D)$$
$$= \frac{P(AB'C'D')}{P(AB'C'D') + P(A'BC'D') + P(A'B'CD') + P(A'B'C'D)}$$

(the probability of the disjunction in the denominator being replaceable by the sum of the probabilities of the disjuncts since these are mutually exclusive). But to get from this to De Morgan's

$$\frac{\mu\nu'\rho'\sigma'}{\mu\nu'\rho'\sigma' + \mu'\nu\rho'\sigma' + \mu'\nu'\rho\sigma' + \mu'\nu'\rho'\sigma}$$

would require stochastic independence of the A, B, C, D, a necessary requirement not mentioned in his statement of the problem. He goes on to discuss the other varieties. We shall forego examining these since their treatment is very similar, though not without remarking, as an aside for the history of logic, on the impressive degree of sophistication which is apparent in De Morgan's handling of the propositional logic involved in his solutions.

The next *Problem* in De Morgan's list is the first which is concerned with arguments (*1847b*, 201):

> *Problem* 3. Arguments being supposed logically good, and the probabilities of their proving their conclusions (that is, of all their premises being *true*) being called their validities, let there be a conclusion for which a number of arguments are presented, of validities a, b, c, &c. Required the probability that the conclusion is proved.
>
> This problem differs from those which precede in a material point. Testimonies are all true together or all false together: but one of the arguments may be perfectly sound, though all the rest be preposterous. The question then is, what is the chance that one or more of the arguments proves its conclusion. That all shall fail, the probability is $a'b'c'\ldots$ that all shall not fail, the probability is $1 - a'b'c'\ldots$.

We translate De Morgan's language as follows:

> argument (premise A, conclusion C) $\ldots\ldots\ldots A \to C$
> argument logically good $\ldots\ldots\ldots\ldots A \to C$ necessary
> argument proves its conclusion $\ldots\ldots\ldots\ldots A(A \to C)$

In these terms the problem proposed by De Morgan may be reexpressed as: given n arguments for a conclusion C from arguments with respective premises $A_1, \ldots, A_n$ all arguments being logically good and with respective validities $a_1, \ldots, a_n$ (i.e., $P(A_i) = a_i$), what is the probability that the conclusion is proved?

Since

$$(A_1 \to C)(A_2 \to C)\ldots(A_n \to C)$$

is logically equivalent to

$$(A_1 \vee A_2 \vee \cdots \vee A_n) \to C,$$

and this is logically valid if the $A_i \to C$ are, the question then is equivalent to asking for $P(A_1 \vee A_2 \vee \cdots \vee A_n)$ given that $P(A_i) = a_i$ for $i = 1, \ldots, n$. The result De Morgan gives, namely $1 - (1 - a_1)(1 - a_2)\ldots(1 - a_n)$, is correct if (using A' for $\neg A$)

$$P(A'_1 A'_2 \ldots A'_n) = P(A'_1)P(A'_2)\ldots P(A'_n),$$

that is, *if the $A_1, \ldots A_n$, are stochastically independent.*[10] Failure to appreciate this requirement (De Morgan makes no mention of it) beguiles him into misinterpreting what appears to him to be a counter-intuitive implication of the result. He says (*1847b*, 201):

> Accordingly, if we suppose n equal arguments, each of validity a, the probability that the conclusion is proved is $1 - (1-a)^n$. And, as in page 187, if the odds against each argument be k to 1, then the number of such arguments being as much as $\frac{7}{10}k$, the conclusion is rendered as likely as not.
>
> But are we really to believe, having arguments against the validity of each of which it is 10 to 1, that seven such arguments make the conclusion about as likely to be true as not. If such be the case, the theory, usually so accordant with common notions, is strangely at variance with them. This point will require some further consideration.

The result referred to on his page 187 in the quotation is that, for large k, with $a = 1/(k+1)$ and $n = 0.7k$,

$$(1 - a)^n = (1 - \frac{1}{k+1})^{0.7k} \approx e^{-0.7} = 0.4966.$$

(For the approximation to be good k need not be very large—even for the $k = 10$ of De Morgan's example it gives a value close to the true value, which is $(1 - \frac{1}{11})^7 = 0.5132$.) The arguments referred to in De Morgan's rhetorical question (each of the odds against of 10 to 1) are not weak in the sense of being poorly constructed—indeed, it is assumed that each premise necessarily implies the conclusion—but that each premise has a small probability of being true. Yet, as De Morgan's calculation shows, the probability of a disjunction of such premises can be considerable. His lengthy and discursive 'further discussion' of the difficulty takes leave of pure logic and concludes that non-formal matters have to be taken into account (*1847b*, 203):

> ... In enumerating the arguments, then, for or against a proposition, those must be included, if any, which arise out of the nature, mode of production, or producers, of any among them. And then until this has been properly done, we are not in a condition to apply the methods of the present chapter.

But the aspect of the *Problem* 3 result which seems paradoxical to De Morgan stems, we believe, from a lack of experience with such inferences, namely those which involve stochastically independent premises

---

10. Independence of $A_1, \ldots, A_n$ implies that of $A'_1, \ldots, A'_n$. See Theorem 5.25 below.

with known, or estimatable, probabilities. Such situations are rare in argumentations, but modern technology, with its demand for high reliability, furnishes us with the following nice illustration.

Suppose a gadget C, requiring a part A, operates necessarily when A does. (For example, a switch A being closed and conducting implies that a parallel circuit of the form $A \vee Q$ also conducts.) If we let the letters assume appropriate propositional senses (i.e., A = part A operates, C = gadget C operates) then $A \to C$ is necessary. (We suppose that the proposition C is analyzable into a compound containing A whose value is *true* when A has this value.) If the reliability of A is 99% (failure probability 0.01) then the reliability of C is also 99%. (We assume all other components trouble-free.) To increase the reliability one uses independent back-ups. Let A_1, A_2, A_3 be components identical with A in function and reliability. Introduce them into C in place of A such that they operate in parallel mode, i.e., functioning as $A_1 \vee A_2 \vee A_3$. Then $A_i \to C$ $(i = 1, 2, 3)$ and the failure probability of C is reduced from one chance in a hundred to 0.01^3, i.e., to one in a million.

The next one of De Morgan's *Problems* asks for the probability that a conclusion is proved when there are arguments of given validities on both sides of the question (*1847b*, 203):

> Problem 4. A conclusion and its contradiction [i.e., its denial] being produced, one or the other of which must be true, and arguments being produced on both sides, required the probability that the conclusion is proved, disproved (i.e., the contradiction proved), or left neither proved nor disproved.
>
> Collect all arguments for the conclusion, as in the last problem, and let a be the probability that one or more of them prove the conclusion. Similarly, let b be the probability that one or more of the opposite arguments prove the contradiction. Both these cases cannot be true, though both may be false. The probabilities of the different cases are thus derived. Either the conclusion is proved, and the contradiction not proved, or the conclusion not proved and the contradiction proved, or both are left unproved. The probabilities for these cases are as $a(1-b), b(1-a)$, and $(1-a)(1-b)$, and the probability that the conclusion is *proved* is $a(1-b)$ divided by the sum of the three, and so on.

To discuss this conveniently we let C be the conclusion and A and B the respective combined arguments for C and for C'. Now De Morgan has for the probability that the conclusion is proved

$$\frac{a(1-b)}{a(1-b) + (1-a)b + (1-a)(1-b)},$$

that is
$$\frac{P(A)P(B')}{P(A)P(B') + P(A')P(B) + P(A')P(B')}.$$

Or, *assuming independence of the arguments for and against,*

$$\frac{P(AB')}{P(AB') + P(A'B) + P(A'B')},$$

which is the conditional probability

$$P(AB' \mid (AB)'),$$

and is indeed equal to $P(A(A \rightarrow C))$, i.e., the probability that A proves C, given that $A \rightarrow C$ and $B \rightarrow C'$ are necessary (since $(A \rightarrow C)(B \rightarrow C')$ implies $(AB)'$ and also that $AB' \leftrightarrow AC$). Note that De Morgan makes no mention of the independence requirement.

In connection with this *Problem* 4 De Morgan remarks (in a passage very reminiscent of Bernoulli—see near end of our §1.4) that the probabilities of arguments on the two sides of a question could both be near 1 (*1847b*, 203–204):

> The predominance of one side or the other, as far as arguments only are concerned, depends on which is the greatest, $a(1 - b)$ or $b(1 - a)$, or simply on which is the greatest, a or b. If the arguments on both sides be very strong, or a and b both very near to unity, then, though $a(1 - b)$ and $b(1 - a)$ are both small, yet $(1 - a)(1 - b)$ is very small compared with either. The ratio of $a(1 - b)$ to $b(1 - a)$ on which the degree of predominance depends, may, consistently with this supposition, be anything whatever. But we cannot pretend that, when opposite sides are thus both nearly demonstrated, the mind can take cognizance of the predominance which depends upon the ratio of the small and imperceptible defects from absolute certainty. The necessary consequences is, that the arguments are evenly balanced, and are as if they were equal: There is no sensible notion of predominance. This is the state to which most well conducted oppositions of arguments bring a good many of their followers. They are fairly outwitted by both sides, and unable to answer either, and the conclusion to which they come is determined by their own previous impressions, and by the authorities to which they attach most weight; and these are, of course, those which favour their own previously adopted side of the question.

But De Morgan's assumption that both a and b, i.e., $P(A)$ and $P(B)$, could be near 1 comes into conflict with $A \rightarrow C$ and $B \rightarrow C'$ being good, i.e., necessary. For if $A \to C$ and $B \to C'$ are both good then (Theorem 4.43(a_1) below)

$$P(A) \leq P(C) \quad \text{and} \quad P(B) \leq P(C'),$$

so that

$$P(A) + P(B) \leq 1,$$

which cannot be if both $P(A)$ and $P(B)$ are near 1. Hence only if their probabilities are near $\frac{1}{2}$ can 'good' arguments on the two sides of a question be 'equally strong'.

As his next problem De Morgan has (*1847b*, 204):

> *Problem* 5. Given both testimony and argument to both sides of a contradiction, one side of which must be true, required the probability of the truth of each side.
>
> This is the most important of our cases, as representing all ordinary controversy. Collect all the testimonies, and let their united force for the first side be μ, and, from the nature of this case, $1 - \mu$ for the other side. Let a and b be the probabilities that the first side and the second side are proved by one or more of the arguments in their favour. Now, observe that, for the *truth* of either side, it is not essential that the argument for it should be valid, but only that the argument against it should be invalid. Accordingly, the probabilities of the two sides are in the proportion of $\mu(1 - b)$ and $(1 - \mu)(1 - a)$, and the probabilities of the two sides are represented by
>
> $$\frac{\mu(1 - b)}{\mu(1 - b) + (1 - \mu)(1 - a)} \qquad \frac{(1 - \mu)(1 - a)}{\mu(1 - b) + (1 - \mu)(1 - a)}.$$

As it includes the non-logical notion of testimony De Morgan's *Problem 5* does not come under our conception of (pure) probability logic. Even so, on its own grounds one can see two failings in the solution: (i) it ignores the possibility of the collective testimony being inconsistent (as does happen in court trials) and (ii) it tacitly assumes that the testimony and the arguments are independent.

The final *Problem* in De Morgan's chapter is an unusual one:

> *Problem* 6. Given an assertion, A, which has the probability a; what does that probability become, when it is made known that there is the probability m that B is a necessary consequence of A, B having the probability b? And what does the probability of B then become?

The problem is unusual in that it concerns, apparently, a conditional probability whose condition is itself a probability statement, namely 'the probability that B is a logical consequence of A is m'. Denoting this statement by '$P(A \rightarrow B) = m$' the problem, taken literally, would be phrased as:

Given the prior probabilities $P(A) = a$ and $P(B) = b$,

find $P(A \mid P(A \rightarrow B) = m)$, and $P(B \mid P(A \rightarrow B) = m)$.

One is reminded of Bayes' PROBLEM (§1.6 above) which also involves a probability statement within a conditional probability, but in the 'consequent', or first component, of the conditional probability pair rather than in the second. Moreover Bayes provides, in effect, a probability distribution for his statements. Apparently not realizing the need for it De Morgan procedes without one and gives the following as a solution (*1847b*, 209):

> First, let A and B not be inconsistent. The cases are now as follows, with respect to A. Either A is true, and it is not true that both the connexion exists and B is false: or A is false. This is much to concise a statement for the beginner, except when it is supposed left to him to verify by collecting all the cases. The odds for the truth of A, either as above or by the collection, are $a\{1 - m(1 - b)\}$ to $1 - a$. As to B, either B is true, or B is false and it is not true that A and the connexion are both true. Accordingly, the odds for B are as b to $(1 - b)(1 - ma)$.

We expand this somewhat condensed solution as follows.
Initially the odds for the assertion A are

$$P(A) : P(A').$$

Since A and $A((A \rightarrow B)B')'$ are logically equivalent, the odds are also

$$P(A((A \rightarrow B)B')') : P(A').$$

When it becomes known that $P(A \rightarrow B) = m$, then these odds become

$$a\{1 - m(1 - b)\} : 1 - a,$$

inasmuch as

$$P(A((A \rightarrow B)B')') = P(A)\{1 - P((A \rightarrow B)B')\}$$
$$= P(A)\{1 - (P(A \rightarrow B)(1 - P(B)))\}.$$

In other words, De Morgan is treating A, $A \to B$ and B as stochastically independent events.

Laying aside the unresolved questions as to what it means for 'B is a necessary consequence of A' to have a probability, and having a probability statement as a condition in a conditional probability, we note a tactic which De Morgan here uses: the event whose probability is sought is expressed as a logical compound (De Morgan chooses to use negation and conjunction) of the events whose probabilities are known. *If* these events are stochastically independent, the probability of the compound is readily computed. But what if the compounds are not independent, or if the objective event is not expressible as a logical function of them? We see no evidence that De Morgan considered this general question. It was left to his contemporary George Boole to do so.

§2.4. George Boole—claims for a general method

Boole's *Mathematical Analysis of Logic (1847)* appeared on the same day[11] as De Morgan's *Formal Logic*, the book whose chapter on probable inference we have just examined. Unlike De Morgan's book, Boole's monograph has no mention of probability aside from a simple remark in the last paragraph of a postscript. In his first publication on the subject, *Boole 1851a*, he says that he has had for a considerable period a general method for treating problems [of a logical nature] in probability. But since there is no hint of such a method in *Boole 1847* and, moreover, the method depends on developments to his logical system subsequent to the monograph, we assume that this method could not have antedated 1847. It first appeared in print in his *The Laws of Thought*, of 1854. Before discussing it in our next section we consider several short items of his on probability published prior to it.

Boole's first publication on probability, *1851a*, is noteworthy as having pointed out that 'contraposing' a conditional probability, i.e., the equating of the probability of A, if B, with the probability of not-B, if not-A, is not valid, a result we have already discussed in connection with the Michell problem (§1.7 above). This is no small accomplishment since there was no clear understanding, even by Boole, of the difference between a conditional probability $P(A \mid B)$ and the probability of a logical conditional $P(B \to A)$.

---

11. According to De Morgan—see his editorial footnote to the posthumously published *Boole 1868*.

After its discussion of the Michell question Boole's paper goes on to simply state what his general method would give for P $[= P(\neg A \mid B)]$ in terms of p $[= P(\neg B \mid A)]$, namely

$$P = \frac{c(1 - a)}{c(1 - a) + a(1 - p)},\tag{1}$$

"where c and a are arbitrary constants, whose interpretation is as follows: viz. a is the probability of the condition A, c is the probability that the event B would happen if the condition A were not satisfied". A simple application of Bayes' rule gives

$$P(\neg A \mid B) = \frac{P(B \mid \neg A)P(\neg A)}{P(B \mid \neg A)P(\neg A) + P(B \mid A)P(A)},$$

which coincides with Boole's result (1). Note that Boole's constants a $[=P(A)]$ and c $[= P(B \mid \neg A) = P(\neg AB)/P(\neg A)]$ are not 'arbitrary' independently of each other.

Before continuing it be might of interest to see how conditional probability and the probability of a conditional are related. We have

$$P(A \mid B) = \frac{P(AB)}{P(B)} = \frac{P(B(B \to A))}{P(B)}.$$

Thus the conditional probability of A, given B, is the probability of the conditional, if B then A, when confined to cases for which B holds and, moreover, with this probability normalized to $P(B)$, i.e., with $P(B)$ taken as the unit. Coincidence occurs in a limiting case since (see Theorem 5.36 below) for $P(B) \neq 0$,

$$P(A \mid B) = 1 \;\leftrightarrow\; P(B \to A) = 1.$$

Without a clear understanding of these notions it is easy to see how the verbal form 'probability of A if B', ambiguous as between 'probability of: A if B' and 'probability of A, if B' could lead to confusion.

This first probability paper of Boole's had an immediate follow-up, *1851b*, which cites two additional results on probability involving logical notions, also without presenting the method. We discuss these results not only for the examples themselves but also for the opportunity afforded of raising questions pertaining to Boole's method. He writes (*1851b = 1952*, 264):

> Quitting this problem [the one just discussed], I shall now notice two others, of which solutions have been given, that appear to

me to be defective in generality from the same cause, viz. the non-recognition of the requisite arbitrary constants.

Given p the probability of an event X, and q the probability of the joint concurrence of the event X and Y: required the probability of the event Y.

The solution of this problem afforded by the general method described in my last letter [i.e., *1851a*] is

$$\text{Prob. of } Y = q + c(1 - p),$$

where c represents the unknown probability, that if the event X does not take place the event Y will take place. Hence, it appears that the limiting probabilities of the event Y are q and $1 + q - p$. The result is easily verified.

·The only published solution of this problem with which I am acquainted is

$$\text{Prob. of } Y = \frac{q}{p},$$

a result which involves the supposition that the events X and Y are independent. This supposition is, however, only legitimate when the distinct probabilities of X and Y are afforded in the data of the question.

The 'published solution' to which he is referring is, no doubt, the one in De Morgan's *1847a* (reprinting *1837*), 401. (See Boole's letter 35 in *Smith 1982*, 51.) Boole's result is easily verified by ordinary methods:

$$P(Y) = P(XY) + P(\neg XY)$$
$$= P(XY) + P(Y \mid \neg X)P(\neg X)$$
$$= q + P(Y \mid \neg X)(1 - p).$$

In a narrow sense Boole had solved his problem of expressing $P(Y)$ in terms of p and q. But his 'arbitrary' constant, $P(Y \mid \neg X)$, depends on X and Y, and hence indirectly on p and q. Moreover we don't know how arbitrary it is. Might its range of values depend on the parameters p and q? On the other hand the limiting values for $P(Y)$ which Boole gives, namely q and $q + (1 - p)$, *are* expressed solely in terms of the given parameters of the problem. Boole gets these values by letting his c (i.e., $P(Y \mid \neg X)$) take on the extreme probability values 0 and 1. Could there be a narrower, hence more precise, range of values which would include all possible values for $P(Y)$? Also does this range include q/p, the value obtained in the special case of X and Y being independent? Clearly $q \leq q/p$, but, in the other direction,

$$q/p \leq q + 1 - p,$$

or equivalently,

$$q(1 - p) \leq p(1 - p),$$

holds if and only if $q \leq p$. Now necessarily $P(XY) \leq P(Y)$. Hence unless the data $\{P(XY) = q, \ P(Y) = p\}$, are inconsistent we do have that q/p does lie within Boole's limits. Consistency conditions such as just considered will be, as we shall later see, one of Boole's major concerns in connection with his general method. As to whether the inequality bounds (here q and $q + (1 - p)$) are *best possible* is a question that was only first considered in the 20th century. (See §3.5 below.)

Now for Boole's second result in *1851b* (*1952*, 264–65):

Given the probabilities p and q of the two premises of the syllogism,

All Ys are Xs

All Zs are Ys

required the probability P of the conclusion

All Zs are Xs.

Here, by the probability p of the premise all Ys are Xs, is meant the probability that any individual of the class represented by Y, taken at random, is a member of the class X, and so in the other cases. The resulting probability of the conclusion afforded by the general method is then

$$P = pq + c(1 - q),$$

where c is an arbitrary constant expressing the unknown probability, that if the minor premise is false the conclusion is true. The limiting probabilities of the conclusion are thus

$$pq \quad \text{and} \quad pq + 1 - q.$$

The only published solution of the above problem with which I am acquainted is P$= pq$, a result which manifestly involves the hypothesis that the conclusion cannot be true on any other grounds than are supplied by the premises.

In the last sentence the "published solution" to which Boole refers is again De Morgan's in *1847a*. Note also that Boole is pointing out that the probability of the premises of a conclusion equals that of the conclusion only if premises and conclusion are equivalent. The passage also shows Boole still not clear on probability of a conditional versus conditional probability.

Although he says p and q are probabilities of the premises of a syllogism, it is not

$$P(\forall x(x \in Y \rightarrow x \in X)) = p,$$
$$P(\forall x(x \in Z \rightarrow x \in Y)) = q,$$

which he assumes to get his solution (as shown later in Boole *1854*, 284–85) but rather, for arbitrary x,

$$P(x \in X \mid x \in Y) = p$$
$$P(x \in Y \mid x \in Z) = q.$$

Thus the problem as he actually treats it is (writing 'X' in place of '$x \in X$', etc.)

Given: $P(X \mid Y) = p, \quad P(Y \mid Z) = q$

find: $P(X \mid Z)$.

$$(2)$$

The following is the closest we can come, using standard conditional probability rules, to Boole's solution for (2):

$$
\begin{aligned}
P(X \mid Z) &= P(XY \vee X\neg Y) \mid Z) \\
&= P(XY \mid Z) + P(X\neg Y \mid Z) \\
&= P(X \mid YZ)P(Y \mid Z) + P(X \mid \neg YZ)P(\neg Y \mid Z) \\
&= p'q + P(X \mid \neg YZ)(1 - q),
\end{aligned}
$$

which differs from what Boole obtains by his method by our having $p' = P(X \mid YZ)$ where he has $p = P(X \mid Y)$. Moreover, our $P(X \mid \neg YZ)$ agrees with Boole's c only if we express his "minor premise is false" by '$\neg(Z \rightarrow Y)$', that is by using the ordinary conditional rather than the probability conditional. Apparently Boole's method entails something other than standard probability rules.

An exposition of the method by which Boole arrived at these announced results appeared several years later in his *The Laws of Thought (1854)*. The claims made for it were extraordinary: namely, that given the probabilities of any system of events one could obtain the consequent or derived probability of any other event. In particular, as he remarks after his treatment of the probabilistic hypothetical syllogism (*1854*, 285–286),

> Let it be observed, that the above method is equally applicable to the categorical syllogism, and not to the syllogism only, but to every form of deductive ratiocination. Given the probabilities separately attaching to the premises of any train of argument, it is always

possible by the above method to determine the consequent probability
of the truth of a conclusion legitimately drawn from such premises. It
is not needful to remind the reader, that the truth and the correctness
of a conclusion are different things.

No one seems to have understood the method and, anyhow, it was declared to be erroneous (*Wilbraham 1854, MacColl 1880, Keynes 1921*). Not
until recent times was a rationale for the method proposed and an explanation for its near success. Full details of this may be found in *Hailperin
1976*, second edition *1986*. Here we present a summary of the method and
an explanation, by way of a simple example, of what we believe underlies
it.

§2.5. Examination of Boole's general method

Involved in Boole's method for solving the general probability problem
are two auxiliary and relatively unproblematic techniques:

(i) a method by which, from a given set of logical equations[12]

$$s = S(x,y,\dots), t = T(x,y,\dots),\dots,u = U(x,y,\dots)$$
$$\text{and } w = F(x,y,\dots), \tag{1}$$

one can obtain, in a certain sense, w as a function of $s, t, \dots, u$ (and thence
F in terms of $S, T, \dots, U$); and

(ii) a technique for converting the probability of a logical function of
events into a corresponding algebraic function of probabilities.

When these are supplemented by additional (though *not* unproblematic)
procedures one obtains Boole's general method. We first enlarge upon (i)
and (ii).

With regard to (i). Any set of Boolean equations can be replaced by
an equivalent single equation (e.g., '$f = 0$ and $g = 0$' is equivalent to

---

12. There is no need here for us to get involved in the details of Boole's use of his
algebra for doing logic. It will suffice to consider the lower case letters as Boolean
variables, i.e., variables ranging over the elements of a Boolean algebra. Similarly, we
shall assume that Boole's '+' corresponds to exclusive *or*, juxtaposition to *and*, and
subtraction from 1, or overline (e.g., $\overline{x}$), to negation.

'$fg + f\overline{g} + \overline{f}g = 0$'). Also elimination can be performed on any variable (e.g., $f(x) = 0$, equivalent to $xf(1)+\overline{x}f(0) = 0$, implies $f(1)f(0) = 0$). The given set of equations (1) can then be converted to a single equation and the $x, y, \ldots$ eliminated, resulting in an equation in $s, t, \ldots, u, w$. Applying his peculiar techniques for doing logic Boole solves (by ordinary algebra!) the equation for w, expands the expression obtained by a special technique, and writes the result as

$$w = A + 0B + \frac{0}{0}C + \frac{1}{0}D. \tag{2}$$

Here A, B, C, D are certain (one or more possibly empty) sums of constituents (basic conjunctions) on $s, t, \ldots, u$, and $A + B + C + D = 1$. The form (2) is interpreted by Boole to mean

$$\begin{cases} w = A + \nu C, & \nu \text{ an indefinite class} \\ D = 0. \end{cases} \tag{3}$$

We have shown (*Hailperin 1986*, §2.5) that this really means

$$\begin{cases} (\exists \nu)(w = A + \nu C) \\ D = 0, \end{cases} \tag{4}$$

or, equivalently in terms of inclusion (a notion not used by Boole),

$$\begin{cases} A \subseteq w \subseteq A + C \\ D = 0, \end{cases} \tag{4'}$$

and that (4) does validly follow from the given equations (1). That (4) follows from (1) is not difficult to show, but to explain how and why Boole's mysterious algebraic techniques work is not easy (*Hailperin 1986*, §2.7). Substitution into (4) of the values of $s, t, \ldots, u, w$ given by (1) yields an expression for $F(x, y, \ldots)$ in terms of $S(x, y, \ldots)$, $T(x, y, \ldots)$, $\ldots$, $U(x, y, \ldots)$. This expression will involve an arbitrary element (Boole's indefinite 'ν') when $C \neq 0$. However substitution into (4'), the form not used by Boole, results only in inclusion bounds for $F(x, y, \ldots)$.

With regard to (ii). Boole seems to have been the first to appreciate and utilize the close relationship between the logic of *not*, *and*, and *or*, and the formal probability properties of compound events. He made effective use of this relationship by restricting *or* to the exclusive sense—so that probabilities of *or*-compounds added— and by believing that all events were ultimately expressible in terms of simple independent events—so that the probabilities of an *and*-compound of such events multiplied. Since Boole used '+' for logical *or*, juxtaposition for logical *and*, and subtraction from 1 (i.e., $1 - x$, also written $\bar{x}$) for negation, he could then obtain the probability of a logical function of simple events by letting the symbols for the simple events stand also for their probabilities—provided the logical function was expressed as a logical sum of exclusive terms, each a product of simple events or their negations. Thus if x and y are simple independent events the probability of $x + \bar{x}y$ would be written also as $x + \bar{x}y$.

BOOLE'S GENERAL PROBLEM. We now state the general problem in probability which Boole claimed that his method could solve. Let S [i.e., $S(x, y, \ldots)$], T, $\ldots$, U, F be any (Boolean) functions of $x, y, \ldots$, and $p, q, \ldots, r$ any numerical values in the interval $[0, 1]$.
Given:

$$P(S(x, y, \ldots)) = p, \; P(T(x, y, \ldots)) = q, \ldots, \; P(U(x, y, \ldots)) = r, \quad (5)$$

find [in terms of $p, q, \ldots, r$]:

$$P(F(x, y, \ldots)).$$

We first describe the method (omitting Boole's justification) and then follow with what we consider to be its rationale, that is an explanation of what underlies it in terms of modern concepts of algebra and probability.

Boole introduces letters $s, t, \ldots, u, w$ which he calls *simple events* and which are set equal to the given functions so as to produce the equations

$$s = S(x, y, \ldots), \; t = T(x, y, \ldots), \; \ldots, \; u = U(x, y, \ldots),$$
$$w = F(x, y, \ldots). \quad (6)$$

From these by (i) he obtains

$$\begin{cases} w = A + \nu C, & \nu \text{ an indefinite class} \\ D = 0, \end{cases} \quad (7)$$

where, A, C, D are sums of constituents on $s, t, \ldots, u$ only. As Boole says, A expresses those combinations of $s, t, \ldots, u$ which imply w, $A + C$ those which are implied by w and D those combinations "the happening of which is totally interdicted". Thus the original problem is translated into the language of simple events, namely as

Given: $P(s) = p, \; P(t) = q, \; \ldots, \; P(u) = r$
find: $P(w)$, where $w = A + \nu C,$

all this being subject to the condition that $D = 0$ or, equivalently, V [$= A + B + C$] $= 1$. Boole now invokes a postulate justifying an "ascent" to a new scheme of simple *independent* events $s', t', \ldots, u'$ with respect to probabilities $p', q', \ldots, r'$, these being so determined that when $s', t', \ldots, u'$ are subject to the same condition (i.e., $V = 1$) as $s, t, \ldots, u$ are (i.e.,, to the condition $V' = 1$, where V' is like V but with primes on $s, t, \ldots, u$) they will have the given probabilities $p, q, \ldots, r$. That is to say, Boole employs his specially introduced postulate to justify treating the probabilities of $s, t, \ldots, u$ as being conditional probabilities of some simple independent ("ideal") events. Hence he writes:

$$\frac{P(s'V')}{P(V')} = p, \quad \frac{P(t'V')}{P(V')} = q, \quad \ldots, \quad \frac{P(u'V')}{P(V')} = r. \tag{8}$$

Having served their purpose the primes on the letters will now be dropped (as was Boole's practice). Equations (8) can be written in a still simpler form by making use of the independence of the (ideal) simple events: replace sV by V_s, the sum of those constituents in the expansion of V which have s (and *a fortiori* no $\bar{s}$) present; similarly for $t, \ldots, u$. Then (8) takes the form of a set of algebraic equations

$$\frac{V_s}{V} = p, \quad \frac{V_t}{V} = q, \quad \ldots, \quad \frac{V_r}{V} = r, \tag{9}$$

with the symbols $s, t, \ldots, u$ now standing, by virtue of (ii), for the respective probabilities of the simple and independent events $s, t, \ldots, u$. And since $w = A + \nu C$, expressed in terms of the ideal elements, would be subject to the condition $V = 1$, its probability would be a conditional probability given V. Boole claimed to have shown (this is discussed in *Hailperin 1986*, §5.6) that equations (9) have a unique solution for $s, t, \ldots, u$ in the probability range $[0, 1]$ if and only if the data (5) is a possible set of probability conditions. (For example, $P(x) = p$, $P(xy) = q$ is possible if and only if $0 \leq q \leq p \leq 1$.) Assuming then that the 'conditions of possible experience' for the data hold, the values for $s, t, \ldots, u$ determined by (9) (in terms of $p, q, \ldots, r$) then determine the value of

$$\frac{A(s, t, \ldots, u) + cC(s, t, \ldots, u)}{V(s, t, \ldots, u)}$$

and gives the probability sought, that is, of $F(x, y, \ldots)$. Here c is an arbitrary value in $[0, 1]$ corresponding to the indefinite ν.

Rather than reproducing for the general situation what we have elsewhere (in our *1986*, §5.4) presented as a rationale for the method just described, we shall just run through the main ideas in the case of a simple example, one of Boole's introduced above:

Given: $P(x) = p$, $P(xy) = q$

find: $P(y)$.

To get y in terms of x and xy Boole introduces the equations

$$s = x, \ t = xy, \ w = y \tag{11}$$

and derives by ordinary algebra

$$w = \frac{t}{s},$$

and then by his peculiar technique

$$w = st + 0s\bar{t} + \frac{0}{0}\bar{s}\bar{t} + \frac{1}{0}\bar{s}t,$$

with the interpretation (ν indefinite)

$$\begin{cases} w = st + \nu\bar{s}\bar{t} \\ \bar{s}t = 0, \end{cases} \quad \text{or also} \quad \begin{cases} st \subseteq w \subseteq st + \bar{s}\bar{t} \\ \bar{s}t = 0, \end{cases} \tag{12}$$

corresponding to $w = A + \nu C$, $D = 0$ of the general discussion. (One can also derive (12) from (11) by standard Boolean algebra rules.)

We shall assume that x and y (and hence xy) are elements of a Boolean algebra $\mathfrak{A}$ and that P is any probability function such that $P(x) = p$, $P(xy) = q$, and otherwise unspecified. To introduce what corresponds to Boole's ideal elements let $\mathfrak{B}$ be a Boolean algebra generated by the free generators s and t ('free' meaning not subject to any relationships except those holding for any Boolean algebra). Corresponding to the equations $s = x$, $t = xy$, we introduce a map h from $\mathfrak{B}$ into $\mathfrak{A}$ such that

$$h(s) = x, \ h(t) = xy$$

and then extend the map homomorphically, with kernel $\{\bar{s}t\}$, to all of $\mathfrak{B}$. (This can be done since s and t are free generators of $\mathfrak{B}$.) This results in a pairing of elements of $\mathfrak{B}$ with (some) elements of $\mathfrak{A}$ as shown in columns

1 and 2 of Table 1.

Table 1

$\mathfrak{B}$	$h(\mathfrak{B})$	$\mathfrak{B}\,|\,V$	$\mathfrak{A}^*$
1. 0	0		
2. $\bar{s}t$	0	$(0,\bar{s}t) = 0\,|\,V$	0
3. t	xy		
4. st	xy	$(t,st) = t\,|\,V = st\,|\,V$	xy
5. $s\bar{t}$	$x\bar{y}$		
6. $s\bar{t}+\bar{s}t$	$x\bar{y}$	$(s\bar{t}, s\bar{t}+\bar{s}t) = s\bar{t}\,|\,V$	$x\bar{y}$
7. $\bar{s}$	$\bar{x}$		
8. $\bar{s}\bar{t}$	$\bar{x}$	$(\bar{s},\bar{s}\bar{t}) = \bar{s}\,|\,V = \bar{s}\bar{t}\,|\,V$	$\bar{x}$
9. s	x		
10. $s+\bar{s}t$	x	$(s, s+\bar{s}t) = s\,|\,V$	x
11. $st+\bar{s}\bar{t}$	$\bar{x}+xy$		
12. $t+\bar{s}\bar{t}$	$\bar{x}+xy$	$(st+\bar{s}\bar{t}, t+\bar{s}\bar{t}) = st+\bar{s}\bar{t}\,|\,V$	$\bar{x}+xy$
13. $\bar{t}$	$\bar{x}+x\bar{y}$		
14. $\bar{t}+\bar{s}t$	$\bar{x}+x\bar{y}$	$(\bar{t},\bar{t}+\bar{s}t) = \bar{t}\,|\,V$	$\bar{x}+x\bar{y}$
15. $s+\bar{s}\bar{t}$	1		
16. 1	1	$(1, s+\bar{s}\bar{t}) = 1\,|\,V$	1

Note that $\bar{s}t$, the D in the $D = 0$ equation of (12), is mapped into the 0 of $\mathfrak{A}$. The set of image elements, $h(\mathfrak{B})$, are elements of a proper (in this example) Boolean subalgebra $\mathfrak{A}^*$ of $\mathfrak{A}$. Moreover this subalgebra is isomorphic to an algebra $\mathfrak{B}\,|\,V$ obtained by identifying elements of $\mathfrak{B}$ which prove to be equal on adjoining the condition $D = 0$ (or $V = 1$). Entries in the same row of columns 3 and 4 are the paired elements in the isomorphism of $\mathfrak{B}\,|\,V$ with $\mathfrak{A}^*$. The elements of $\mathfrak{B}\,|\,V$ are (non-uniquely) denoted by $b\,|\,V$ for $b \in \mathfrak{B}$. Column 3 shows the identification resulting for $\bar{s}t = 0$. Since $\mathfrak{B}\,|\,V$ and $\mathfrak{A}^*$ are isomorphic we may assume that any probability function defined on the one can also be defined on the other. Note also that (for this example) the objective function y, whose probability is sought, is not in $\mathfrak{A}^*$.

Continuing with our construction of Boole's ideal elements, let P^0 be a probability function defined on $\mathfrak{B}$ with respect to which s and t are stochastically independent and such that the equations

$$\frac{P^0(sV)}{P^0(V)} = p, \qquad \frac{P^0(tV)}{P^0(V)} = q, \tag{13}$$

are satisfied. For our example these equations are

$$\frac{s}{s+\bar{s}\bar{t}} = p, \qquad \frac{st}{s+\bar{s}\bar{t}} = q,$$

(here $s = P^0(s)$, etc.) and indeed they have the unique solution

$$s = \frac{p - q}{\overline{q}}, \quad t = \frac{q}{p}. \tag{14}$$

(Note that $(p - q)/\overline{q}$ and q/p would not be probability values if p were less than q.) Thus since s and t (the events, that is) are stochastically independent and generate $\mathfrak{B}$, the probabilities of all elements of $\mathfrak{B}$ are determined by (14). The probability measure P^0 is now used to define a measure P^* on $\mathfrak{B}|V$ by defining

$$P^*(b|V) = \frac{P^0(bV)}{P^0(V)},$$

that is, as the conditional probability of b given V. Then (13) may be written as

$$P^*(s|V) = p, \quad P^*(t|V) = q. \tag{15}$$

Now by (11) and (12)—using the inclusion form in (12) to avoid unimportant complications—we have (dropping the useless $\overline{x}(xy) = 0$)

$$x(xy) \subseteq y \subseteq x(xy) + \overline{x}(\overline{xy}).$$

Thus for any P defined on $\mathfrak{A}$ and such that $P(x) = p$, $P(xy) = q$,

$$P(x(xy)) \leq P(y) \leq P(x(xy)) + P(\overline{x}(\overline{xy})) \tag{16}$$

(In the general case such a P may not be unique.[13]) Boole, not distinguishing $\mathfrak{A}$ and $\mathfrak{A}^*$, identifies P with P^* so that from (16)

$$P^*((s|V)(t|V)) \leq P(y) \leq P^*((s|V)(t|V)) + P^*((\overline{s}|V)(\overline{t}|V)). \tag{17}$$

(We do not replace '$P(y)$' by '$P^*(y)$' since, in this example, y is not an element of $\mathfrak{A}^*$.) Since the map $b \to b|V$ is a homomorphism, (17) can be written

$$P^*(st|V) \leq P(y) \leq P^*(st|V) + P^*(\overline{st}|V),$$

or by definition of P^*,

$$\frac{P^0(st)}{P^0(V)} \leq P(y) \leq \frac{P^0(st)}{P^0(V)} + \frac{P^0(\overline{st}|V)}{P^0(V)} \tag{18}$$

---

13. As a simple example, let $\mathfrak{A}$ be generated by x and y and let $P(x) = p$. Then $P(\overline{x}y)$ and $P(\overline{x}\,\overline{y})$ can be any non-negative values so long as they add up to $1 - p$.

Since s and t, elements of B, are by assumption stochastically independent the quantities bounding $P(y)$ can (by (ii)) be expressed as algebraic functions of $s = P^0(s)$ and $t = P^0(t)$ and then by (14) as functions of the given parameters p and q. Although in this example, and a large number of Boole's worked examples, one does come out with correct values for the objective probability, examples exist where one does not—at least by currently accepted methods (*Wilbraham 1854, Hailperin 1986*, 249, 253). The source of the difference with standard probability theory is Boole's assumption that $P^*(b\,|\,V)$ can be $P^0(bV)/P^0(V)$, with P^0 a probability function on $\mathfrak{B}$ with respect to which the algebraically independent generators s, t are *also* stochastically independent. While it is the case that $\mathfrak{B}\,|\,V$ is isomorphic to $\mathfrak{A}^*$, so that one can impose on $\mathfrak{B}\,|\,V$ a probability function matching that on $\mathfrak{A}^*$, and also the case that such a probability function can be taken as a conditional probability (with condition V) on $\mathfrak{B}$, it is not necessary that this probability function on $\mathfrak{B}$ be the particular P^0 described, i.e., the one that treats s and t as stochastically independent—there can be others. Take for example,

$$\text{Given:} \quad P(x) = p, \quad P(y) = q,$$
$$\text{find:} \quad P(xy). \tag{19}$$

Boole's method would set $s = x$, $t = y$, so that $V = 1$ and then have $P^*(s\,|\,V) = P^0(s) = p$, $P^*(t\,|\,V) = P^0(t) = q$. Here assuming that s and t are stochastically independent, i.e., that $P^0(st) = P^0(s)P^0(t)$, is equivalent to assuming that x and y are. As early as 1867 C. S. Peirce, commenting on this feature of Boole's ideas said, "... there can be no doubt that an insurance company, for example, which assumed that events were independent without any reason to think that they really were so, would be subject to a great hazard." Contemporary opinion sides with Peirce and against Boole. (See Theorem 4.52(a) below which shows that, on the given data in (19), $P(xy)$ can be any value between $\max(0, p + q - 1)$ and $\min(p, q)$.)

Boole believed that his method also worked for data which included conditional probability statements as, for example (in present day notation), data of the form

$$P(S\,|\,T) = p, \quad \text{that is,} \quad \frac{P(ST)}{P(T)} = p.$$

To reduce this to the already treated case, Boole wrote it as two equations

$$P(ST) = ap, \quad P(T) = a,$$

where a is a new constant—in essence changing the problem by introduction of another parameter. Thus his treatment of the probabilistic hypothetical syllogism begins with (*1854*, 284):

Let the syllogism in its naked form be as follows:

Major premiss: If the proposition Y is true X is true.
Minor premiss: If the proposition Z is true Y is true.
Conclusion: If the proposition Z is true, X is true.

Suppose the probability of the major premiss to be p, that of the minor premiss q.

The data then are as follow, representing the proposition X by x, etc., and assuming c and c' as arbitrary constants:

$$\text{Prob. } y = c, \quad \text{Prob. } xy = cp;$$
$$\text{Prob. } z = c', \quad \text{Prob. } yz = c'q;$$

from which we are to determine,

$$\frac{\text{Prob. } xz}{\text{Prob. } z} \quad \text{or} \quad \frac{\text{Prob. } xz}{c'}.$$

This translates the two premises into four equations with four parameters, the two given probabilities and two additional probability values. *Hailperin 1986*, §6.7, shows how such problems can be treated without the introduction of additional parameters; the method is elaborated in §5.4 below. (See Example 6.72 of our *1986*, or Theorem 5.455 below.)

There is an aspect of Boole's probability method which we have so far neglected in our discussion, namely the determination of the 'conditions of possible experience' which he found were necessary and sufficient for the validity of the method. He devised three different approaches to determine these conditions. The one he settled on—a form of Fourier's method (§0.5) for solving systems of inequations by elimination—also furnished a solution to his general problem, though not in the sense in which he conceived of it. It can be shown that this Fourier technique applied to Boole's general problem furnishes functions of the parameters which are best possible upper and lower bounds on the sought probability (*Hailperin 1965; 1986* §5.7). We shall be discussing all of this in detail later (§3.4). Here we only illustrate the ideas using a simple example, the one used above: given $P(x) = p$, $P(xy) = q$, find $P(y)$.

Introducing quantities k_1, k_2, k_3, k_4 standing for the probabilities of the four constituents xy, $x\overline{y}$, $\overline{x}y$, $\overline{x}\,\overline{y}$, one can formulate the problem as follow:

Given the system of equations and inequations

$$k_1 + k_2 = p$$
$$k_1 = q$$
$$k_1 + k_2 + k_3 + k_4 = 1, \quad k_i \geq 0 \ (i = 1, 2, 3, 4) \tag{20}$$
$$w = k_1 + k_3$$

determine what conditions on p and q are implied by the system independently of the values of k_1, k_2, k_3, k_4.

Viewing (20) as a system of non-strict inequations (an equation, $a = b$, is equivalent to $a \leq b$ and $b \leq a$) one can apply Fourier's technique (Boole makes no mention of Fourier) to eliminate all variables but w. This results in the inequalities

$$0 \leq q,$$
$$q \leq p \leq 1,$$
$$q \leq w, \tag{21}$$
$$w \leq q + 1 - p.$$

The first two lines give Boole's 'conditions of possible experience' for the given problem and the last two lines the 'limits within which a solution for w must lie'. But this is really equivalent to the solution Boole obtained by his 'ideal elements' methods, i.e.,

$$w = q + c(1 - p),$$

with $c \in [0, 1]$. For the value of c is not determined except for being in the interval $[0, 1]$, and hence this solution says no more than that w lies between q and $q + (1 - p)$.

Boole must have thought that his ideal elements method to be more general in that it produced specific values for the probability within, and in some cases strictly within, the limits provided by the elimination technique. He did not realize that it actually was less general in that its restrictive features could reduce the size of the solution set. Moreover, as Wilbraham pointed out, the solution (set) produced by Boole's method may not be one of particular interest. While he was not entirely successful we must give credit to Boole for envisioning and clearly formulating the general problem which, among other things, provides a basis for a general probabilistic inference scheme.

§2.6. On combining 'evidence'

Beginning with Bernoulli and continuing on through the eighteenth and first half of the nineteenth century there was extensive interest in the general topic of combining 'arguments', 'testimonies', or 'evidence' in a probabilistic setting. It engaged the attention of philosophically minded mathematicians, mathematically minded theologians, and mathematicians eager to rationalize the practice of jurisprudence. (For a detailed and interesting historical account see *Daston 1988*, §6.3.) Nowadays the topic is not even mentioned in standard texts on applications of probability.[14] In addition to the unclarity in its material aspects (e.g., the notion of 'credibility of a witness') the treatments often involved errors or confusions of a logico-probabilistic nature. It is this latter aspect which will be the primary interest of this section. Our sample examples will be from the nineteenth century, two being presented here and a third in our next section. We open with one from a paper by Bishop Terrot, 'On the Possibility of Combining two or more Probabilities of the same Event, so as to form one Definite Probability' (*Terrot 1857*).[15]

Terrot's paper begins with an excerpt from "the very popular Treatise on Logic by Dr. Whately, now Archbishop of Dublin", which he will be criticizing (*Whately 1844*, 211):

> As in the case of two probable premises, the conclusion is not established except upon the supposition of their being *both* true, so in the case of two (and the like holds good for any number) distinct and independent indications of the truth of some proposition, unless *both* of them *fail*, the proposition must be true: we therefore multiply together the fractions indicating the probability of the failure of each—the chances against it—and the result being the total chances against the establishment of the conclusion by these arguments, this fraction being deducted from unity, the remainder gives the probability for it.
>
> e.g., A certain book is conjectured to be by such and such an author, partly 1st, from its resemblance in style to his known works; partly, 2nd, from its being attributed to him by someone likely to be pretty well informed. Let the probability of the conclusion, as deduced from these arguments by itself, be supposed $\frac{2}{5}$, and in the other case $\frac{3}{7}$; then the *opposite* probabilities will be respectively $\frac{3}{5}$ and $\frac{4}{7}$, which

14. We are specifically referring to these eighteenth and nineteenth century treatments. In the twentieth century there has been a revival of interest in constructing a mathematical theory of combining evidence, but from a quite different standpoint. In *Shafer 1976*, for example, the subject is based on what are called "belief functions", which are not the same as the standard probability functions used in our chapters 4 and 5.

15. Charles Hughes Terrot (1790–1872) was appointed Bishop of Edinburgh and Pantonian professor in 1841. As a student at Trinity College, Cambridge, from 1808 to 1812, he was an associate of Whewell, Peacock, and Mill. He graduated B.A. with mathematical honors and, despite excursions into literature and extensive writings in theology, managed to maintain an active interest in mathematics, contributing numerous papers on the subject to the *Transactions of the Royal Society of Edinburgh*, of which Society he was a Fellow.

multiplied together give $\frac{12}{35}$ as the probability against the conclusion; i.e., the chance that the work may *not* be his, notwithstanding the reasons for believing that it is; and consequently, the probability in *favour* of the conclusion will be $\frac{23}{35}$, or nearly $\frac{2}{3}$.

Whately's derivation in the first paragraph of this quotation is essentially the same as that of De Morgan's *Problem* 3 (see our §2.3), except that Whately does not neglect to say that the two 'indications' of the truth of the conclusion are 'distinct and independent'. If we let A_1 and A_2 be the two independent "indications of the truth of a proposition" C, then "unless both of them fail, the proposition must be true" translates to '$(A_1 \lor A_2) \to C$'. Then, assuming the probability of the conclusion to be that of the premise, we have Whately's

$$P(C) = P(A_1 \lor A_2)$$
$$= 1 - P(\neg A_1 \neg A_2)$$
$$= 1 - P(\neg A_1)P(\neg A_2).$$

However Terrot claims that Whately's reasoning must be erroneous as it leads to inconsistency. He goes on to show that using it one can arrive at the result that the probabilities for and against the conclusion in Whately's example (in the second paragraph of the quotation) do not add up to 1 (*Terrot 1857, Boole 1952*, 488):

> Let us then take as the proposition whose probability is to be found, the negative—he did not write it—the partial probabilities *for* which are by the data $\frac{3}{5}$ and $\frac{4}{7}$. The opposite probabilities are now $\frac{2}{5}$ and $\frac{3}{7}$, and their product is $\frac{6}{35}$, the probability *against* the conclusion whose probability we are now seeking. Consequently, $1 - \frac{6}{35} = \frac{29}{35}$ is the probability *for* our conclusion, namely that he did not write the book. But by the former calculation, the probability of the same conclusion was found to be $\frac{12}{35}$: and, as these incompatible results follow from the same principle and method, the principle and method must be erroneous.

After discussion, and finding similar fault with a solution by De Morgan (discussed by us in §2.2 above), Terrot goes on to present his own solution—but based on a significantly altered reformulation (p. 490):

> Let us now consider the Problem under the following form. *A, whose veracity is undoubted, states that, from his knowledge of the facts of the case, the probability of the event E is p/q. B, under the same conditions states, that it is r/s. Supposing the facts known*

by each to be altogether distinct, what is the proper measure of the
expectation formed in a third mind by these two statements?

Ignoring the epistemic dress of this formulation we note that it is not
about the probabilities of arguments (premises) which 'indicate' a conclusion but, rather, about the probabilities of an event, given certain facts;
in other words Terrot's version involves conditional probabilities. This is
borne out by the attempt at a solution he presents (which we shall not
reproduce). It leads him to "an infinite series of infinite series" which he
cannot sum: "If they can be summed their sum divided by the infinite of
the second order n^2, is the probability required."

Boole makes a contribution to the discussion at hand in an appendix
to his Keith Prize Memoir (*Boole 1857*). After quoting the passage from
Whately (given by us at the beginning of this section) he says (*1952*, 385):

> A confusion may here be noted between the probability that a
> conclusion is proved, and the probability in favour of a conclusion
> furnished by evidence which does not prove it. In the proof and
> statement of his rule, Archbishop Whately adopts the former view of
> the nature of the probabilities concerned in the data. In the exempli-
> fication of it, he adopts the latter. He thus applies the rule to a case
> for which it was not intended and to which it is, in fact, inapplicable.

In other words, Boole is bringing out the distinction between the probability of C when it is the case that C is a consequence of A, and the
(conditional) probability of C supposing that A. He then points out that
the probabilities in Whately's numerical example are conditional probabilities and that a result he (Boole) has obtained earlier in his paper can
produce a solution to the problem when the probabilities are so interpreted
(*1952*, 384):

> Taking the problem in its intended meaning, each of the frac-
> tions, $\frac{2}{5}$, $\frac{3}{7}$, measuring, not the probability of the truth of certain
> premises, but the probability drawn from these premises, as condi-
> tions, in favour of a certain supposition (I use this word in preference
> to conclusion), we are no longer permitted to apply the formula above
> determined. And we are not permitted to do so, because the proba-
> bilities with which we are concerned are conditional, and their posses-
> sion of this character greatly increases the difficulty of the problem.
> Its rigorous formal solution is given in Art. 34, and shows that the
> probability sought is, generally speaking, indefinite,—a result which
> agrees with the conclusions of Bishop Terrot, by whom the error to
> which attention as been directed was first point out, Transactions of
> the Royal Society of Edinburgh, vol. xxi., p. 369.

It seems to us that Boole is overly generous in his remarks with regard to Terrot's accomplishment, for it was he himself who had clearly identified the cause of the inconsistency which Terrot had produced. Moreover, Terrot had not *shown* that the probability sought is in general indefinite, but merely asserted "In no case, except when $p/q = r/s$, so far as I can see, can the sum of sums, or the whole probability, be determinately expressed". Boole's Art. 34 solution is markedly different from Terrot's. This is how Boole formulates the problem (*1952*, 355):

> Required the probability of an event z, when two circumstances x and y are known to be present—the probability of the event z, when we only know of the existence of the circumstance x being p,—and its probability when we only know of the existence of y being q.

In modern notation this is

$$\text{Given:} \quad P(z\,|\,x) = p,\ P(z\,|\,y) = q$$
$$\text{find:} \quad P(z\,|\,xy). \tag{1}$$

Note that the question about how to combine the two conditional probabilities of the same event is resolved by Boole's taking the conjunction xy as the condition. However, he does not solve this problem, i.e., (1), but replaces it by another in which there are no longer conditional probabilities, these being eliminated by the introduction of new parameters c and c', where $c = P(x)$ and $c' = P(y)$. The result which he gives—too complicated to be reproduced here (see our *1986*, §6.5)—suffers from the general defects of his method described above in §2.5. Our solution to this 4-parameter version is Example 6.74 on page 404 of *Hailperin 1986*. For a solution of the problem as originally stated see Theorem 5.45 below.

Archbishop Whately's popular treatise on logic was not the only one which failed to distinguish between conditional probabilities and the probability of a conditional. The equally popular J. S. Mill's *A System of Logic*, first published in 1843 and going through many editions in his lifetime, also fails to make the distinction. Its Volume II contains a chapter on 'approximate generalizations'. These are statements of the form 'nine out of every 10 A are B', and have for Mill a probabilistic meaning since he says of it (i.e., the example) "there will be one chance in ten of error in assuming that any A, not individually known to us, is a B." We shall render this as a conditional probability: for x arbitrary, $P(x \in B\,|\,x \in A) = \frac{9}{10}$. Mill, as with Whately, wishes to combine separate probabilities for a conclusion into a single probability (*Mill 1879*, 147–48):[16]

---

16. We quote the more accessible 1879 edition. The body of it differs from the original *1843* only in some minor stylistic changes but, significantly, in the inclusion of a lengthy footnote to be presently mentioned.

When approximate generalizations are joined by way of addition
we may deduce from the theory of probabilities laid down in a former
chapter, in what manner each of them adds to the probability of a
conclusion which has the warrant of them all.

But rather than a general demonstration Mill presents the calculation
for a specific instance (*1879*, 135–36):

> If, on an average, two of every three As are Bs, and three of every
> four Cs are Bs, the probability that something which is both an A
> and a C is a B, will be more than two in three, or than three in four.
> Of every twelve things which are As, all except four are Bs by the
> supposition; and if the whole twelve, and consequently those four,
> have the characters of C likewise, three of these will be Bs on that
> ground. Therefore, out of twelve which are both As and Cs, eleven
> are Bs.

The inference can be given the symbolic rendition:

$$P(x \in B \mid x \in A) = \tfrac{2}{3}$$
$$P(x \in B \mid x \in C) = \tfrac{3}{4}$$
$$[P(x \in B \mid x \in A \wedge x \in C) \geq \tfrac{2}{3}, \; \geq \tfrac{3}{4}]$$
$$\text{therefore} \quad P(x \in B \mid x \in A \wedge x \in C) = \tfrac{11}{12}.$$

The demonstration Mill offers is faulty. He assumes he has 12 things
which are A and C. Of these 12 four (1/3 of 12) are, by virtue of their
being As, not Bs. But then he says that by virtue of their being Cs three
of these four (already said to be not Bs!) are Bs. Hence leaving 11 of the
12 as Bs. Mill goes on to present a second demonstration (p. 136):

> The argument would in the language of the doctrine of chances be
> thus expressed: the chance that an A is not a B is 1/3, the chance
> that a C is not a B is 1/4; hence if the thing be both an A and a C,
> the chance is 1/3 of 1/4 = 1/12.*

The footnote * (omitted here) goes on at length—for the better part of
two half pages of fine print—revealing Mill's difficulties with this inference
form, but which he now thinks he has overcome. That his inference, now
in complementary form,

$$P(x \in \overline{B} \mid x \in A) = \tfrac{1}{3}$$
$$P(x \in \overline{B} \mid x \in C) = \tfrac{1}{4} \tag{2}$$
$$\text{therefore} \quad P(x \in \overline{B} \mid x \in A \wedge x \in C) = \tfrac{1}{3} \cdot \tfrac{1}{4} = \tfrac{1}{12},$$

is incorrect can be shown by Terrot's counterexample: for if the inference form (2) were correct replacing $\overline{B}$ by B, and 1/3 by 2/3, and 1/4 by 3/4, would result in premises equivalent to the original premises. But the resulting conclusion, $P(x \in B \mid x \in A \wedge x \in C) = \frac{2}{3} \cdot \frac{3}{4} = \frac{1}{2}$, is incompatible with that of (2), the two probabilities not adding to 1.

Mill goes on to explain that "This argument [for the inference form] presupposes (as the reader will doubtless have remarked) that the probabilities arising from A and C are independent of one another." But even adjoining this additional hypotheses does not render a valid inference.

We believe Mill's difficulties stem from a faulty interpretation of his notion of 'approximate generalization', which is supposed to generalize 'All As are Bs'. He thinks that 'a fraction p of the As are Bs'—rendered by us as $P(x \in B \mid x \in A) = p$—will do it. But, as we have seen, this cannot be correct. If instead of the conditional probability one uses the probability of the conditional, i.e., $P(x \in A \rightarrow x \in B) = p$, then the calculations come out in agreement with Mill's. For, given

$$P(x \in A \rightarrow x \in B) = p \quad \text{and} \quad P(x \in C \rightarrow x \in B) = q,$$

and, supposing that $x \in A\overline{B}$ and $x \in C\overline{B}$ are independent, then

$$
\begin{aligned}
(1 - p)(1 - q) &= P(x \in A \wedge x \notin B)P(x \in C \wedge x \notin B) \\
&= P(x \in A\overline{B})P(x \in C\overline{B}) \\
&= P(x \in A\overline{B} \wedge x \in C\overline{B}) \\
&= P(x \in A\overline{B}C) = P(x \in AC \wedge x \in \overline{B});
\end{aligned}
$$

so that, on taking complements,

$$P(x \in AC \rightarrow x \in B) = 1 - (1 - p)(1 - q).$$

(Note that Mill's assumption that A and C are independent is not sufficient; one needs, rather, the independence of $A\overline{B}$ and $C\overline{B}$.) A contributing factor to Mill's confusion could be that $P(B \mid A) = p$ and $P(A \rightarrow B) = p$ *are* equivalent when $p = 1$. Mill takes the former to be his 'approximate generalization' but calculates with it as if it were the latter.

§2.7. Peirce and MacColl—Conditional probability symbolized

In a very early paper of C. S. Peirce's (On an Improvement in Boole's Calculus of Logic, *1867*) occur several items on probability which should be included in our study. One of these—historically the earliest—is his introduction of a symbol for conditional probability, and a list of some of its formal properties. Another is his criticism of Boole's (mild) confusion, mentioned above, between the probability of a conditional and conditional probability. In connection with this Peirce presents his solution of the problem of the probability of a 'hypothetical syllogism' in which the premises and conclusion are conditional probabilities. A third is his mention of an error in Venn's *Logic of Chance (1866)*. We shall reproduce Peirce's remarks on the error as it involves, once again, the problem of combining 'evidence' discussed in our preceding section. We shall need a brief introduction to Peirce's notation and conceptions, the two not being readily discussed separately.

As we know, Boole used arithmetical symbols to represent class operations. One of Peirce's 'improvements' on Boole's calculus was to replace its peculiar logical addition by class union. (Apparently this was independent of Jevons' introduction of it a few years earlier.) Peirce also departed notationally from Boole by using a comma underneath an arithmetic operation or relation symbol (plus, times, equality, ...) to indicate that these are being used in a logical rather than an arithmetical sense. This accords with the sound principle of not using the same symbol with different meanings except in distinct contexts. But then he violates this principle when introducing probability, which he identifies with relative frequency (*1867; 1984*, 18):

> Let every expression for a class have a second meaning, which is its meaning in an equation. Namely, let it denote the proportion of individuals of that class to be found among all the individuals examined in the long run.
> Then we have

(27)   If $a = b$ [then] $a = b$

(28) $a + b = (a + b) + (a, b).$

With some care the scheme can be made to work for explicitly given expressions. Thus in Peirce's (28) the first occurrences of a and b are taken numerically because of the infixed '+', but then the second occurrences are to be taken as logical classes (infixed '+' with a comma underneath);

however the result of this class union, and the class intersection (a, b), are to be taken numerically. Boole, in contrast, would have written (28) as

$$n(a) + n(b) = n(a + \bar{a}b) + n(ab).$$

Although Boole uses '+' here in two senses, clear distinction is maintained by use of the numerical operator 'n' (See *Boole 1854*, chapter XIX, Of Statistical Conditions). As we have seen, Boole makes the transition from the logical to the numerical (or algebraic) form of an equation by arranging to have his logical equations use *or* only disjunctively and also arranging that it only connects terms which are conjunctions of stochastically independent events. Peirce may have had a similar end result in mind, but without assuming stochastic independence unless there are positive (material) reasons for it—an aspect of Boole's theory he was highly critical of. His (28) establishes the connection between addition of probabilities and non-exclusive *or*, but for logical conjunction he needs a supplementary notion. This notion corresponds to conditional probability, but for Peirce it is a distinctive notion, not only denoting relative frequency but also having an associated class meaning (*1867; 1984*, 18):

> Let b_a denote the frequency of b's among the a's. Then considered as a class, if a and b are events, b_a denotes the fact that if a happens b happens.

(29) $$ab_a = a, b.$$

Peirce doesn't explain what class it is that b_a is being considered as, or how it 'denotes' a fact. He goes on (p. 18):

> It will be convenient to set down some obvious and fundamental properties of the function b_a.

(30) $$ab_a = ba_b$$

(31) $$\phi(b_a \text{ and } c_a) = (\phi(b \text{ and } c))_a$$

(32) $$(1 - b)_a = 1 - b_a$$

(33) $$b_a = \frac{b}{a} + b_{(1-a)}(1 - \frac{1}{a})$$

(34) $$a_b = 1 - \frac{1-a}{b}b_{(1-a)}$$

(35) $$(\phi a)_a = (\phi(1))_a.$$

Since Peirce here uses '=' without a comma the terms on the right and left hand sides are to be taken in the numerical sense. Formula (30) is indeed obvious from (29) and the commutivity of logical multiplication (i.e., $(a, b) \doteq (b, a)$). Skipping over (31) for the moment, the formulas (32)–(34) are evident on the basis of relative frequency, especially if (33) and (34) are written in the more perspicuous algebraically equivalent symmetric forms,

$$ab_a + \bar{a}b_{\bar{a}} = b \quad \text{and} \quad ba_b + \bar{b}a_{\bar{b}} = a.$$

Formula (35) is evident by (29) and the logical identity $(a, \phi(a)) \doteq (a, \phi(1))$. It is not clear how to make sense of (31). We first note that his '$\phi(b$ and $c)$' represents a logical function of the two arguments b and c—presumably Peirce uses 'and' in place of the customary comma separating function arguments so as to avoid confusion with his symbol for class intersection (i.e., a comma under the empty space for juxtaposition). Thus the right hand side of (31) has a clear meaning which, in modern notation, would be written $P(\phi(b, c) \,|\, a)$. But what does the left hand side mean? If, for example, $\phi(b$ and $c)$ were to be class union $b + c$, then the right hand side is $(b + c)_a$. But then the left hand side is $b_a + c_a$, to which Peirce has not assigned any numerical meaning.

Despite the unintelligible (31) Peirce is quite clear on the nature of conditional probability and that it differs from probability of a conditional. Indeed, he severely criticizes Boole for confusing the two, in particular Boole's use of conditional probabilities in a probabilistic version of the hypothetical syllogism (discussed by us in §2.5 above). Moreover, he contends that Boole's solution to the hypothetical syllogism problem (that is, Boole's value for the probability of the conclusion) is anyhow incorrect. He then goes on to present his solution. It contains seven introduced parameters and is incomprehensible to us.

A concluding paragraph in Peirce's 1867 paper refers to a mistake Venn makes in answering a question which Venn had concocted in his *1881* to illustrate a point. The question is: Suppose an insurance company knows that nine-tenths of the Englishmen who go to Madeira die, and nine-tenths of consumptives who go there get well. How should the insurance company treat consumptive Englishmen?

The question involves conditional probabilities

$$P(D \,|\, E), \quad P(D \,|\, C), \quad P(D \,|\, CE)$$

and is thus a logico-probability problem of the kind discussed in our preceding section. In his *Logic of Chance*, wherein the question is posed, Venn argues that the "Office could safely carry out its proceedings" either insuring consumptive Englishmen as consumptives or as Englishmen. Referring to this solution in his review of Venn's book, Peirce says (*1867b*, 319 footnote):

> *This is an error. For supposing every man to be insured for the same amount, which we may take as our unit of value, and adopting the notation,
>
> (c, e) = number of consumptive Englishmen insured.
>
> $(c, \overline{e})$ =   ”   consumptives not English   ”
>
> $(\overline{c}, e)$ =   ”   not consumptive English   ”
>
> x = unknown ratio of consumptive English who *do not die* in the first year.
>
> The amount paid out yearly by the company would be, in the long run,
>
> $$\tfrac{1}{10}(c, \overline{e}) + \tfrac{9}{10}(\overline{c}, e) + x(c, e),$$
>
> and x is unknown. This objection to Venn's theory may, however, be waived.

Peirce gives no indication of how he obtained this result. In a Note to this remark (*Peirce 1984*, 514) the editor says that both Venn's and Peirce's analysis of the problem are wrong, that the actual payment would be "$x(c, \overline{e}) + y(\overline{c}, e) + z(c, e)$ and x, y and z are unknown". If Peirce's formula were correct, the note continues, then "x would have to be both $\tfrac{1}{10}$ and $\tfrac{9}{10}$". Apparently the editor thinks that Peirce's terms represent the [expected] number dying in each of the classes $(c, \overline{e})$, $(\overline{c}, e)$ and (c, e). But the coefficient $\tfrac{1}{10}$, for example, is not the probability of a member of $(c, \overline{e})$ dying but that of a member of c (the class of consumptives). Hence Peirce's $\tfrac{1}{10}(c, \overline{e})$ is not the number of the $(c, \overline{e})$ who die. We believe Peirce's formula with the one parameter is correct, and for the following reason. Our explanation also gives a characterization of the x in Peirce's formula.

Assuming the universe of discourse to be $C \vee E$, so that $D = DC \vee DE$, one readily obtains by elementary probability rules:

$$P(D) = P(D \mid C)P(C) + P(D \mid E)P(E) - P(D \mid CE)P(CE).$$

Replacing $P(C)$ by $P(CE) + P(C\overline{E})$, $P(E)$ by $P(CE) + P(\overline{C}E)$, $P(D \mid CE)$ by $1 - P(\overline{D} \mid CE)$, and using Venn's data $P(D \mid C) = \tfrac{1}{10}$, $P(D \mid E) = \tfrac{9}{10}$, one finds

$$P(D) = \tfrac{1}{10}P(C\overline{E}) + \tfrac{9}{10}P(\overline{C}E) + P(\overline{D} \mid CE)P(CE),$$

where $P(\overline{D}\,|\,CE)$ corresponds to Peirce's x. Although he says that x is unknown, it can be shown that x, i.e., $P(\overline{D}\,|\,CE)$, can have any value in the unit interval (Theorem 5.45, below). Thus as Peirce implies (though doesn't show), it could be disadvantageous to the insurance company to include CE with either $C\overline{E}$ or $\overline{C}E$ since that would mean replacing x by either $\frac{1}{10}$ or $\frac{9}{10}$, whereas it might be more than $\frac{9}{10}$.

Somewhat later and independently of Peirce, Hugh MacColl had introduced a symbol for conditional probability and investigated its properties in some detail. We conclude our chapter with a summary of his work as it relates to our topic.

With Boole propositional logic was a special application of his class calculus algebra (for the details see our *1986*, chapter 3). It is noteworthy that, contemporaneously with Frege, MacColl had developed propositional logic as a branch of logic independent of the class calculus (or term logic of the traditional syllogism). In his 'The Calculus of Equivalent Statements and Integration Limits' (*1877*) MacColl presents a "new analytic method" which is clearly a propositional logic of *not*, *and* and *or*. His view of its future usefulness seems to us (with hindsight) quite restrained: "The chief use of the method, as far as I have yet carried it, is to determine the new limits of integration when we change the order of integration or the variables in a multiple integral,..." A second follow-up paper (*1878*) brings in the conditional ("implication") as a logical connective and its chief properties delineated. According to MacColl, it was his interest in solving probability problems (of the puzzle variety) which stimulated him to develope his form of symbolic logic. It also led him to the introduction of the symbol for conditional probability. We quote from his *1880*, 113:

> In applying symbolic logic to probability, the following notation will, I think, be found useful.
> Def. 1.—The symbol x_a denotes the *chance* that the statement x is true *on the assumption that the statement* a *is true*.

Although outwardly MacColl's notation for conditional probability is the same as that of Peirce, conceptually it differs from it in that the two components are statements rather than classes. Moreover, MacColl attaches no logical, i.e., non-numerical, meaning to his x_a. The absolute (i.e., unconditional) probability of a statement x is written 'x_ϵ': "The symbol ϵ is an equivalent for the [logical] symbol 1, and denotes any statement whose truth is taken for granted *throughout the whole of an investigation*." Thus, unlike with Boole and Peirce, MacColl's notation clearly distinguishes between an argument (event, proposition, class) and its probability.

We present a summary of the conditional probability items in MacColl's *1880*. He has no formal development—demonstrations are based on informally understood meanings and by the use of examples in which random points are selected from geometrical regions with probabilities assumed to be in proportion to the areas of the regions from which the selections are made.

1. It is noted that an implication, a implies x, is equivalent to the probability statement $x_a = 1$.

2. He defines a and x to be *independent* if any of the four conditions holds (the prime denotes negation):

$$x_a = x_\epsilon, \quad a_x = a_\epsilon, \quad x_a = x_{a'}, \quad a_x = a_{x'}.$$

These are all provably equivalent to the usual product formulation of independence which, in MacColl's notation, would be

$$(ax)_\epsilon = a_\epsilon x_\epsilon.$$

3. Results of the following kind are produced:

$$x_a = \frac{(ax)_\epsilon}{a_\epsilon}$$
$$(a+b)_x = a_x + b_x - (ab)_x,$$

the first of these being the usual present-day definition of conditional probability.

4. Some results, e.g.,

$$x_{x+y} = \frac{x_\epsilon}{(x+y)_\epsilon}, \quad (xy)_x = y_x,$$

involve a bit of logic.

5. The 'fundamental rule in The Inverse Method of Probability' is easily derived from his rules in the form

$$a_x = \frac{a_\epsilon x_a}{\alpha_\epsilon x_\alpha + \beta_\epsilon x_\beta + \gamma_\epsilon x_\gamma + \ldots},$$

on the assumption that x implies $\alpha + \beta + \gamma + \ldots$.

All these results, while not particularly deep, do show the gain in clarity, precision, and generality, resulting from the introduction of a suitable notation for conditional probability.

MacColl illustrates his rules by solving two problems from Boole's *Laws of Thought*, one of them being Boole's 'Challenge Problem'—so dubbed since it was proposed by Boole as a test of the "received" theory of probability as against his. (For the history see our *1986*, §6.2.) The problem is the following:

An event E can happen only as a result of either of two 'causes' A_1 or A_2. The probability of A_1 is c_1 and tht of A_2 is c_2. Also the probability of E, if A_1, is p_1 and that of E, if A_2, is p_2. Required the probability of E. In modern notation

$$\text{Given:} \quad P(A_1) = c_1, \ P(A_2) = c_2$$
$$P(E \mid A_1) = p_1, \ P(E \mid A_2) = p_2,$$
$$E \to A_1 \vee A_2,$$
$$\text{find:} \quad P(E).$$

MacColl's solution runs as follows. Since, by hypothesis, $E \leftrightarrow E(A_1 \vee A_2)$,

$$
\begin{aligned}
P(E) &= P(EA_1) + P(EA_2) - P(EA_1 A_2) \\
&= P(A_1)P(E \mid A_1) + P(A_2)P(E \mid A_2) - P(A_1 A_2)P(E \mid A_1 A_2) \\
&= c_1 p_1 + c_2 p_2 - P(A_1 A_2)P(E \mid A_1 A_2). \quad (1)
\end{aligned}
$$

Boole's method yielded a solution depending only on the four parameters c_1, c_2, p_1, p_2, namely: the unique value of u in the probability range satisfying the equation

$$
\frac{(u - c_1 p_1)(u - c_2 p_2)}{c_1 p_1 + c_2 p_2 - u} = \frac{(1 - c_1(1 - p_1) - u)(1 - c_2(1 - p_2) - u)}{1 - u} \quad (2)
$$

MacColl attempts to prove Boole's solution to be wrong, but we have found his argument faulty (*Hailperin 1986*, 366–68). In a subsequent paper, *MacColl 1897*, he recurs to the problem and this time notes that the third term on the right in (1) "may have numberless values, all consistent with the data" whereas Boole's equation (2) produces only one value (in the probability range). He asserts: "The fallacy that vitiates Boole's whole reasoning on probability is to be found in his definition of independent events". This isn't entirely correct: the source of Boole's deviancy from "received" probability theory is his assumption that one can "ascend" to simple independent events representing the compound events of the problem, and by appropriately conditioning based on the logical relations of these compounds, reproduce the problem in terms of such conditioned events. When there are no logical relations, then the events are treated as independent (see §2.5 above).

Our solution to Boole's Challenge Problem is presented in §5.5 below.

Chapter 3

The Twentieth Century

§3.1. Confirmation—from Keynes to Carnap

Beginning with Francis Bacon (1561–1626), and continuing on up to the present, an array of writers concerned themselves with the question of inductive inference. The absence of a rational basis for such inferences was pointedly brought out by Hume (1711–1776). And, once it became clear that inductive inferences could not attain certainty, it was natural to think of introducing probabilistic notions. Perhaps the earliest attempts at linking probability with contingent inferences of this nature were those associated with Bayes' theorem and inverse probability. Here, as indicated in §1.6 above, the interest is in the probability of an event occurring *in a single trial* (Bayes) or on the *next* occasion (Laplace) if it has occured n times previously. This is to be contrasted with the more usual inductive inference which has a universal generalization for its conclusion. (In this connection it might be noted that the problem of how to associate probability with universally quantified propositions has, as yet, no recognized satisfactory solution.)

With regard to generalized inductive inferences we have the pessimistic opinion of C. D. Broad who in a paper on the 'logic of inductive inference' wrote (*1918*, 26):

> ... (1) that unless inductive conclusions be expressed in terms of probability all inductive inference involves a formal fallacy; (2) that the degree of belief which we actually attach to the conclusions of well-established inductions cannot be justified by any known principle of probability, unless some further premise about the physical world be assumed; and (3) that it is extremely difficult to state this premise so that it shall be at once plausible and non-tautologous.

Notwithstanding, there were, in the first half of the twentieth century, at least three efforts at a grand synthesis of foundations of probability and inductive inference. We shall only briefly discuss them—two in this section and the third in the next—confining our attention just to aspects relevant to our topic; we wish to distinguish our concept of probability logic from that of these efforts, which are often described as being a "logic of probable inference". To avoid possible misunderstanding, we emphasize that we are not judging the larger aims of these studies.

Keynes' *A Treatise on Probability (1921)* opens with a quotation from Leibniz which expresses the hope for a new kind of logic that would treat of degrees of probability. Apparently fulfilment of this hope was for Keynes a principal aim. But a great part of his *Treatise* is devoted to matters of philosophical analysis, to criticism of previous writings, to applications of probability, and to a justification of induction, little of this being related our study. Only slightly more than two of its 33 chapters have a connection with what we consider to be probability logic. Keynes characterizes this portion as follows (*1921*, 133):

> The object of this [chapter XII] and the chapters following is to show that all the usually assumed conclusions in the fundamental logic of inference and probability follow rigorously from a few axioms, in accordance with the fundamental conceptions expounded in Part I. This body of axioms and theorems corresponds, I think, to what logicians have termed the *Laws of Thought*, when they have meant by this something narrower than the whole system of formal truth. But it goes beyond what has been usual, in dealing at the same time with the laws of probable, as well as of necessary, inference.

What Keynes here refers to as "the fundamental conceptions expounded in Part I" may be briefly outlined as follows. It is asserted that all propositions are true or false.[1] Probability (less than certainty) only arises when there is an argument for a conclusion which does not follow demonstratively from the premises. If knowledge of h justifies a rational belief in a of degree α, it is said that there is a probability relation of degree α between a and h, and this is written '$a/h = \alpha$'.[2] In some cases this degree may be numerical, but need not be in general. Keynes does not consider

---

1. This assertion was disputed even in ancient times. See, for example, the summary of the early history of this doctrine in the Appendix 'On the history of the law of bivalence' of *Lukasiewicz 1930* (English translation in *Lukasiewicz 1970*, 176–78). In modern times quantum mechanics provides some grounds for believing that bivalence of propositions is not necessarily applicable to the physical world.

2. Admitting to its notational deficiencies Keynes nevertheless adopts it instead of '$P(a/h) = \alpha$' as being 'less cumbrous'.

the question of reconciling uniqueness of the degree of rational belief with the possibility of there being many *different* non-demonstrative arguments from one and the same given set of premises to a given conclusion—for example, the one-step inference '*h* to *a*' is itself a non-demonstrative argument (except if the *a* happens to be a logical consequence of *h*). His (i.), p. 135, postulates uniqueness if *h* is not an inconsistent conjunction.

Many subsequent writers (though not all) refer to Keynes' concept as 'degree of confirmation' rather than as probability; or, as in Carnap (*1950*, 2nd edition *1962*) as a special kind of probability—'probability$_1$', in contrast to 'probability$_2$' which pertains to relative frequency. Sometimes it is referred to as 'epistemic' or as 'inductive' probability. Keynes' belief that the notion is fundamentally non-numerical has had little following and many writers drop this feature; similarly for his epistemic-mentalistic description in terms of knowledge and belief. Nicod (*1930*, 207), for example, simply refers to a Keynes relation of probability as a *probable inference*. We discuss his few brief remarks about it.

The material occurs at the beginning of Nicod's lengthy essay on induction, which is taken as a species of inference. Two kinds of inference are distinguished (*Nicod 1930*, 207):

> We first regard inference as the perception of a connection between the premises and conclusion which asserts that the conclusion is true if the premises are true. This connection is implication, and we shall say that an inference grounded on it is a *certain inference*. But there are weaker connections which are also the basis of inferences. They have not until recently received any universal name. Let us call them with Mr. Keynes *relations of probability* (*A Treatise on Probability*, London, 1921, ch. 1.). The presence of one of these relations among the group A of propositions and the proposition B indicates that in the absence of any other information, if A is true, B is probable to a degree *p*. A is still a group of premises, B is still a conclusion, and the perception of such a relation between A and B is still an inference: let us call this second kind of inference *probable inference*.

Nicod's phrase "in the absence of any other information" we take to mean that only the premises in *A* (or *A*, if a single premise) are to be considered. This is the normal way one thinks of inference. His understanding of a Keynes probability relation, $P(B/A) = p$, is that 'if *A* is true, *B* is probable to a degree *p*'. We take 'indicates' to mean 'implies'. In his discussion immediately following, the '*A* is true' becomes '*A* is certain'. Nicod is thus affirming the correctness of the inference form

$$\frac{P(B/A) = p, \ A \text{ is certain}}{B \text{ is probable to a degree } p}$$

or, if we write it to accord with Keynes' view that probability is a two-argument notion,

$$\frac{P(B/A) = p, \; P(A/t) = 1}{P(B/t) = p}$$

where 't' stands for a logically true sentence. But affirming the correctness of this inference form and nothing else, tells us little about $P(B/A) = p$. (Compare, for example, a proposed explanation of the ordinary conditional $C(A, B)$ in propositional logic which says that it indicates

$$\frac{C(A, B), \; A}{B}.$$

Here either AB, $A \rightarrow B$, or $A \leftrightarrow B$ are all interpretations for $C(A, B)$ which produce a valid inference form.)

Nicod then goes on to the case of probable premises (*1930*, 208):

> We have so far considered only premises that are certain. But any inference which yields something, starting from premises taken only as certain, still yields something starting from premises taken only as probable, and this holds for both types of inferences, certain and probable. We can even assert that starting from premises which, taken together have a probability p, a certain inference will confer on its conclusion the same probability p; and a probable *inference*, which would confer on its conclusion the probability q if its premises were certain, will confer on its conclusion the probability pq.

Both of the assertions in this quotation are incorrect. In the first place a certain (i.e., necessary) inference having a premise with probability p,[3] confers on its conclusion not probability p but probability *not less that* p. (For a partial history of this error and its correction by Bolzano, see *Hailperin 1988b*, §§4–5, also §2.1 above). As for the second assertion, the claim is

$$\frac{P(B/A) = q, \; P(A/t) = p}{P(B/t) = pq}.$$

The conclusion here can't be valid since by the multiplication rule (*Keynes 1921*, 135),

$$pq = P(B/At)(P(A/t) = P(AB/t).$$

However, since $P(B/t) \geq P(AB/t)$, it would be valid if the conclusion were

$$P(B/t) \geq pq.$$

---

3. This could mean that the premises of the inference are either '$P(B/A) = 1$, $P(A) = p$', or '$P(A \rightarrow B) = 1$, $P(A) = p$'. The difference is immaterial since, for $p > 0$, '$P(B/A) = 1$' is equivalent to '$P(A \rightarrow B) = 1$'. (See Theorem 5.36 below.)

We return to Keynes' *Treatise*. In chapter 12 he presents an axiomatic formulation for his a/h notion. But what one finds there could not pass muster by current standards. For example, there is no listing of undefined notions, no specification of proper syntactic structure, and no explicit rules of inference; further, some definitions are conditional (so that eliminability is in doubt), while some lack axioms to make them operative. Rather than introducing this system—which would require extensive emendatory discussion—we shall instead present what we believe to be an adequate surrogate, namely the axiomatization of 'confirmation' due to J. Hosiasson(-Lindenbaum) (*1940*, 133):

> Let a, b, and c be variable names of sentences belonging to a certain class,[1] the operations $a.b$, $a + b$, and $\bar{a}$, the (syntactical) product, the sum and the negation of them. Let us further assume the existence of a real non-negative function $c(a, b)$ of a and b, when b is not self-contradictory. Let us read '$c(a, b)$' 'degree of confirmation of a with respect to b' and take the following axioms:
>
> Axiom I.   If a is a consequence of b, $c(a, b) = 1$.
> Axiom II.   If $\overline{a.b}$ is a consequence of c,
>
> $$c(a + b, c) = c(a, c) + c(b, c).$$
>
> Axiom III. $c(a.b, c) = c(a, c) \cdot c(b, a.c)$.
> Axiom IV.   If b is equivalent to c, $c(a, b) = c(a, c)$.[2]
>
> As may easily be seen, the interval of variation of c is $(0, 1)$; this is quite conventional.
>
> [1] The class must be broad enough to include all sentences for which we desire to speak about confirmation.
>
> [2] These axioms are analogous to St. Mazurkiewicz's system of axioms for probabilities (see *Zur Axiomatik der Wahrscheinlichkeitsrechnung, Comptes rendus des séances de la Société des Sciences et des lettres de Varsovie*, vol 25 (1932)).

What is here referred to as 'a certain class' is not narrowly specified, but by virtue of the presence of logical sum, product, and negation in the axioms it is clear that the class would have to be closed under these operations. Conceivably the sentences of the class could involve additional logical structure (e.g., quantifiers) but this can play no role in confirmation theory (as based on these axioms) since the axioms show no interaction of the c function except with propositional connectives. Although, by Axiom IV, provision is made for replacement in the second argument place of the c function of one (logically) equivalent sentence by another, there is none justifying replacement in the first position. We take this to be an oversight and shall assume that Axiom IV is modified so as to include this provision. Subsequent to this axiomatization by Hosiasson other formulations

of confirmation have appeared. However, the differences between these and
Hosiasson's are minor and have no relevance to our study.

Let t be some fixed logically true sentence. Then $c(a, t)$ is a one-place
function of the sentence variable 'a'; if $c(a, t) \neq 0$ one readily proves from
axioms III, I, and IV, that

$$c(b, a) = c(a.b, t)/c(a, t). \tag{1}$$

Thus the two-place $c(b, a)$ is expressible in terms of a one-place function. It
is readily seen that when the formal $c(a, t)$ is interpreted as a probability,
$P(a)$, and (1) taken to be a definition of conditional probability, $P(b \mid a)$, the
above axioms for confirmation become correct assertions about conditional
probability. Thus the axioms of Hosiasson are insufficient to characterize
anything different from conditional probability. If, as in the opinion of some
writers, the confirmation notion is central to inductive logic, something
more needs to be added.

During the 1940s two separate approaches were initiated to define confirmation as a syntactic notion so as to provide a basis for inductive logic. For
our purposes it will suffice to sketch one of these—that due to Carnap—as
it is the one which was extensively developed and is widely known. (For
the other, see *Hempel 1943*.)

Unlike Keynes and Hosiasson who, in connection with confirmation, rec-
ognize that sentences are combinable by sentential connectives but say
nothing further about the syntax of the language, Carnap assumes that he
has a specific (type of) language for which his explicatum for the informal
probability$_1$, generically denoted '$c(h, e)$', is to be defined. He contends
that the notion should be 'L-determinate', i.e., depend only on the logical
structure of h and e. Initially only simple languages are considered. The
ultimate goal is a definition of the notion in a language adequate for science.
In what follows we shall only be referring to his $\mathcal{L}_N$, a predicate language
over a universe of N individuals, hence effectively without quantifiers. The
language $\mathcal{L}_N$ has the N individual symbols $a_1, \ldots, a_N$.

Carnap acknowledges that an explicatum for probability$_1$ should have
the commonly accepted general properties of confirmation (as given, for
example, in Hosiasson's axioms). A function having these properties he
calls a regular c-function. However, he argues that a much narrower notion
is needed for inductive logic (*Carnap 1950*, or *1962*, 344):

> If the axioms of a system hold for all regular c-functions, then that
> system represents only a very small part of the theory of probability$_1$.
> This part, it is true, is of great importance because it contains the
> fundamental relations between c-values. But its weakness becomes
> apparent from the following facts. Let e and h be factual sentences

in $\mathcal{L}_N$ such that e L-implies neither h nor $\sim h$. Then a theory of the kind mentioned does not determine the value of $c(h, e)$. Moreover, it does not even impose any restricting conditions upon this value: the assignment of any arbitrarily chosen real number between 0 and 1 is compatible with the theory. ... A theory of this kind states merely relations between c-values; thus if some c-values are given, others can be computed with the help of the theorems. There is an analogous restriction in the theory of probability$_2$; however, here the restriction is necessary. The statement of a particular value of probability$_2$ for two given properties is, in general, a factual statement (10B). Therefore, a logicomathematical theory of probability$_2$ cannot yield statements of this kind but must restrict itself to stating relations between probability$_2$ values. On the other hand, in the case of a theory of probability$_1$, there is no reason for this restriction. A sentence of the form '$c(h, e) = r$' is not factual but L-determinate. Therefore, a logicomathematical theory of probability$_1$, in other words, a system of inductive logic, can state sentences of this form. The fact that the axiom systems for probability$_2$ restrict themselves to statements which hold for all regular c-functions make these systems unnecessarily weak.

At the time he wrote this Carnap thought that an adequate explicatum for probability$_1$ would be something like his function c^*, described in the appendix to his *1950*. He later acknowledged that this was not satisfactory, that c^* was only one in a whole continuum of possible c-functions and then, later, that even these were not entirely satisfactory as a basis for inductive logic (*Carnap and Jeffreys 1971*, 1). Nevertheless, by describing his c^* (for $\mathcal{L}_N$) we can illustrate what sort of conception Carnap had in mind for his inductive logic, and to compare it with our notion of probability logic.

For Carnap a c-function is defined in terms of a measure m over sentences of the language, namely by setting

$$c(h, e) = m(e.h)/m(e) \ \text{ for } m(e) \neq 0.$$

For simple languages $\mathcal{L}_N$ specifying the values m for state-descriptions (basic conjunctions of atoms and negated atoms) is sufficient, these values being positive real numbers whose sum taken over all the state-descriptions equals 1. All other values are then determined via additivity of the measure. To obtain his c^* Carnap specifies a particular measure, m^*, which is (i) 'symmetric' and (ii) has the same value for all 'structure-descriptions' of $\mathcal{L}_N$. A symmetric m function is one which assigns equal values to isomorphic state-descriptions—two state-descriptions being isomorphic if one is obtainable from the other by a permutation of the individual symbols, plus rearranging of conjunctions. A structure-description (determined by a state-description) is the disjunction of all state-descriptions isomorphic to it, arranged in lexicographical order. The value of m^* for a state-description S_i is then given by

$$m^*(S_i) = 1/\tau\zeta_i$$

when τ is the number of state-descriptions of $\mathcal{L}_N$ and ζ_i the number of state-descriptions isomorphic to S_i.

For our purposes we need not entertain this much complexity—it will suffice to limit ourselves to an $\mathcal{L}_1$, a language with one individual, in which case permuting individual symbols is irrelevant and a structure-description is the same as a state-description. We take this $\mathcal{L}_1$ as having the indivdual symbol 'a' and three one-place predicate symbols 'Q_1', 'Q_2', and 'Q_3'. Let $A_i = Q_i(a)$, $(i = 1, 2, 3)$. Then there are only 2^8 logically inequivalent sentences constructible for this language, namely those obtained by taking a disjunction of some, none, or all of the eight state-descriptions

$$A_1A_2A_3, \ \overline{A}_1A_2A_3, \ A_1\overline{A}_2A_3, \ \ldots, \ \overline{A}_1\overline{A}_2\overline{A}_3. \tag{2}$$

To each of these state-descriptions Carnap's theory would assign an equal measure $1/8$. The one-argument $c^*(h, t)$, which by definition is $m^*(h)$, would equal $n(h)/8$, where $n(h)$ is the number of these which imply h. This value is independent of any interpretation of a, Q_1, Q_2, Q_3 and accords with Carnap's conception of (inductive) probability as a measure of closeness to validity based on the syntactic structure of sentence in a given language.

In contrast, probability logic as we are conceiving it is semantically rather than syntactically based. Generality and independence of interpretation are obtained by allowing any assignment of real numbers $k_1, \ldots, k_8$ to the state-descriptions (2), subject only to the requirement that they be non-negative and sum to 1. Specific application determines what particular set of real numbers to choose. On what basis one chooses these eight values, be it statistical, subjective, or merely hypothetical, is not a relevant matter for (pure) probability logic, any more than the basis on which one chooses to assign truth values to sentences when applying verity logic. Carnap would refer to our notion as being 'factual'. We think 'semantic' would be more appropriate as we believe '$P(h \mid e) = p$' to be analogous to the semantic sentence "The truth value of 'Snow is white' is truth" rather than to the factual sentence 'Snow is white'. Our ideas of a probability logic are more fully described in chapter 4 below.

We turn now to a distinctively different approach which also aims at justifying induction via a theory of probability and which its originator also referred to as a 'probability logic'.

§3.2. Reichenbach: probability as multivalued logic

Beginning with his dissertation of 1915, and continuing on through to the 1950s, H. Reichenbach wrote extensively on philosophical and technical matters connected with foundations of probability. Our references will be confined to his *The Theory of Probability (1949)*, a compendium of his settled views.

Reichenbach's notion of probability is, like Keynes', relational though conceptually different. Whereas Keynes thinks of it as a two-argument relation having a (possibly but not necessarily numerical) degree, Reichenbach takes it to be a three-argument relation with two of these arguments being classes of a special kind and the third a real number. For his 'fundamental probability relation' he writes

$$(i)(x_i \in A \underset{p}{\supset} y_i \in B), \tag{1}$$

where '$\underset{p}{\supset}$' is referred to as 'probability implication'. The classes A and B are *thought of* as having members $x_1, x_2, \ldots, x_i, \ldots$ and $y_1, y_2, \ldots, y_i, \ldots$ respectively. These are assumed to be denumerably ordered, and (1) is taken to mean that a probability implication relation holds for all corresponding pairs (x_i, y_i). It is also read as 'if an $x_i \in A$, then $y_i \in B$ with probability p'. Since the universal quantifier '(i)' and the inner structures of A and B play no role in his listing of the formal properties of '$\underset{p}{\supset}$' Reichenbach simplifies (1) to

$$A \underset{p}{\supset} B$$

or, later on, it is written

$$P(A, B) = p.$$

This latter expression is presented as a whole without any meaning attached to the separate parts, i.e., to '$=$', or '(A, B)', or '$P(A, B)$'. To avoid the typographical inconvenience of having complex numerical expressions such as '$\phi(q, r)$' underneath the (probability) implication symbol Reichenbach uses the form

$$(\exists p)(A \underset{p}{\supset} B)(p = \phi(q, r)).$$

(This is how Reichenbach writes it; for a correct rendering the scope of the quantifier should be the entire conjunction, not just the first member.)

We shall first state Reichenbach's axioms and then briefly describe their content (*1949*, 54–62):

Axiom I. $p \neq q \supset [(A \underset{p}{\supset} B).(A \underset{q}{\supset} B) \equiv (\overline{A})]$

Axiom II1. $(A \supset B) \supset (\exists p)(A \underset{p}{\supset} B).(p = 1)$

Axiom II2. $\overline{(\overline{A})}.(A \underset{p}{\supset} B) \supset p \geq 0.$

Axiom III. $(A \underset{p}{\supset} B).(A \underset{q}{\supset} C).(A.B \supset \overline{C}) \supset (\exists r)(A \underset{r}{\supset} B \vee C).(r = p + q)$

Axiom IV. $(A \underset{p}{\supset} B).(A.B \underset{u}{\supset} C) \supset (\exists w)(A \underset{w}{\supset} B.C).(w = p \cdot u)$

According to Axiom I the same probability implication from A to B holds for two distinct reals if and only if A is empty. (By his conventions '$(\overline{A})$' means '$(i)(x_i \in \overline{A})$', i.e., that A has no members). By Axiom II1, the p for $A \underset{p}{\supset} B$ is 1 if A is contained in (implies) B; and by II2, $p \geq 0$ if it is not the case that A is empty. Axioms III and IV are the usual addition and multiplication rules, appropriately framed for the context.

In chapter 4 of his book Reichenbach 'constructs' a frequency interpretation for his $A \underset{p}{\supset} B$ in terms of postulated properties of sequences associated with A and B. However one can give a more general and simpler interpretation: Let P be a probability function defined on a Boolean algebra which is strictly positive (as well as normed to 1 and additive). One readily verifies that Reichenbach's axioms I–IV are satisfied if A and B are arbitrary elements of the Boolean algebra and '$A \underset{p}{\supset} B$' means

$$P(A.B) = p \cdot P(A),$$

i.e., that the conditional probability of B given A is p, but in this variant form rather than the more usual quotient form. (In Axiom I, since in our interpretation P is assumed to be strictly positive, the clause '$(\overline{A})$', i.e., A is empty, is equivalent to '$P(A) = 0$'.) Thus Reichenbach's axioms for his probability implication, as Hosiasson's for confirmation, need supplementation to characterize a notion which is distinguishable from conditional probability based on a one-argument probability function.

Of particular interest for our study is Reichenbach's claim (*1949*, chapter 10) to have constructed a probability logic with continuously many truth values. We believe this claim to be unsubstantiated. Our reasons are listed below after the following discussion.

Reichenbach introduces the notation 'hz_i' for '$z_i \in A$' and 'fx_i' for '$x_i \in B$' and converts his probability form

$$(i)(z_i \in A \underset{p}{\supset} x_i \in B)$$

to

$$(i)(hz_i \underset{p}{\supset} fx_i), \tag{2}$$

which is then written

$$P(hz_i, fx_i) = p. \tag{3}$$

But if read literally this latter form, containing free 'i', would be saying that P maps the i-th pair of corresponding members of the two sequences into p, which is clearly not intended since Reichenbach's interpretation for $P(hx_i, fy_i)$ involves the entire sequences (see (8) below.) Hence a more appropriate notation would be

$$P(\hat{h}, \hat{f}) = p, \tag{3'}$$

where $\hat{h} = \{ hx_i : i \in \omega \}$ and $\hat{f} = \{ fy_i : i \in \omega \}$ are the entire sequences; this would accord with his earlier notation $P(A, B) = p$.

Viewing (3) [better would be (3′)] as an assertion about the two propositional sequences Reichenbach says (p. 395): "The frequency interpretation [of (2) or (3)] is contructed by counting the number of true propositions 'fx_i' within the subsequence selected by the true propositions 'hz_i'." When hz_i is true for each i, Reichenbach writes

$$P(fx_i) = p \tag{4}$$

with frequency interpretation

$$P(fx_i) = \lim_{m \to \infty} \frac{1}{m} \mathop{\mathbf{N}}_{i=1}^{m} \{V[fx_i] = 1\}, \tag{5}$$

whose meaning is that $P(fx_i)$ is the limit of the relative frequency of the number of true occurrences of fx_i in the sequence. (Note that the right-hand side of (5) has no free 'i'.)

Since the probability logic Reichenbach intends to present deals with propositional sequences rather than propositions, he wishes to define operations for propositional sequences analogous to those for propositions. All goes well for those that correspond to the usual connectives. Using parentheses about the general term of a sequence to denote the sequence (hence binding its free 'i') he defines (p. 398):

$$(fx_i) \vee (gy_i) =_{Df} (fx_i \vee gy_i)$$
$$(fx_i).(gy_i) =_{Df} (fx_i.gy_i) \tag{6}$$
$$\overline{(fx_i)} =_{Df} (\overline{fx_i}).$$

These are legitimate definitions of operations on propositional sequences which result in propositional sequences and their P values would be given by (5). (These values ought to be written '$P((fx_i))$'.) But then he goes on with (p. 399, display numberings changed so as to mesh with ours):

We must now introduce a new propositional operation that will allow us to write degrees of relative probabilities. Since probabilities of this kind are written in the form $P(fx_i, gy_i)$, the content of the parentheses in this expression can be regarded as a compound proposition, resulting from the two components by a propositional operation, denoted by the comma. We shall call it *operation of selection*, or *comma operation*. We can also put the comma between propositional sequences. We then define, by analogy with those of (6),

$$(fx_i), (gy_i) =_{Df} (fx_i, gy_i). \tag{7}$$

The frequency interpretation of $P(fx_i, gy_i)$ is given by the expression

$$P(fx_i, gy_i) = \lim_{n \to \infty} \frac{\overset{n}{\underset{i=1}{N}}\{V[fx_i.gy_i] = 1\}}{\overset{n}{\underset{i=1}{N}}\{V[fx_i] = 1\}}. \tag{8}$$

This form makes clear why we speak of the operation of selection. The proposition fx_i selects the subsequence in which we count the frequency of gy_i. Since (7) allows us to regard the comma on the left side of (8) as standing between the propositional sequences, we can also say that the comma operation represents a selection from one sequence by another sequence.

The argument here lacks cogency. Reichenbach says that the content of the parentheses in '$P(fx_i, gy_i)$' can be regarded as a compound proposition. This seems to indicate that he is thinking of the 'fx_i' and 'gy_i' appearing therein as propositions, i.e., as instances of the propositional sequences. But, as we have earlier noted and as (8) shows, they have to be propositional sequences, not propositions. Furthermore, he seems to think that the selection idea provides him with a sequence whose general term is 'fx_i, gy_i'. But the selection only provides a term for *some* values of i, namely those for which fx_i is true, and nothing when fx_i is false. Hence no meaning for 'fx_i, gy_i' for all i if fx_i has false instances.

In his truth tables Reichenbach uses probability values in place of the two truth values of ordinary two-valued logic. These are probabilities of propositional sequences. However, unlike for the two-valued logic, where the value of a compound depends on the values of the two components, say fx_i and gy_i, he uses the three values

$$P(fx_i), \quad P(gy_i), \quad P(fx_i, gy_i),$$

to specify the value of a compound of fx_i and gy_i. (We are following Reichenbach's convention of omitting the parentheses around a general term of a propositional sequence allowing the general term to stand for the sequence.) This is how his truth tables appear (p. 400), abbreviated by the omission of the columns for '⊃' and '≡'):

Table1. Truth tables of probability logic

A. Negation

$P(f)$	$P(\overline{f})$
p	$1-p$

Restrictive Conditions:

1. $\frac{p+q-1}{p} \le u \le \frac{q}{p}$

2. $P(f, f) = 1$

B. Binary Operations

$P(f)$	$P(g)$	$P(f,g)$	$P(f \vee g)$	$P(f.g)$	$P(g,f)$
p	q	u	$p + q - pu$	pu	pu/q

Reichenbach considers these tables to be a sufficient basis for a probability logic. But let us compare them with the analogous ones for two-valued logic. We use 'V' for 'the truth value of'.

$V(\phi)$	$V(\psi)$	$V(\overline{\phi})$	$V(\phi \vee \psi)$	$V(\phi.\psi)$
p	q	$1-p$	$\max\{p, q\}$	pq

Here p and q are variables ranging over the set $\{0, 1\}$, with 0 and 1 having their usual arithmetic properties. Another way of expressing this is by the equations

$$V(\overline{\phi}) = 1 - V(\phi)$$
$$V(\phi \vee \psi) = \max\{V(\phi), V(\psi)\} \qquad (9)$$
$$V(\phi.\psi) = V(\phi) \cdot V(\psi).$$

Since a (two-valued) logical expression is constructed out of finitely many basic elements (atomic propositions) its V value is determined in terms of those of the basic elements by a recursive use of (9). This enables two-valued logic to have a (an effective) notion of logical consequence. That a formula ψ is a logical consequence of a set of premises $\phi_1, \ldots, \phi_n$ can be determined by examining all possible assignments of 0 or 1 to the basic elements common to $\phi_1, \ldots, \phi_n, \psi$. But Reichenbach's tables provide no means of reducing P values of expressions to P values of simpler expressions nor, for that matter, is there a clear specification of the syntax of the language. We do not even know if, for example, '$\phi \lor (\psi, \chi)$' or '$(\phi, \psi).\chi$' are meaningful.

Since Reichenbach has

— no adequate justification for including 'fx_i, gy_i' in his language
— no complete specification of the syntax of his language
— no provision for general determination of the probability value of an expression in terms of its components, and
— no definition of logical consequence

we do not, on the basis of our understanding of the concept, accept his claim to have constructed a probability logic.

§3.3. Probabilistic inference revived

Despite De Morgan's cogently argued advocacy for probable inference as a part of formal logic, and Boole's conceiving of, and if not succeeding at least showing the feasibility of, a general method for obtaining such inferences, little of substance ensued. As evidenced by the writers we have mentioned in our preceding two sections, the interest was in justifying methodological principles of science. Not until we come to Suppes' *1966* is there something for our historical account. And even in his paper the topic of probabilistic inference was not central; it arose in the context of a criticism of employing the concept of 'total evidence' as a means of resolving the 'statistical paradox'. The paradox is stated in Suppes' opening paragraph (*1966*, 49):

> My purpose is to examine a cluster of issues centering around the so-called statistical syllogism and the concept of total evidence. The kind of paradox that is alleged to arise from uninhibited use of the statistical syllogism is of the following sort.
>
> > (1) The probability that Jones will live at least fifteen years given that he is now between fifty and sixty years of age is r. Jones is now between fifty and sixty years of age. Therefore, the probability that Jones will live at least fifteen years is r.
> >
> > (2) The probability that Jones will live at least fifteen years given that he is now between fifty-five and sixty-five years of age is s. Jones is now between fifty-five and sixty-five years of age. Therefore, the probability that Jones will live at least fifteen years is s.

The paradox arises from the additional reasonable assertion that $r \neq s$, or more particularly that $r > s$.

If we write the inference forms involved in Suppes' statement as

$$
\frac{\begin{array}{c} P(A\,|\,B) = r \\ B \end{array}}{P(A) = r} \qquad (1)
\qquad\qquad
\frac{\begin{array}{c} P(A\,|\,C) = s \\ C \end{array}}{P(A) = s} \qquad (2)
$$

then the rule of total evidence considers it illegitimate to use B and C separately as evidence when it is know that $B \wedge C$ is the case.

Aside from the difficulties associated with the rule of total evidence (see the criticism by Suppes in his paper on p. 50) there is the question as to what an inference form such as (1) means. As we have seen in our §3.1, Nicod considered (1)—with the major premise taken as a statement of a Keynes probability relation—to be a way of explaining Keynes' notion, thus implying that (1) was for him clearly valid. On the other hand Suppes says (p. 58):

> ... Now let us schematize this inference [i.e., (1)] in terms of *hypothesis* and *evidence* as these notions occur in Bayes' theorem
>
> $$
> \frac{\begin{array}{l} P(\text{hypothesis}\,|\,\text{evidence}) = r \\ \text{evidence} \end{array}}{\therefore P(\text{hypothesis}) = r},
> $$
>
> and the incorrect character of this inference is clear. From the standpoint of Bayes' theorem it asserts that once we know the evidence, the posterior probability $P(H\,|\,E)$ is equal to the prior probability $P(H)$, and this is patently false.

We have then two diametrically opposed opinions on the validity of (1). This inference form—with the $P(A\,|\,B)$ taken in the general sense of confirmation—was the subject of a lengthy essay by H. Kyburg. The essay was followed by a discussion involving seven participants of the colloquium at which it was presented (*Lakatos 1968*, 98–165). The positions on the inference were so widely diverse that we may reasonably conclude that it has no clear meaning.

Replacing the second premise in (1) by the probability statement '$P(B) = 1$' does produce the valid form

$$
\frac{\begin{array}{c} P(A\,|\,B) = r \\ P(B) = 1 \end{array}}{\therefore P(A) = r} \qquad (3)
$$

But (3) is not usable for inferences of the kind in the statistical paradox, for '$P(B) = 1$' is true if and only if B is the sure event, and false otherwise.

We turn to some probabilistic inferences which Suppes presents. These are all clearly valid forms, since they are consequences of basic probability principles.

In place of the disputed (1) Suppes considers the natural generalization of the rule of detachment for probable inference to be

$$\frac{\begin{aligned} P(A \mid B) &= r \\ P(B) &= \rho \end{aligned}}{\therefore P(A) \geq r\rho} \tag{4}$$

or, more generally,

$$\frac{\begin{aligned} P(A \mid B) &\geq r \\ P(B) &\geq \rho \end{aligned}}{\therefore P(A) \geq r\rho} \tag{5}$$

He contrasts this with the one which uses the ordinary conditional

$$\frac{\begin{aligned} P(B \to A) &\geq r \\ P(B) &\geq \rho \end{aligned}}{\therefore P(A) \geq r + \rho - 1} \tag{6}$$

These results are then reexpressed so as to feature the nearness of the probability to 1:

$$\frac{\begin{aligned} P(B \to A) &\geq 1 - \epsilon \\ P(B) &\geq 1 - \epsilon \end{aligned}}{\therefore P(A) \geq 1 - 2\epsilon} \tag{7} \qquad \frac{\begin{aligned} P(A \mid B) &\geq 1 - \epsilon \\ P(B) &\geq 1 - \epsilon \end{aligned}}{\therefore P(A) \geq (1 - \epsilon)^2} \tag{8}$$

Comparing these with necessary (verity) inference forms, Suppes observes that rules involving two premises (e.g., A, $A \to B$ $\therefore$ B) have corresponding probabilistic forms (e.g., (7)) in which the probability bound on the conclusion (of the necessary form) is less that that of the premises, though for single premise inferences (e.g., A $\therefore$ $A \lor B$) this bound remains unchanged. Stated formally (*Suppes 1966*, 54):

Theorem 1. *If $P(A) \geq 1 - \epsilon$ and A logically implies B, then $P(B) \geq 1 - \epsilon$.*

Theorem 2. *If each of the premises $A_1, \ldots, A_n$ has probability at least $1 - \epsilon$ and these premises [conjointly] logically imply B, then $P(B) \geq 1 - n\epsilon$.*
Moreover, in general the lower bound $1 - n\epsilon$ cannot be improved on, i.e., equality holds in some cases whenever $1 - n\epsilon \geq 0$.

Proofs of these results are quite straightforward except for the 'moreover' part of Theorem 2. This is the example which Suppes gives to show that the lower bound for the conclusion in Theorem 2 can be attained (*1966*, 55):

The example I use is most naturally thought of as a temporal sequence of events $A_1, \ldots, A_n$. Initially we assign

$$P(A_1) = 1 - \epsilon$$
$$P(\overline{A_1}) = \epsilon.$$

Then [assigning]

$$P(A_2 \mid A_1) = \frac{1 - 2\epsilon}{1 - \epsilon}$$
$$P(A_1 \mid \overline{A_1}) = 1,$$

and more generally

$$P(A_n \mid A_{n-1}A_{n-2} \ldots A_1) = \frac{1 - n\epsilon}{1 - (n-1)\epsilon}$$
$$P(A_n \mid A_{n-1}A_{n-2} \ldots \overline{A_1}) = 1$$

$$\vdots$$

$$P(A_n \mid \overline{A_{n-1}}\,\overline{A_{n-2}} \ldots \overline{A_1}) = 1,$$

in other words for any combination of preceding events on trials 1 to $n - 1$ the conditional probability of A_n is 1, except for the case $A_{n-1}A_{n-2} \ldots A_1$.

Suppes then shows by induction that $P(A_nA_{n-1} \ldots A_1) = 1 - n\epsilon$. Two comments are here in order. First, Suppes neglects to show that an assignment of probabilities such as he describes is a possible one. Secondly, a much simpler example can be given. Consider the probability 'space' which is the unit square with probability 'being' the area of a subregion. Assume $0 < \epsilon < \frac{1}{n}$ and let the square be divided so that there are n mutually exclusive subregions $\overline{A_1}, \ldots, \overline{A_n}$, each with probability equal to ϵ. Then

$$P(A_1A_2 \ldots A_n) = 1 - P(\overline{A_1} \vee \overline{A_2} \vee \cdots \vee \overline{A_n})$$
$$= 1 - \sum_{i=1}^{n} P(\overline{A_i})$$
$$= 1 - n\epsilon.$$

Looking back 119 years to De Morgan's *Formal Logic*, chapter 10, (discussed in §2.2 above) and comparing De Morgan's examples with those of

Suppes' we note, not surprisingly, that advances have been made: in clarity with regard to the nature of logic, probability and their interrelations; in separation of logical from non-logical subject matter; in the introduction of inequality relations, and in formal correctness. Except for forms involving conditional probability Suppes limits himself, as did De Morgan, to inference forms in which the formula involved in the conclusion is a (verity) necessary consequence of those involved in the premises. Results are derived as in a mathematical theory and there is no semantic criterion as to what constitutes a valid inference. We would describe what Suppes presents as some applications of probability notions to ordinary (two-valued) logic, not a probability *logic*.

In connection with his Theorem 2 we have noted Suppes' interest in a bound being best possible ("cannot be improved on"). Apparently he was unaware that Boole had, long before, investigated the subject from a general viewpoint. This is the topic of our next section.

§3.4. Bounds on probability—early history

The determination of probability bounds is in its own right a topic of interest. We shall later see, in chapter 6, applications to estimation in statistics, to fault analysis in combinational (switching) circuits, and to network reliability. But it has additional significance for our study in that it plays a fundamental role in the construction of a probability logic— meaning by this a genuine logic and not simply an application of probability to (verity) logic. We shall see that the notion of *best possible probability bounds* which we arrive at is independent of any particular events or any particular probability space. This is analogous to the property of validity in verity logic, which is independent of any particular propositions or models.

Our historical account of probability bounds requires back-tracking to the nineteenth century as the topic was initiated and quite extensively developed by Boole, chiefly in his *Laws of Thought*. We begin the discussion by quoting the opening paragraphs from chapter XIX, Of Statistical Conditions (*1854*, 295):

> 1. By the term statistical conditions, I mean those conditions which must connect the numerical data of a problem in order that those data may be consistent with each other, and therefore such as statistical observations might actually have furnished. The determination of such conditions constitutes an important problem, the

solution of which, to an extent sufficient at least for the requirements
of this work, I purpose to undertake in the present chapter, regarding
it partly as an independent object of speculation, but partly also as
a necessary supplement to the theory of probabilities already in some
degree exemplified.

He then goes on to state the nature of the connection:

> 2. There are innumerable instances, and one of the kind presented
> itself in the last chapter, Ex. 7, [or (2), in §2.7 above] in which the
> solution of a question in the theory of probabilities is finally dependent
> upon the solution of an algebraic equation of an elevated degree. In
> such cases the selection of the proper root must be determined by
> certain conditions, partly relating to the numerical values assigned in
> the data, partly to the due limitation of the element required. The
> discovery of such conditions may sometimes be effected by unaided
> reasoning. For instance, if there is a probability p of the occurrence
> of an event A, and a probability q of the concurrence of the said event
> A, and another event B, it is evident that we must have

$$p \overset{=}{>} q.$$

> But for the general determination of such relations, a distinct method
> is required, and this we proceed to establish.

As this quotation indicates, Boole's interest in the topic arose from a
specific need connected with his probability method, but he notes its importance for obtaining consistency conditions on probability data. (In later
writings he replaces the phrase 'statistical conditions' by 'conditions of possible experience'.)

When Boole speaks of the "general determination of such relations"
he is apparently referring to a uniform procedure which, when applied in
a given instance, results in consistency conditions. Also involved, though
not explicitly mentioned, is another type of generality: since the conditions
are derived using only general principles of probability theory they hold
good for any probability situation, i.e., putting it in a formal setting, they
hold good for any probability algebra (or space). And, as these conditions
are expressed in terms of inequations, they are directly connected with
the question of bounds. Thus, for the simple example Boole mentions,
if $P(A) = p$, then 0 and p are, respectively, lower and upper bounds on
$P(AB)$ good no matter what A and B may be. As another example, if
$P(A) = p$ and $P(B) = q$, then for the conjunction AB,

$$0 \leq P(AB) \leq \text{ minimum of } p \text{ and } q.$$

Thus 0 and $\min(p,q)$ are lower and upper bounds on $P(AB)$. We proceed to the details of Boole's procedures. We say procedure*s* for he has more than one.

The approach he uses in *Laws of Thought* is not based directly on the notion of probability but indirectly through the operator 'n', where '$n(x)$' stands for the number of elements in a class x. Relations among such values lead first to relations among relative frequencies by dividing through by $n(1)$, the number of elements in the universal class 1, and then to probabilities by passing to the limit. (Boole is quite vague about this transition from relative frequencies to probabilities.) In the interests of brevity we shall be referring chiefly to upper bounds or, as Boole calls them, 'major limits' of a class. Minor limits are obtained by subtraction from 1 of major limits of the complementary class.

Boole first considers the problem of obtaining major and minor limits of a (Boolean) class expression in terms of the number of elements in the classes represented by the class symbols present in the expression. Thus for the class expression 'xy' the major limit is 'the least of $n(x)$ and $n(y)$'; and for any 'constituent' (i.e., basic conjunction) the major limit is the least of the values for the individual conjuncts. For the general case of any class expression Boole replaces the class expression by an equivalent logical sum of constituents (i.e., by a disjunctive normal form). The problem is stated as a proposition (*1854*, 300):

PROPOSITION II.

6. To determine the major numerical limit of a class expressed by a series [i.e., logical sum] of constituents of the symbols x, y, z, &c., the values of $n(x), n(y), n(z)$, &c., and $n(1)$, being given.

The solution is given as a rule (*1854*, 301):

RULE.—Take one factor from each constituent, and prefix to it the symbol n, add the several terms or results thus formed together, rejecting all repetitions of the same term; the sum thus obtained will be a major limit of the expression, and the least of all such sums will be the major limit to be employed.

Thus as an example Boole lists for $n(xy + x(1-y)z)$ the sums

1. $n(x)$	4. $n(y) + n(x)$
2. $n(x) + n(1-y)$	5. $n(y) + n(1-y)$
3. $n(x) + n(z)$	6. $n(y) + n(z)$

One readily verifies the correctness of the rule as giving *an* upper bound. But note the absence of a claim for its being the best, i.e., narrowest that could be obtained—Boole's concluding sentence in the RULE simply states that the least of all such sums "will be the major limit to be employed [for purposes of his probability method]". An example showing that Boole's RULE does not yield the optimal bound first appeared in *Hailperin 1965*, §4.

Boole goes on to consider the most general situation stating it as a problem, somewhat unclearly, as follows (*1854*, 304):

PROPOSITION IV

Given the respective number of individuals composed in any classes s, t, &c., logically defined, to deduce a system of numerical limits of any other class w, also logically defined.

By 'logically defined' Boole means that there are (Boolean) equations

$$s = \phi(x, y, \ldots, z), \quad t = \psi(x, y, \ldots, z), \quad \ldots, \quad w = F(x, y, \ldots, z) \quad (1)$$

defining $s, t, \ldots, w$ in terms of a set of (unspecified) class symbols $x, y, \ldots, z$. By his general method in logic Boole deduces from a system such as (1) that

$$w = 1\,A + 0\,B + \frac{0}{0}\,C + \frac{1}{0}\,D, \quad (2)$$

where A, B, C, D are constituents on $s, t, \ldots$ determined by the method. The interpretation which Boole gives to this peculiar equation involves use of his indefinite class symbol 'v' but can be seen to be equivalent to

$$\begin{cases} A \subseteq w \subseteq A + C \\ D = 0. \end{cases} \quad (3)$$

Here $D = 0$ is the necessary condition that there be the (two-sided inclusion) solution for w given by the first equation. It is the case that (3) is always a correct consequence of (1). (See *Hailperin 1986*, §2.6.) For the simple example, introduced in §2.5, where

$$s = x, \quad t = xy, \quad w = y, \quad (4)$$

we have seen that Boole's technique gives

$$w = \frac{t}{s} = st + 0\,s\bar{t} + \frac{0}{0}\,\bar{s}t + \frac{1}{0}\,\bar{s}\bar{t}$$

with the meaning

$$\begin{cases} st \subseteq w \subseteq st + \bar{s}t \\ \bar{s}\bar{t} = 0. \end{cases}$$

Returning to the general case we state Boole's solution to the problem proposed in his PROPOSITION IV: the major limits of w are given by the major limits of $A + C$, these limits being expressed in terms of the given (numerical) values of $s, t, \ldots$. The equation $D = 0$, from which w is absent, provides the necessary condition for the existence of a solution, and implies that all minor limits of D be ≤ 0. Thus for the simple example (4) for which $A + C$ is $st + \overline{s}\overline{t}$, the major limits for w are, by Boole's RULE,

$$n(s) + n(\overline{t}), \quad n(t) + n(\overline{s}). \tag{5}$$

The minor limit of $\overline{s}t$, the D for this example, is $n(\overline{s}) + n(t) - 1$, (i.e., $1 -$ major limit of the complement $s + \overline{s}t$) which gives

$$n(\overline{s}) + n(t) - 1 \leq 0,$$

or

$$n(t) \leq n(s), \tag{6}$$

as the necessary condition for a solution. (Clearly correct for $s = x$ and $t = xy$. Note that by adding $n(\overline{t})$ to both sides of (6) the first limit in (5) is not less that 1 and hence can be dropped.)

With regard to this method, later referred to by Boole as the 'prior' method, he has the disclaimer (*1854*, 310):

> 13. It is to be observed, that the method developed above does not always assign the narrowest limits which it is possible to determine. But in all cases, I believe, sufficiently limits the solutions of questions [of the general kind he has proposed] in the theory of probabilities.

No example is given to justify the assertion in the first sentence. But since the method depends on Boole's above cited RULE, the example in our *1965*, §4, referred to above, does provide one. Most interestingly, Boole then goes on to outline an entirely different method (*1854*, 310–11):

> The problem of the determination of the narrowest limits of numerical extension of a class is, however, always reducible to a purely algebraical form.* Thus, resuming the equation
>
> $$w = A + 0B + \frac{0}{0}C + \frac{1}{0}C,$$
>
> let the highest inferior numerical limit of w be represented by the formula $a\,n(s) + b\,n(t) \ldots + d\,n(1)$, wherein $a, b, c, \ldots, d$ are numerical

constants to be determined, and s, t, &c., the logical symbols of which A, B, C, D are constituents. Then

$$a\,n(s) + b\,n(t) \ldots + d\,n(1)$$
$$= \text{minor limit of } A \text{ subject to the condition } D = 0.$$

He then describes how to find the highest inferior limit:

Hence if we develop the function

$$as + bt \ldots + d,$$

reject from the result all constituents which are found in D, the co-efficients of those constitutents which remain, and are found also in A, ought not individually to exceed unity in value, and the coefficients of those constituents which remain, and which are not found in A, should individually not exceed 0 in value. Hence we shall have a series of inequalities of the form $f \overset{=}{<} 1$, and another series of the form $g \overset{=}{<} 0$, f and g being linear functions of a, b, c, &c. Then those values of $a, b..d$, which, while satisfying the above conditions, give to the function

$$a\,n(s) + b\,n(t) \ldots + d\,n(1),$$

its highest value must be determined, and the highest value in question will be the highest minor limit of w. To the above we may add the relations similarly formed for the determination of the relations among the given constants $n(s), n(t), \ldots n(1)$.

Boole's footnote (here omitted) refers to a lost manuscript of which he recollects only the impression that the principal of it was the same as just described, that it was developed in considerable detail, and its sufficiency was formed. (But see our *1965*, §7, where we express the opinion that his recollection concerning its sufficiency was in error.)

This sketch of a 'purely algebraic form' which Boole presents merits special mention: it is an early historical example of a linear programming problem, as well as a correct method of finding optimal probability bounds. However, aside from the obscurity due to his peculiar logical methods, there are three major gaps that need to be filled so as to make it cogent. One is a justification for using A, a sum of constituents on $s, t, \ldots$, in place of $F(x, y, z, \ldots)$. For, when the variables $s, t, \ldots$ in A are replaced by the functions of $x, y, z, \ldots$ as given in (1), the resulting expression, while implying $F(x, y, z, \ldots)$, need not be equivalent to it. (See our simple example (4) above). Another gap is the assumption that the best lower bound is a *linear* function of the given values $n(s), n(t), \ldots$. Finally, there is no proof that the bounds are *best* possible.

Assuming that the gaps are filled one can justify Boole's assertion that the best possible lower bound ('highest minor limits of w') is obtainable as described, namely by finding the maximum value of the linear form

$$a\,n(s) + b\,n(t) + \cdots + d\,n(1) \tag{7}$$

subject to the linear inequation system which he gives. In showing this Boole resorts to an expansion of the numerico-logical expression

$$as + bt + \cdots + d, \tag{8}$$

where, $a, b, \ldots, d$ are numerical variables and $s, t, \ldots$ are logical symbols. Such an expression, which seems queer to us, is a feature of his algebraic method of doing logic. However one can also obtain the result in a less mysterious manner. Let

$$\bigvee_s K_i, \quad \bigvee_t K_i, \quad \ldots, \quad \bigvee_1 K_i$$

be respectively the expansions of $s, t, \ldots, 1$ as sums of constituents on $s, t, \ldots$ (The expansion of 1 has all possible constituents on $s, t, \ldots$) Since the operator n distributes over a sum of mutually exclusive classes, we have

$$n(s) = \sum_s n(K_i), \quad n(t) = \sum_t n(K_i), \quad \ldots, \quad n(1) = \sum_1 n(K_i).$$

By (i) multiplying these equations respectively by $a, b, \ldots, d$, (ii) adding to form a single equation, and (iii) collecting terms on the right, we obtain

$$a\,n(s) + b\,n(t) + \cdots + d\,n(1) = \sum_i h_i\, n(K_i), \tag{9}$$

where the h_i are linear combinations of $a, b, \ldots, d$. The indicated sum in (9) extends over all possible constituents K_i on $s, t, \ldots$. Since we seek the maximum value of (7) subject to the condition $D = 0$ we can delete any term $h_i\, n(K_i)$ in this sum if the K_i occurs in D—for if $D = 0$, then $n(K_i) = 0$ for any K_i in D. If (7) is to represent a lower bound for $n(A)$ subject to $D = 0$, then for each K_i which is present in A its coefficient h_i on the right in (9) must not exceed 1, and for each K_i not present in A, its coefficient h_i must not exceed 0. Hence if a lower bound for $n(A)$ is representable in the form (7) then its maximum value, subject to the described linear inequalities, is the best possible lower bound. (Boole says nothing about mathematical techniques for finding the maximum of a linear form under linear constraints—a subject of paramount interest in current linear programming theory.) To illustrate with a simple example what we have just described in general, let $s = x, t = xy$; we wish to obtain the highest minor limit of w where $w = y$. Here (see (4) above) $A = st$ and $D = \bar{s}t$. Set $L = a\,n(s) + b\,n(t) + c\,n(1)$, a linear form in variables a, b, c. Then

$$
\begin{aligned}
L = {} & a(n(st) + n(s\bar{t})) \\
& + b(n(st) + n(\bar{s}t)) \\
& + c(n(st) + n(s\bar{t}) + n(\bar{s}t) + n(\bar{s}\bar{t}))
\end{aligned}
$$

which then gives

$$
\begin{aligned}
L = {} & (a + b + c)n(st) \\
& + (a + 0 + c)n(s\bar{t}) \\
& + (0 + b + c)n(\bar{s}t) \\
& + (0 + 0 + c)n(\bar{s}\bar{t}).
\end{aligned}
$$

Since $\bar{s}t = D$ we delete the term $n(\bar{s}t)$ and, since st is the only constituent of A, we set

$$
\begin{aligned}
a + b + c &\le 1 \\
a + c &\le 0 \tag{10} \\
c &\le 0.
\end{aligned}
$$

The maximum of L subject to these constraints is (by linear programming theory) attained at a vertex of the polytope specified by (10). Here there is only one vertex, namely $(0, 1, 0)$. Hence $0\,n(s) + 1\,n(t) + 0\,n(1) = n(t)$ is the minor limit of w $(= y)$.

Perhaps the easiest and most informative way of filling the gaps mentioned above so as to make Boole's argument cogent, is by way of the duality theorem of linear programming. As we shall be referring to it in our next section it will be convenient to postpone completion of the discussion of Boole's 'purely algebraic form' until then.

In a subsequent paper appearing in the *Philosophical Magazine* Boole presents a distinctively different method of obtaining consistency conditions and bounds on the probability of a logical function. His opening paragraph declares it to be "an easy and general method of determining such conditions" and, referring to his earlier method in the *Laws of Thought*, "... the method there developed is difficult of application and I am not sure that it is equally general with the one I am about to explain."

In this paper (*1854*a) Boole drops the use of '*n*' and expresses everything in terms of probability. The general problem proposed in PROPOSITION IV quoted above is expressed via the equations

$$P(\phi(x, y, \ldots, z)) = p, \quad P(\psi(x, y, \ldots, z)) = q, \quad \ldots,$$
$$P(F(x, y, \ldots, z)) = w \qquad (11)$$

where $\phi, \psi, \ldots, F$ are given Boolean functions, with $\phi(x, y, \ldots, z)$, $\psi(x, y, \ldots, z), \ldots$ having respective probabilities $p, q, \ldots$. The object is to find bounds on w, the probability of $F(x, y, \ldots, z)$. Boole's first step is to convert equations (11) to a set of linear equations in the probabilities of the constituents on $x, y, \ldots, z$. This is accomplished by replacing each logical function by its expansion in terms of constituents on $x, y, \ldots, z$ and distributing P over the sums of constituents. If we denote constituent probabilities generically by k_i the resulting equations take the form

$$\sum_\phi k_i = p, \quad \sum_\psi k_i = q, \quad \ldots, \quad \sum_F k_i = w. \qquad (12)$$

To these equations Boole adjoins the normative conditions that the sum over all k_i be 1, and that for each i, $k_i \geq 0$. Finding the minimum value of the linear form $\sum_F k_i$ subject to these linear constraints is a linear programming problem. Boole took it for granted that the minimum value thus obtained would be 'best possible'. A mathematical proof may be found in our *1965*, §3. In effect Boole treats the problem as a *parametric* linear program and expresses his result in terms of the parameters $p, q, \ldots$ as follows.

First he shows how to successively eliminate variables one by one from a system of linear (equations and) inequations. Though he makes no mention of it, the method had been described earlier by Fourier in the 1820s. (See *Dantzig 1963*, 84) Next he applies (Fourier) elimination (of the k_i) to the combined system, i.e., to (12) plus the normative conditions, with w taken as a parameter along with $p, q, \ldots$. This ultimately results in a series of equations

$$L_1 \leq w, \quad L_2 \leq w, \quad \ldots$$
$$w \leq U_1, \quad w \leq U_2, \quad \ldots \qquad (13)$$

in which the L_i and U_i are linear forms in the parameters $p, q, \ldots$. It is clear that the L_i are minor limits, and the U_i major limits, of $P(F(x, y, \ldots, z))$. The largest of the L_i and the smallest of the U_i are what Boole takes as the narrowest limits on $P(F(x, y, \ldots, z))$ that are expressible in terms of $p, q, \ldots$. Finally, eliminating w from (13) results in a set of inequalities on $p, q, \ldots$ which are the necessary conditions that equations (11) have a solution for w.

When Boole describes this method as 'easy and general' as compared with the 'purely algebraic form' sketched in his *Laws of Thought* he underestimates the computational difficulties involved, for Fourier elimination can result in dauntingly large (exponentially exponential) numbers of inequalities.

In concluding this section we should like to mention two papers from the Russian literature—one from the 19th and the other from the 20th century—both presenting essentially the same (incomplete) solution that Boole gave to his general probability problem. The only significant respect in which they differ from Boole's is in the use of inclusion instead of Boole's indefinite class symbol to indicate a logical range of values. Boole's general problem has been discussed by us in our *1988b*, §9 (and in much fuller detail in our *1986*, chapters 4 and 5). The matter is related to the question of bounds in that Boole's solution often leads to a range of values as the answer. But this range need not be the same as the full range of possible values allowed by the given conditions of the problem, i.e., his method need not give a complete solution. This same failing applies to the two Russian papers which we now briefly discuss.

Poretzky's *1887* provides five worked examples of which four are taken without change from *Boole 1854*. These examples of Boole's have all been discussed in our *1986*. Poretzky's fifth example is of no interest here as it is a simple one with an exact value as a solution. *Čirkov 1971* has two worked examples of which the first has an exact value as a solution, but the second results in a range of values.

The problem is stated by Čirkov in technical applications-like form referring to 'automata', '0,1 signals', etc. However in purely mathematical terms it is as follows:

Given events A_1, A_2, A_3, A_4 of which it is only known that

$$P(A_1) = a_1, \quad P(A_2) = a_2, \quad P(A_3) = a_3, \quad P(\overline{A}_1\overline{A}_2\overline{A}_3\overline{A}_4) = b,$$

find $P(A_4)$.

Letting r be any value for $P(A_4)$ which satisfies the hypothesis Čirkov finds that the bounds on r are given by

$$\frac{(1 - b - a_1)(1 - b - a_2)(1 - b - a_3)}{(1 - b)^2} \leq r \leq 1 - b.$$

One can construct counter-examples that show that the lower limit is not best possible.[4] Application of the technique described below in §4.5

---

4. Take A_1, A_2, A_3 to be mutually exclusive and $P(A_1) = P(A_2) = P(A_3) = P(\overline{A}_1\overline{A}_2\overline{A}_3\overline{A}_4) = 1/4$. Then Čirkov's lower bound is $\frac{1/2 \cdot 1/2 \cdot 1/2}{(3/4)^2}$, which is greater than 0, a value $P(A_4)$ can take on.

yields the following optimal bounds:

$$\text{Lower} = L = \max(0,\ 1 - (a_1 + a_2 + a_3 + b))$$
$$\text{Upper} = U = \min(1,\ 1 - b),$$

with

$$0 \le L \le U \le 1$$

being the consistency condition on the parameters. Čirkov's problem is a slight generalization of one of Boole's (*1854*, 281) to which it reduces by setting $a_3 = 0$. The result then is the same as Boole's.

§3.5. Best possible probability bounds

After Boole's 1854 work on the probability of logical functions there seems to be nothing of note on this topic until Fréchet's *1935*. Although in this paper Fréchet cites Boole for a simple inequality, there is no mention of Boole's general results described in our preceding section.

Fréchet addresses himself to the following problem. What is the best that can be said about the probability P_a of an alternation of events $H_1 \vee H_2 \vee \cdots \vee H_n$ if all one knows is that the events have, respectively, probabilities $p_1, \ldots, p_n$? He readily shows that

$$p_1 + \cdots + p_n - (n - 1) \le P_a \le p_1 + \cdots + p_n$$

and hence, setting

$$\omega(p_1, \ldots, p_n) = \max\{0,\ p_1 + \cdots + p_n - (n - 1)\}$$
$$\Pi(p_1, \ldots, p_n) = \min\{1,\ p_1 + \cdots + p_n\},$$

that

$$\omega(p_1, \ldots, p_n) \le P_a \le \Pi(p_1, \ldots, p_n). \tag{1}$$

Fréchet derives the inequalities (1) by simple direct means applied to the specific compound event involved. Boole on the other hand had, as we have seen in the preceding section, presented general methods for obtaining bounds on the probability of an arbitrary (logically expressed) compound event; moreover, there was the additional generality that the components

H_i of the compound event may be subject to additional, logically expressed, conditions. Though Boole declares that his method yields the 'narrowest limits' no justification of this is given, an essential for which is a criterion for judging when a bound is best possible. Fréchet's paper does implicitly provide one for his particular case, namely: the functions $\omega(p_1, \ldots, p_n)$ and $\Pi(p_1, \ldots, p_n)$ are *best possible bounds* for $P_a = P(H_1 \vee \cdots \vee H_n)$ if, for any pair of functions $f(p_1, \ldots, p_n)$ and $F(p_1, \ldots, p_n)$ such that for all possible values $p_1, \ldots, p_n$ in the unit interval and chance events $H_1, \ldots, H_n$ [in a probability space] for which $p_1, \ldots, p_n$ are their respective probabilities and

$$f(p_1, \ldots, p_n) \leq P \leq F(p_1, \ldots, p_n),$$

one has [for all $p_1, \ldots, p_n$ in the unit interval]

$$f(p_1, \ldots, p_n) \leq \omega(p_1, \ldots, p_n) \leq \Pi(p_1, \ldots, p_n) \leq F(p_1, \ldots, p_n).$$

It is clear that a comparable criterion applies to any compound event and not just this particular one—provided that its components (the H_i) have unrestricted probabilities. This would not be the case if the H_i were subject to restrictions.

Fréchet employs his criterion to show that the functions ω and Π are the best that one can do. Considering the lower bound in (1), he asserts that for any set of reals $p_1, \ldots, p_n$ in the unit interval one can define a chance event G of probability $p' = \omega(p_1, \ldots, p_n)$, then events H_j with respective probabilities p_j $(j = 1, \ldots, n)$ which take place only if G does, so that $P_a \leq p'$. For these events

$$f(p_1, \ldots, p_n) \leq P_a = p' = \omega(p_1, \ldots, p_n),$$

thus establishing the result for the lower bound.

Some 30 years later, in *Hailperin 1965*, attention was called to the three methods Boole had developed for obtaining probability bounds: (i) the 'prior' method of his chapter XIX in *1854*, (ii) the 'purely algebraic form' which he claimed did give narrowest limits, and (iii) the 'easy and general' method of his *1854a* of which he was unsure that it was equivalent to the purely algebraic form. Additionally, our *1965* gave an example to show that the 'prior' method did not in all cases provide a best possible bound; also, that methods (ii) and (iii) were primal and dual—hence equivalent—forms of the same linear program, and that the solution of the linear program did give the best possible bounds. Implicit in our paper, though not explicitly stated until *1976*, §5.6, was the conclusion that the solution to Boole's General Probability Problem consisted in specifying the best possible upper and lower bounds on the probability sought. It will be instructive to describe these ideas in the context of a specific worked example. The description will refer to the notion of a probability algebra, or space—as in our *1965*—rather than to our subsequently developed notion of a probability logic.

The simple example we select is the one already referred to in §2.5: given $P(A_1 A_2) = p$, $P(A_2) = q$, find $P(A_1)$. Since, as is the case here, the given conditions need not determine a unique value for the unknown, we seek best possible functions of p and q which between them include any possible value for $P(A_1)$. The qualification 'best possible' is important since it is not sufficient simply to derive, by probability principles or axioms of probability algebras, a conclusion about $P(A_1)$. We wouldn't be sure that the conclusion was the most general result.

Let $\alpha = \alpha(p, q)$ be the set of all possible values of w such that there is a probability algebra $< \mathfrak{B}, P >$ with $A_1, A_2 \in \mathfrak{B}$, $P(A_1 A_2) = p, P(A_2) = q$ and $w = P(A_1)$. If α is non-empty (it could be empty if, for example, q were less than p), then as a non-empty bounded set of real numbers it has a least upper bound $U(p, q)$. By virtue of its definition $U(p, q)$, as a function of p and q, is *the best possible upper bound function for $P(A_1)$ subject to the given conditions*. Similarly $L(p, q)$, defined as the greatest lower bound of α, is the best possible lower bound. But these definitions are non-effective, furnishing no means for computing values of $U(p, q)$ and $L(p, q)$. To obtain computable forms we approach the problem from another direction.

Suppose we have a probability algebra $< \mathfrak{B}, P >$ with $A_1, A_2 \in \mathfrak{B}$. Set $k_1 = P(A_1 A_2)$, $k_2 = P(A_1 \overline{A}_2)$, $k_3 = P(\overline{A}_1 A_2)$, $k_4 = P(\overline{A}_1 \overline{A}_2)$, so that $P(A_1) = k_1 + k_2 (= w)$. As Boole does in his 'easy and general' method we express the given conditions in the form

$$1k_1 + 0k_2 + 0k_3 + 0k_4 = p$$
$$1k_1 + 0k_2 + 1k_3 + 0k_4 = q \qquad (2)$$
$$1k_1 + 1k_2 + 1k_3 + 1k_4 = 1$$
$$k_1, k_2, k_3, k_4, \geq 0,$$

where the first two equations are the explicit conditions, and the remaining lines are implicit ones. In matrix notation these conditions become

$$\begin{bmatrix} 1 & 0 & 0 & 0 \\ 1 & 0 & 1 & 0 \\ 1 & 1 & 1 & 1 \end{bmatrix} \begin{bmatrix} k_1 \\ k_2 \\ k_3 \\ k_4 \end{bmatrix} = \begin{bmatrix} p \\ q \\ 1 \end{bmatrix} \qquad (3)$$

$$[k_1 \quad k_2 \quad k_3 \quad k_4]^{\mathrm{T}} \geq \mathbf{0}.$$

Now define $\beta(p, q)$ to be the set of values $w = k_1 + k_2$ where (k_1, k_2, k_3, k_4) is any quadruple which satisfies the system (3), considered simply as an algebraic system. One can show that the sets $\alpha(p,q)$ and $\beta(p,q)$ are the same and that $U(p,q)$ and $L(p,q)$ are its maximum and minimum. Hence finding best possible bounds for $P(A_1)$ is equivalent to solving the linear programming problem: find the max (and min) of the linear form $k_1 + k_2$ subject to the linear constraints (3). If, following Boole, we apply Fourier elimination (of k_1, k_2, k_3, k_4) to the system we obtain formed by adjoining the equation $w = k_1 + k_2$ to (3), we obtain

$$
\begin{aligned}
0 &\leq q \\
q &\leq p \leq 1 \\
q &\leq w \\
w &\leq q + 1 - p \\
0 &\leq w, \ w \leq 1
\end{aligned}
\tag{4}
$$

so that from the last four inequations we have,

$$
\begin{aligned}
\max w &= \min\{q + 1 - p, \ 1\} \\
\min w &= \max\{0, \ q\},
\end{aligned}
\tag{5}
$$

and from the first two the consistency condition,

$$
0 \leq q \leq p \leq 1,
\tag{6}
$$

for the existence of the solution.

By virtue of the duality theorem of linear programming an equivalent problem to finding the minimum of $(w =) \ 1k_1 + 1k_2 + 0k_3 + 0k_4$, subject to (3), is finding the maximum of $(u =) \ px_1 + qx_2 + 1x_3$, subject to

$$
\begin{bmatrix}
1 & 1 & 1 \\
0 & 0 & 1 \\
0 & 1 & 1 \\
0 & 0 & 1
\end{bmatrix}
\begin{bmatrix}
x_1 \\
x_2 \\
x_3
\end{bmatrix}
\leq
\begin{bmatrix}
1 \\
1 \\
0 \\
0
\end{bmatrix}
\tag{7}
$$

with the x_i unrestricted. Note that the coefficients in the objective function w of the primal form, i.e., $1, 1, 0, 0$, now are the constant terms on the right side of (7), that the constant terms on the right side of the equations in (3), i.e., $p, q, 1$, now are the coefficients of the objective function u of the dual form, and that the matrix of 0s and 1s in (7) making up the coefficients of the x_i, is the transpose of the matrix of coefficients in (3). This dual form of the linear programming problem has the significant feature that the parameters p, q now are no longer involved in specifying the region of feasible solutions, and hence the polytope specified by the inequalitites

(7) is expressed in purely numerical terms. The theory tells us that the optimal value of the objective function is obtained at a vertex, i.e., at a basic feasible solution $(x_1^{(i)}, x_2^{(i)}, x_3^{(i)})$. For our example these corner points are $(0,0,0)$ and $(0,1,0)$, so that the maximum value for u is the maximum of $p0 + q0 + 0$ and $p0 + q1 + 0$, that is, $\max\{0, q\}$.

We return to the unfinished business of §3.4 concerning Boole's use of $A(s, t, \dots)$ in place of $F(x, y, \dots)$ to obtain the optimal lower bound. We use our ongoing simple example to illustrate the ideas. Set $s = xy$, $t = y$, and $w = x$. Then Boole's solution for w in terms of s and t is (using inclusion instead of Boole's indefinite symbol):

$$\begin{cases} st \subseteq w \subseteq st + \bar{s}\bar{t} \\ \bar{s}t = 0, \end{cases}$$

and the method entails obtaining the optimum value for $P(st)$ so as to provide a lower bound for $P(w)$. (In this example $A(s, t, \dots)$ is st, and $F(x, y, \dots)$ is $W[= x]$.) In the table below we have, as a brief study will show, a combined representation of the relationships between x, y and s, t, w as well as the conditions expressed by (7).

$K_i(x,y)$	$s(= xy)$	$t(= y)$	1		st		$w(= x)$
xy	1	1	1		1		1
$x\bar{y}$	0	0	1	$(\leq)$	0	$(\leq)$	1
$\bar{x}y$	0	1	1		0		0
$\bar{x}\bar{y}$	0	0	1		0		0

(Table 1)

In this table the '1' or '0' in a column indicates the presence or absence of the constituent (of column 1) in the expansion of the term (as a function of x, y) heading the column. (Lines 2 and 4 of the table show that w can't be a truth-function of s and t since it is assigned different values for the same s and t values.) Ignoring columns 1 and 5 gives us the system (7), with $s, t, 1$ playing the role of x_1, x_2, x_3. Note, in (7), that any set of values (x_1, x_2, x_3) making its line 4 true also makes its line 2 true since $0x_1 + 0x_2 + 1x_3 \leq 0$ implies $0x_1 + 0x_2 + 1x_3 \leq 1$. Thus if in Table 1 we replace, under w, the '1' in line 2 by '0'—which then becomes the same as column 5—we lose nothing from the set of feasible points containing the optimal value. Thus the optimal value for $P(st)$ will be the same as that for $P(w)$. For the general case there will be a table similar to Table 1:

$K_i(x, y, \ldots)$	s t . . . u	1		w
K_1		1		w_1
.	(1's and 0's)	1		w_2
.		.	$(\leq)$	.
.		.		.
.		.		.
.		.		.
K_{2^n}		1		w_{2^n}

If for any pair $K_i, K_j, (i \neq j)$ the values under $s, t, \ldots, u$ are the same in both rows i and j but there are different values in these rows under w, then the value under w which is '1' can be replace by '0' without changing the feasible set (since ≤ 0 implies ≤ 1). We carry out this operation until no change is possible and let the resulting column be headed by 'w^*'. But these values under w^* are the same as that for a function of $s, t, \ldots, u$ obtained by taking the logical sum of constituents on $s, t, \ldots, u$ that is maximal with respect to implying w, and this is Boole's $A(s, t, \ldots, u)$.[5]

Note that the described procedure reduces the number of distinct rows in the matrix inequality and hence, after duplicate rows are deleted, reduces the amount of computation needed to find the vertices of the polytope.

The linear programming approach to logico-probability questions pioneered by Boole can be used to derive a large variety of probability identities, as was noted by Rényi in his *1970*, §2.6. Rényi considers linear relations of the form

$$\sum_{i=1}^{N} c_i P(F_i) = 0 \quad (\text{or, also, } \leq 0, \geq 0) \tag{8}$$

where the c_i are real constants and the $F_i = F_i(A_1, \ldots, A_n)$ are Boolean polynomials of the events $A_1, \ldots, A_n$. By direct argument Rényi shows (Theorem 2.61, p. 64) that (8) is an identity, i.e., is valid in all probability spaces $S = (\Omega, \mathcal{A}, P)$, if it is valid in the 2^n special cases when each of the $A_1, \ldots, A_n$ is allowed to be either 0 or Ω, that is, if it is valid in the trivial space with just the two events 0 and Ω. After proving the theorem Rényi remarks (p. 68 with a reference to *Ha[i]lperin 1965*) that one may also view the result from the viewpoint of linear programming, but says nothing further. It is a rather nice application of the technique. We shall be presenting it in chapter 6.

---

5. Proof is given in (i) on page 159 of *Hailperin 1986*.

After our *1965*, four papers appeared which were concerned with logical-probability or probability-logical questions and which used linear programming. Apparently all four were independent developments since none of them mentions earlier or contemporaneous sources. Two of these will be described in chapter 6. The other two, which have logic as their primary interest, are discussed in the next two sections.

§3.6. Transmission of uncertainties in inferences

Featured in *Adams-Levine 1975* is an investigation of how uncertainties are transmitted from premises to conclusion in deductive inferences. This notion of inference is a generalized one and need not be necessary, i.e., deductively sound: any ordered pair $<< \phi_1, \ldots, \phi_n >, \psi >$ consisting of an n-tuple $< \phi_1, \ldots, \phi_n >$ of sentences (called the 'premises') and a sentence ψ (called the 'conclusion') is referred to as an *inference*. The *uncertainty* of a sentence is (by definition) the probability of its denial. The central question considered is:

If the premises ϕ_i of an inference have uncertainties that, respectively, do not exceed ϵ_i $(i = 1, \ldots, n)$ what is the maximum uncertainty of the conclusion ψ?

Since $P(\overline{\phi}) \leq \epsilon$ is equivalent to $\epsilon \leq P(\phi) \leq 1$ the notion of uncertainty of a sentence not exceeding ϵ is equivalent to that of its being within ϵ of 1. We have noted (in §3.3 above) that Suppes had presented some inferences featuring premises with probabilities near 1. What is different here is that with Adams-Levine the conclusion formula need not be a necessary consequence of the premise formulas. Moreover, they address themselves to a general class of inferences and not just specific inferences.

Their paper states that the sentences can be from an arbitrary first-order language. But many logical notions that are appealed to (e.g., logical implication, logical equivalence, consistency) would in general be non-effective if applied to first-order languages; and, moreover, nowhere is any use made of quantifier structure. Accordingly we shall assume that only the sentential (propositional, truth-functional) structure of sentences is relevant. With this understanding we see that the Adams-Levine question is a special subcase of Boole's 'general' probability problem, as extended to inequalities in

Hailperin 1965 §6, namely:

$$Given: \quad P(\bar{\phi}_i) \leq \epsilon_i, \quad (i = 1, \ldots, n)$$

$$find: \quad \text{the best possible upper bound for } P(\bar{\psi})$$

(the bound to be in terms of $\epsilon_1, \ldots, \epsilon_n$). The method which Adams-Levine use to solve this problem is quite different from the (extended) Boole approach and merits discussion.

Though the inferences Adams-Levine consider are not required to be deductively sound, nor need the premises and conclusion be related, the method is unnecessarily encumbered (we believe) with logical distinctions such as whether the premises are consistent or inconsistent,[6] which subsets of the premises imply or do not imply the conclusion, and others. These distinctions, to be sure, are involved later on in the paper, for example in tracing the effects of particular subsets of the premises on the conclusion's uncertainty. But, as we shall see, the maximum uncertainty of the conclusion is obtainable from a linear program which can be set up without appeal to these logical distinctions; and once such a linear program has been set up one can read off from the algebraic form the various logical items which the Adams-Levine method would have needed as prerequisite to its employment. An example or two will make this clear. But first an outline of the paper's procedure.

An initial step is the replacement of the conclusion ψ by another sentence defined in terms of the premises (the sentence may be ψ itself). The sentence replacing ψ for purposes of finding maximum uncertainty, which can take on either of two equivalent forms, me(I) or ms(I), is obtained as follows.

A subset P' of the set of premises P of an inference $I = < \phi_1, \ldots, \phi_n, \psi >$ is said to be *essential* for I if the complimentary set $P - P'$ does not imply ψ. An essential (for I) set is minimal if no proper subset is essential. Let $E_1, \ldots, E_s$ be minimal essential subsets for I. Then, where $\bigvee E_i$ is the disjunction of members of E_i (or the logically false sentence if E_i is empty), me(I) is defined to be $(\bigvee E_1) \wedge (\bigvee E_2) \wedge \cdots \wedge (\bigvee E_s)$ (or the logically true sentence if there are no minimal essential sets). A premise subset P' of P is *sufficient* (for I) if it implies ψ, and is minimal if no proper subset of it does. If the minimal sufficient subsets are $S_1, \ldots, S_r$ then ms(I) is defined to be $(\bigwedge S_1) \vee (\bigwedge S_2) \vee \cdots \vee (\bigwedge S_r)$ where $\bigwedge S_i$ is the conjunction of elements of S_i (or the logically true sentence if S_i is empty). If there are no minimal sufficient subsets then ms(I) is defined to be the logically false sentence. Thus, as an example, the inference

$$I = < A, B, -(AB), A \leftrightarrow -B >,$$

---

6. Inconsistent premises, e.g., ϕ and $\bar{\phi}$, need not be excluded—one can have $P(\bar{\phi}) \leq \epsilon_1$ and $P(\bar{\bar{\phi}}) \leq \epsilon_2$ so long as $1 \leq \epsilon_1 + \epsilon_2 \leq 2$.

with conclusion $A \leftrightarrow -B$, has the two minimal essential subsets

$$E_1 = \{A, B\} \text{ and } E_2 = \{-(AB)\}.$$

Hence

$$\text{me}(I) = (A \vee B) \wedge (-(AB)) = A\bar{B} \vee \bar{A}B,$$

which happens in this case to be equivalent to the conclusion.

It is asserted in the paper that me(I) and ms(I) are equivalent and that either can replace ψ for the purpose of finding the maximum uncertainty of ψ. Proving the first of these assertions is straightforward, but not the second. It is further stated (p. 433):

> The reduction [replacement of ψ by ms(I)] can be carried one step farther. If the premises are consistent then each premise can be replaced by a distinct atomic letter with the same replacements being made in ms(I) or me(I), and the new inference will be equivalent to the original with respect to uncertainty maximization. The same reduction can also be carried out when the premises are inconsistent, except that in this case it is necessary to add non-logical axioms to the language which specify in effect that sets of atomic formulas which correspond to inconsistent premise sets are inconsistent.

Note the correspondence of Adams-Levine's introducing atomic letters with Boole's introducing the letters $s, t, \ldots, u$ in place of the given functions of $x, y, \ldots, z$ (§§2.5 and 3.4 above). We now look at Adams-Levine's method of obtaining the linear inquality systems expressing $P(\bar{\phi}_i) \leq \epsilon_i$, and the linear form for $P(\overline{ms(I)})$. This requires examining the premises and determining the minimal essential subsets and the minimal negatively sufficient subsets of the premises. The first of these notions we have already introduced. As for the second , a subset of an inconsistent set of premises is *negatively sufficient* if its complement (in the set of premises) is consistent and sufficient (to imply the conclusion); if the premises are consistent, the sets are empty. Using these notions Adams-Levine construct a *minimal falsification matrix* from which *minimal falsification functions* are obtained to provide the linear forms in terms of which the linear program is stated. We shall not state the general definition of these two concepts but they may be inferred from a simple example we take from *Adams-Levine 1975* p. 435 (in this example ms(I) is equivalent to ψ):

> A simple illustration is the inference of the conclusion '$A \leftrightarrow -B$' from the three premises 'A', 'B', and '$-(A\&B)$' (the premises are both redundant and inconsistent). In this example there are two minimal sufficient premise sets, $S_1 = \{A, -(A\&B)\}$ and $S_2 =$

$\{B, -(A\&B)\}$, two minimal essential premise sets $E_1 = \{A, B\}$ and $E_2 = \{-(A\&B)\}$, and two negatively sufficient premise sets, $NS_1 = \{A\}$, and $NS_2 = \{B\}$.

For this example (call it EXAMPLE 1) their minimal falsification matrix is:

| | premises: | | | conclusion : | |
	A	B	$-(A\&B)$	$A \leftrightarrow -B$	
$E_1 = \{A, B\}$	1	1	0	1	(1)
$E_2 = \{-(A\&B)\}$	0	0	1	1	
$NS_1 = \{A\}$	1	0	0	0	
$NS_2 = \{B\}$	0	1	0	0	

In this matrix the rows are values of minimal falsification functions (one for each row) and are weighted—in the case of essential rows (here the first two) by non-negative reals $p_1, \ldots, p_s$ (in the example $s = 2$). and for the negatively sufficient rows (here the last two) by non-negative reals $q_1, \ldots, q_t$ (in the example $t = 2$) and the sum of all the weights is set equal to 1. Probabilities are then assigned by summing weighted *column* entries. Thus:

$$P(\bar{A}) = 1 \cdot p_1 + 0 \cdot p_2 + 1 \cdot q_1 + 0 \cdot q_2$$
$$P(\bar{B}) = 1 \cdot p_1 + 0 \cdot p_2 + 0 \cdot q_1 + 1 \cdot q_2$$
$$P(\overline{AB}) = 0 \cdot p_1 + 1 \cdot p_2 + 0 \cdot q_1 + 0 \cdot q_2$$

and $$P(\overline{A \leftrightarrow \bar{B}}) = 1 \cdot p_1 + 1 \cdot p_2.$$

This yields the linear program:

$$\text{maximize } p_1 + p_2$$

subject to

$$\begin{bmatrix} 1 & 0 & 1 & 0 \\ 1 & 0 & 0 & 1 \\ 0 & 1 & 0 & 0 \end{bmatrix} \begin{bmatrix} p_1 \\ p_2 \\ q_1 \\ q_2 \end{bmatrix} \leq \begin{bmatrix} \epsilon_1 \\ \epsilon_2 \\ \epsilon_3 \end{bmatrix}$$

$$p_1 + p_2 + q_1 + q_2 = 1, \quad p_1, p_2, q_1, q_2 \geq 0.$$

We now show how Adams-Levine's minimal falsification matrix (1) can be obtained by a straightforward (to us, less obscure) method, Boole's 'purely algebraic form' described above in §3.4.

Introduce *letters* $\phi_1, \phi_2, \phi_3, \psi$ to represent the logical functions, i.e., set

$$\phi_1 \leftrightarrow A, \quad \phi_2 \leftrightarrow B, \quad \phi_3 \leftrightarrow \overline{AB}, \text{ and } \psi \leftrightarrow A\bar{B} \vee \bar{A}B. \tag{2}$$

Eliminate A and B from these equivalences by any one of a number of means, e.g., by computing all 2^3 basic conjunctions on ϕ_1, ϕ_2, ϕ_3, or by Boole's algebraic techniques (mentioned in §2.5 and §3.4 above). Then one obtains

$$\phi_1 \phi_2 \bar{\phi}_3 \vee \bar{\phi}_1 \bar{\phi}_2 \phi_3 \leftrightarrow \bar{\psi}$$
$$\bar{\phi}_1 \bar{\phi}_2 \bar{\phi}_3 \vee \bar{\phi}_1 \phi_2 \bar{\phi}_3 \vee \phi_1 \bar{\phi}_2 \bar{\phi}_3 \vee \phi_1 \phi_2 \phi_3 \leftrightarrow f, \tag{3}$$

where f is the logically false sentence. (In general, the first line could be a two-sided inclusion of $\bar{\psi}$ by sums of basic conjunctions and not an equivalence.) As Boole would have put it, the basic conjunctions in the second line (corresponding to his $D = 0$ equation) are those combinations which are 'impossible' or excluded by the data. It is convenient to display this information in a table. Here $\frac{1}{0}$ stands for 'impossible' and we enter the table with values of $\bar{\phi}_i$ and $\bar{\psi}$ so as to match up with Adams-Levine's scheme.

	$\bar{\phi}_1$	$\bar{\phi}_2$	$\bar{\phi}_3$	$K_i(\bar{\phi}_1, \bar{\phi}_2, \bar{\phi}_3)$	$\bar{\psi}$	
$i = 1.$	1	1	1	$\bar{\phi}_1 \bar{\phi}_2 \bar{\phi}_3$	$\frac{1}{0}$	
2.	1	1	0	$\bar{\phi}_1 \bar{\phi}_2 \phi_3$	1	
3.	1	0	1	$\bar{\phi}_1 \phi_2 \bar{\phi}_3$	$\frac{1}{0}$	
4.	1	0	0	$\bar{\phi}_1 \phi_2 \phi_3$	0	(Table A)
5.	0	1	1	$\phi_1 \bar{\phi}_2 \bar{\phi}_3$	$\frac{1}{0}$	
6.	0	1	0	$\phi_1 \bar{\phi}_2 \phi_3$	0	
7.	0	0	1	$\phi_1 \phi_2 \bar{\phi}_3$	1	
8.	0	0	0	$\phi_1 \phi_2 \phi_3$	$\frac{1}{0}$	

Deleting rows containing $\frac{1}{0}$ (unrealizable, hence contributing no probability) produces

	$\bar{\phi}_1$	$\bar{\phi}_2$	$\bar{\phi}_3$	$K_i(\bar{\phi}_1, \bar{\phi}_2, \bar{\phi}_3)$	$\bar{\psi}$	
$i = 2.$	1	1	0	$\bar{\phi}_1 \bar{\phi}_2 \bar{\phi}_3$	1	
4.	1	0	0	$\bar{\phi}_1 \phi_2 \phi_3$	0	(Table B)
6.	0	1	0	$\phi_1 \bar{\phi}_2 \phi_3$	0	
7.	0	0	1	$\phi_1 \phi_2 \bar{\phi}_3$	1	

This is identical with Adams-Levine's minimal falsification matrix (1) except that our columns are headed by the negated letters, and there is a different arrangement of the rows. Looking at these rows for which there is a 1 under $\overline{\psi}$ enables us to pick out the essential sets of premises (in the example they happen also to be minimal) and for the rows for which there is a 0 under $\overline{\psi}$ we obtain the negatively sufficient sets of premises. From Table B we see that the entire probability space consists of the four constituents listed under $K_i(\overline{\phi}_1, \overline{\phi}_2, \overline{\phi}_3)$. Assigning probabilities ('weights') p_1, p_2, q_1, q_2 to these constituents (letters 'p' for those corresponding to essential premise sets, and letters 'q' for those corresponding to negatively sufficient ones) enables us to write

$$P(\overline{A}) = p(\overline{\phi}_1) = P(\overline{\phi}_1\overline{\phi}_2\overline{\phi}_3 \vee \overline{\phi}_1\phi_2\phi_3) = 1p_1 + 0p_2 + 1q_1 + 0q_2,$$

and similarly for the other formulas of the inference, which reproduces the Adams-Levine linear program for the inference.

Since this EXAMPLE 1 doesn't show all the features of the Adams-Levine approach we present another one, one in which the premises are consistent and sufficient. This is taken from *Adams-Levine 1975* p. 44. Call it EXAMPLE 2. The premises are A, B, C and the conclusion, $A(B \vee C)$, is equivalent to ms(I). Thus we wish to find the maximum of $P(\overline{A} \vee \overline{BC})$ subject to $P(\overline{A}) \leq \epsilon_1$, $P(\overline{B}) \leq \epsilon_2$, $P(\overline{C}) \leq \epsilon_3$. In this case $\phi_1 \leftrightarrow A$, $\phi_2 \leftrightarrow B$, $\phi_3 \leftrightarrow C$ and there is no need to introduce the ϕ_i letters. Moreover there are no excluded combinations implied by the premises. Hence

	$\overline{A}$	$\overline{B}$	$\overline{C}$	$K_i(\overline{A},\overline{B},\overline{C})$	$\overline{A} \vee \overline{BC}$	
$i = 1.$	1	1	1	$\overline{A}\,\overline{B}\,\overline{C}$	1	
2.	1	1	0	$\overline{A}\,\overline{B}\,C$	1	
3.	1	0	1	$\overline{A}\,B\,\overline{C}$	1	
4.	1	0	0	$\overline{A}\,B\,C$	1	(Table C)
5.	0	1	1	$A\,\overline{B}\,\overline{C}$	1	
6.	0	1	0	$A\,\overline{B}\,C$	0	
7.	0	0	1	$A\,B\,\overline{C}$	0	
8.	0	0	0	$A\,B\,C$	0	

Lines 1–5 indicate which premise sets are the essential ones and lines 6–8 the negatively sufficient ones. Here there are minimal sets, line 4 and line 5 for the essential ones, and line 8 for the negatively sufficient ones. Thus we obtain Adams-Levine's minimal falsification matrix for this problem, namely

	$\overline{A}$	$\overline{B}$	$\overline{C}$	$\overline{A} \vee \overline{BC}$
$E_1 = \{A\}$	1	0	0	1
$E_2 = \{B,C\}$	0	1	1	1
$NS_1 = \{\}$	0	0	0	0

An equivalent reduction of Table C can be obtained without appeal to any of these special notions. We expand all functions as logical sums of basic conjunction on A, B, C and use the probabilities of the conjunctions as variables. Then the constraint system for the linear program is

k_1	k_2	k_3	k_4	k_5	k_6	k_7	k_8		
1	1	1	1	0	0	0	0	$\leq \epsilon_1$	
1	1	0	0	1	1	0	0	$\leq \epsilon_2$	(Table D)
1	0	1	0	1	0	1	0	$\leq \epsilon_3$	
1	1	1	1	1	1	1	1	$= 1$	

(The detached coefficient format for the system is used to conserve space.) It can be readily seen that columns headed by k_1, k_2, k_3, k_6, and k_7 can be dropped and one has an equivalent system. Even more readily can this be seen from the dual system, namely:

	w_1	w_2	w_3	v		
1.	1	1	1	1	≥ 1	
2.	1	1	0	1	≥ 1	
3.	1	0	1	1	≥ 1	
4.	1	0	0	1	≥ 1	
5.	0	1	1	1	≥ 1	(Table E)
6.	0	1	0	1	≥ 0	
7.	0	0	1	1	≥ 0	
8.	0	0	0	1	≥ 0	

$$w_1, w_2, w_3 \geq 0, \ v \text{ unrestricted}$$

We observe here that, since for each i, $w_i \geq 0$, any set of values w_1, w_2, w_3, v making the inequality in line 4 true also makes those in lines 1, 2, and 3 true. Thus these three lines—corresponding to non-minimal essential sets of premises—are (algebraically) redundant and can be deleted. Similarly line 8 implies lines 6 and 7 and they too can be deleted.

As in *Hailperin 1965*, §4, Adams-Levine also go over to a dual formulation so as to have, as illustrated by Table E, a constraint system which is independent of parameters (here the ϵ_i). These constraints can then be treated purely numerically. The parameters ϵ_i then appear only in the linear objective function—here $w_1\epsilon_1 + w_2\epsilon_2 + w_3\epsilon_3 + v$—which is to be minimized.

We continue with the second of the two independently conceived papers applying linear programming methods to logico-probability questions.

§3.7. Nilsson's probabilistic logic

Unlike the paper discussed in our preceding section, which views the topic as an application of probability to ordinary logic, Nilsson lays claim to a 'probabilistic logic' (*1986*, 72):

> In this paper we present a semantical generalization of ordinary first-order logic in which the truth values of sentences can range between 0 and 1. The truth value of a sentence in *probabilistic logic* is taken to be the *probability* of that sentence in ordinary first-order logic. We make precise the notion of the probability of a sentence through a possible-worlds analysis. Our generalization applies to any logical system for which the consistency of a finite set of sentences can be established.

We single out three items to be noted in this quotation:

(i) that a semantic generalization of ordinary first-order logic is presented,

(ii) that the notion of the probability of a sentence is made precise through a possible-worlds analysis and

(iii) that the generalization applies to any logical system for which the consistency of a finite set of sentences can be established.

We begin our discussion with (ii), making the notion of the probability of a sentence precise. Nilsson's analysis of "the probability of a sentence" starts out with (p. 72):

> To define what we mean by the *probability of a sentence* we must start with a sample space over which to define probabilities (as is customary in probability theory). A sentence S can be either *true* or *false*. If we were concerned about just the one sentence S, we could imagine two sets of *possible-worlds*—one, say $\mathcal{W}_1$, containing worlds in which S was *true* and one, say $\mathcal{W}_2$, containing worlds in which S was *false*. The actual world, the world we are actually in, must be in one of these two sets, but we might not know which one. We can model our uncertainty about the actual world by imagining that it is in $\mathcal{W}_1$ with probability p_1 and is in $\mathcal{W}_2$ with some probability $p_2 = 1 - p_1$. In this sense we can say that the probability of S (being true) is p_1.

Nilsson believes that he has to have a sample space whose subsets are assigned probabilities, and chooses for this space the set of all possible worlds. Then "the sentence S has probability p_1" is represented by

$$p(R \in \mathcal{W}_1(S)) = p_1, \tag{1}$$

where R is the real world and $\mathcal{W}_1(S)$ is the subset of all possible worlds in which S is true. ('$\mathcal{W}_1(S)$' is used rather than '$\mathcal{W}_1$' to indicate its dependence on S.) We do not see the advantage in modelling S's uncertainty in this way. Why not simply

$$p(S) = p_1? \tag{2}$$

This would serve the same purpose, requiring only readjustment in one's thinking so as to be able to attribute probability to sentences,[7] and not necessarily to sets. The gain with (2) over (1) is the elimination of the ontological notions of 'real world' and 'set of possible worlds'.

Nilsson also introduces 'impossible worlds' (*1986*, 72):

> If we have L sentences, we might have as many as 2^L sets of possible worlds. Typically though, we will have fewer than the maximum number because some combinations of *true* and *false* values for our L sentences will be logically inconsistent. We cannot, for example, imagine a world in which S_1 is *false*, S_2 is *true* and $S_1 \wedge S_2$ is *true*. That is, some of the sets of the 2^L worlds might contain only impossible worlds.

He illustrates this with an example of three sentences $\{P, \ P \supset Q, \ Q\}$:

> The consistent sets of truth values for these three sentences are given by the columns in the following table:

P	*true*	*true*	*false*	*false*
$P \supset Q$	*true*	*false*	*true*	*true*
Q	*true*	*false*	*true*	*false*

$$\tag{3}$$

In this case, there are four sets of possible worlds each one corresponding to one of these four sets of truth values.

These sets of consistent truth value assignments—which 'correspond to' sets of possible worlds—are then taken as elements of a sample space on which to define a probability distribution. But for this there is no need for the possible-impossible worlds notion—one can obtain these four consistent assignments to which probability is attached just by an examination of the syntactic structure of the sentences (assuming, as Nilsson does, that the sentences are expressed as logical functions of a set of atomic sentences). Equivalently, and somewhat more conveniently, one can use constituents (basic conjunctions) on $S_1 = P$, $S_2 = P \supset Q$, $S_3 = Q$, that is *sentences*,

---

7. A basic tenet of this monograph. In §4.7 below we discuss the relationship between the customary set-theoretic and our logic-theoretic approach to probability.

to represent the consistent truth value assignments (the four columns), namely the four sentences

$$S_1 S_2 S_3, \ S_1 \overline{S}_2 \overline{S}_3, \ \overline{S}_1 S_2 S_3, \ \overline{S}_1 S_2 \overline{S}_3. \tag{4}$$

These are obtained by writing down all possible constituents on S_1, S_2, S_3 and deleting those which become inconsistent when S_1 is replaced by P, S_2 by $P \supset Q$ and S_3 by Q. The information contained in (3) is reproduced by an 'incidence matrix':

	$S_1 S_2 S_3$	$S_1 \overline{S}_2 \overline{S}_3$	$\overline{S}_1 S_2 S_3$	$\overline{S}_1 S_2 \overline{S}_3$	
$S_1(=P)$	1	1	0	0	(5)
$S_2(=P \supset Q)$	1	0	1	1	
$S_3(=Q)$	1	0	1	0	

where a 1 in the body of the table indicates that the formula to the left in that row appears unnegated, and a 0 that it appears negated, in the basic conjunction at the head of the column. Table (5) establishes a one-to-one correspondence between consistent assignments of truth values to S_1, S_2, S_3 and constituents on these letters. We can reproduce (5) in a slightly different way which serves to bring out the connection with Adams-Levine (§3.6) and Boole (§2.5). Consider the truth table on sentence variables S_1, S_2, S_3:

			S_1	S_2	S_3
$i = 1.$		$S_1 S_2 S_3$	1	1	1
	2.	$S_1 S_2 \overline{S}_3$	1	1	0
	3.	$S_1 \overline{S}_2 S_3$	1	0	1
	4.	$S_1 \overline{S}_2 \overline{S}_3$	1	0	0
	5.	$\overline{S}_1 S_2 S_3$	0	1	1
	6.	$\overline{S}_1 S_2 \overline{S}_3$	0	1	0
	7.	$\overline{S}_1 \overline{S}_2 S_3$	0	0	1
	8.	$\overline{S}_1 \overline{S}_2 \overline{S}_3$	0	0	0

If we set $S_1 \leftrightarrow P$, $S_2 \leftrightarrow P \supset Q$, $S_3 \leftrightarrow Q$ and 'algebraically' eliminate P and Q we obtain the analogue of Boole's $D = 0$ equation, namely

$$S_1 S_2 \overline{S}_3 \vee S_1 \overline{S}_2 S_3 \vee \overline{S}_1 \overline{S}_2 S_3 \vee \overline{S}_1 \overline{S}_2 \overline{S}_3 \leftrightarrow f,$$

representing those alternands on S_1, S_2, S_3, which are impossible (by *logic*) on the data, hence contributing no probability. Striking out the corresponding rows, i.e., rows 2, 3, 7 and 8, produces

	S_1	S_2	S_3
$S_1 S_2 S_3$	1	1	1
$S_1 \overline{S}_2 \overline{S}_3$	1	0	0
$\overline{S}_1 S_2 S_3$	0	1	1
$\overline{S}_1 S_2 \overline{S}_3$	0	1	0

which is the transpose of (5).

Since Nilsson's development is based on an analysis such as is illustrated in (3), and since we have seen that one can arrive at it by ordinary logical means, here too we believe the notion of 'worlds' is not essential. In fact, they disappear for Nilsson when he identifies (p. 73) sets of possible worlds with sets of truth values for sentences. We suggest going one step further and replace the sets of truth values by constituents (on the S_1, S_2, S_3), which are sentences.

We turn to the other items singled out from our initial quotation. With regard to item (i) the description of his 'probabilistic logic' is quite brief (*Nilsson 1986*, 73) :

> ... Since we typically do not know the ordinary (*true/false*) truth value of S in the actual world, it is convenient to imagine a logic that has truth values intermediate between *true* and *false* and, in this logic, define the truth value of S to be the probability of S. In the context of discussing uncertain beliefs, we use the phrases *the probability of S* and *the (probabilistic logic) truth value of S* interchangeably.

However the details of the 'logic' which Nilsson says it is convenient to imagine are not spelled out. All we are told is that it has ('truth') values between *true* and *false* and that the truth value of a sentence S is defined to be the probability of S. We have the following comments:

1. A (syntax) language for the logic is not explicitly stated. Nilsson implies (see our opening quotation from his paper) that the sentences are those of ordinary first-order logic.
2. There is no full account of the semantics and its relationship to the syntax. How does the value (probability) of a compound sentence relate to those of its components? And what if the sentence is a quantified expression?
3. There is no explicit definition of logical consequence. However Nilsson does introduce a procedure he calls "probabilistic entailment" which concerns "determining the probability of an arbitrary sentence given a set B of sentences and their probabilities".

If we restrict Nilsson's "logical systems" to sentential logic then his "determining the probability of an arbitrary sentence given a set B of sentences and their probabilities" is exactly Boole's General Probability Problem (see §§2.4, 2.5 above). Apparently when Nilsson wrote his paper he was unaware of any previous history, as his only relevant reference is to *Adams-Levine 1975* which, likewise, was unaware of earlier work. It is of interest to look at the problem from Nilsson's perspective, as we now do.

Suppose we are given a set S of sentences and a probability distribution over the sets of 'possible worlds' for S (i.e., a probability distribution over the consistent basic conjunctions of elements of S). The combined relationship that obtains between the probabilities that accrue to the sentences of S by virtue of their syntactic structure and the probabilities of the distribution, is expressed by Nilsson as a matrix equation

$$\Pi = \mathbf{VP}, \tag{6}$$

where Π is a column matrix whose entries are the probabilities of the members of S, where $\mathbf{V}$ is the matrix whose columns are the sets of truth values (in the form of 1's and 0's) which the sentences have in the possible worlds, and where $\mathbf{P}$ is the column matrix of the probability distribution values. Thus for the example $S = \{P, \ P \supset Q, \ Q\}$ equation (6) is

$$\begin{bmatrix} p(P) \\ p(P \supset Q) \\ p(Q) \end{bmatrix} = \begin{bmatrix} 1 & 1 & 0 & 0 \\ 1 & 0 & 1 & 1 \\ 1 & 0 & 1 & 0 \end{bmatrix} \begin{bmatrix} p_1 \\ p_2 \\ p_3 \\ p_4 \end{bmatrix} \tag{7}$$

where the matrix $\mathbf{V}$ of 1's and 0's is the incidence matrix (5). To (6) and (7) one needs to add the conditions $p_i \geq 0$ and $\sum_i p_i = 1$. (This latter condition is replaceable by the addition of a logically true sentence to S which would then add a row of 1's to $\mathbf{V}$ and also a 1 to Π.) In the example (7) the first equation,

$$p(P) = p_1 + p_2 + 0p_3 + 0p_4,$$

'says' that the probability $p(P)$ is composed of the p_1-th part contributed by $S_1 S_2 S_3$ and the p_2-th part contributed fy $S_1 \overline{S}_2 \overline{S}_3$, and similarly for the other three equations. An analysis such as this first occurs in Boole *1854*, chapter XIX (modernized in *Hailperin 1965*).

The procedure which Nilsson employs to determine the probabilistic entailment of sentence S by a set of sentences $\mathcal{B}$ ('beliefs') with given probabilities, is to adjoin S to the set $\mathcal{B}$ and then for this enlarged set determine the region of possible probability values for S in the Π-space, subject to the contraint $\Pi = \mathbf{VP}$ (for the enlarged $\mathcal{B}$). Examination of this region gives Nilsson the allowable probability values for the sentence S. Thus, for the example $S = Q, \mathcal{B} = \{P, P \supset Q\}$, the matrix equation is (7). Because the

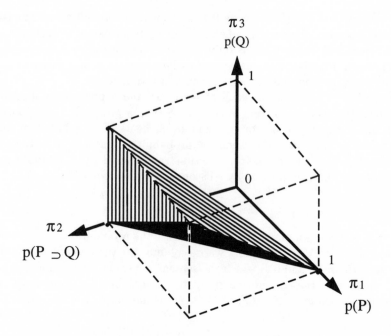

FIG. 3.71. The convex region of consistent probability values for P, $P \supset Q$, and Q.

small number of sentences involved Nilsson can view the problem geometrically. The region specified by the matrix equation is depicted in Figure 3.71 (taken from Nilsson, p. 76).

The solid has two non-vertical faces and two vertical faces (one hidden from view). From the equations of the non-vertical planes one readily deduces that if $(p(P),\ p(P \supset Q),\ p(Q))$ is a point of the solid then

$$p(P \supset Q) + p(P) - 1 \leq p(Q) \leq p(P \supset Q).$$

This same result, as an example of logical consequence in probability logic (generalized *modus ponens*), appeared in *Hailperin 1984*, 209–210, and is Theorem 4.51 below.

Nilsson observes that inconsistent probabilities could be assigned to premises (elements of $\mathcal{B}$). Any vertical line in the figure not intersecting the solid would intersect the $\pi_1\pi_2$ plane in a point $(p(P),\ p(P \supset Q))$ which furnishes no value for $p(Q)$. He also points out that the probability $p(Q)$ is not determined uniquely but only within bounds. He does not consider the question as to whether these bounds are best possible, i.e., give the strongest entailment result.

Our next comment concerns the third item of note listed by us at the beginning of this section. Nilsson states that the "semantical generalization of ordinary first-order logic" applies to any logical system for which the consistency of a finite set of sentences can be established. We take it that "established" means by an effective procedure. Since consistency for arbitrary finite sets of sentences of first-order logic is not an effective notion some restriction, e.g., to sentential logic or to monadic predicate logic, is needed; and, in the latter case, as we shall now see, no more than sentential logic is used.

The example of Nilsson's we wish to discuss is referred to as a "simple problem in first-order logic". It concerns the probabilistic entailment of $\exists z Q(z)$ by $\exists y P(y)$ and $\forall x(P(x) \supset Q(x))$. Using Skolem functions he finds the consistent assignments for the three sentences to be (using our form)

	$S_1 S_2 S_3$	$S_1 \overline{S}_2 S_3$	$S_1 \overline{S}_2 \overline{S}_3$	$\overline{S}_1 S_2 S_3$	$\overline{S}_1 S_2 \overline{S}_3$
$S_1(\exists y P(y))$	1	1	1	0	0
$S_2(\forall x(P(x) \supset Q(x)))$	1	0	0	1	1
$S_3(\exists z Q(z))$	1	1	0	1	0

from which he derives

$$p(\exists y P(y)) + p(\forall x(P(x) \supset Q(x))) - 1 \leq p(\exists z Q(z)) \leq 1. \tag{8}$$

However the problem is readily expressible on the sentential level. We rewrite the sentences (using obvious abbreviations) as follows:

$$\exists y P(y) = \exists x(P(x)Q(x) \lor P(x)\overline{Q(x)})$$
$$= \exists(PQ) \lor \exists(P\overline{Q}) = A \lor B$$
$$\forall x(P(x) \supset Q(x)) = \neg\exists x(P(x)\overline{Q(x)})$$
$$= \neg\exists(P\overline{Q}) = \neg B$$
$$\exists z Q(z) = \exists x(P(x)Q(x) \lor \overline{P(x)}Q(x))$$
$$= \exists(PQ) \lor \exists(\overline{P}Q) = A \lor C.$$

The set is then of the form

$$S = \{A \lor B, \ \neg B, \ A \lor C\}$$

and produces the result (8) without reference to the quantifier structure of A, B, or C.

What we have shown here in this particular case is true in general for monadic predicate sentences. For such sentences consistency can be determined in a domain of individuals having no more that 2^n individuals, where n is the number of distinct predicates present. (See, e.g., *Hilbert-Bernays 1968*, 194.) Thus all quantification sentences can be replaced by finite conjunctions or alternations, which then converts the problem to the sentential (i.e., non-first-order) level. (Systematic methods for carrying out the reduction—as in our working of Nilsson's example—can be gleaned from *Quine 1972*, §§18–24.)

Although realizing that the problem of probabilistic entailment can be treated by linear programming methods, Nilsson nevertheless doesn't make use of the results and methods of this subject. For example, he remarks (p. 86) that his method can be extended to include ("belief") sentences with given upper and lower bound probabilities and not just exact values. But rather than including these as inequality constraints (as in our *1965*, §6) he suggests instead calculating bounds for the entailed S by first using one set of extreme values, and then again for the other set of extreme values.

There is a brief discussion of conditional probabilities in probabilistic logic. The method of treatment suggested is to write the conditional probability as a quotient of unconditional probabilities and to find the bounds for each separately. We contend this cannot give best results in all cases. See, for example, the criticism of a similar procedure of Boole's in our *1986*, 371–372; it contains in its §6.7 a quite different method, which doesn't separate the numerator and denominator probabilities but treats the quotient as a whole. Our chapter 5 below has a detailed treatment of the method.

A number of papers based on Nilsson's subsequently appeared in various journals devoted to operations reserach, artificial intelligence, expert systems and the like. We cite only *Georgakopoulos, Kavvadias and Papadimitriou 1988* and *Jaumard, Hansen and Poggi de Aragão 1991*, these being specially concerned with the large scale computations that arise when there are many variables and conditions. The latter of these two papers includes in its bibliography references to other papers that were spawned by Nilsson's.

§3.8. Probability logic of Scott and Krauss (*1966*)

The paper we discuss in this section antedated both that of Adams-Levine (§3.6) and that of Nilsson (§3.7). According to the authors it was inspired by *Gaifman 1964*, which introduced and investigated probability measures on (finitary) first-order languages. Scott and Krauss frame their results for *infinitary* first-order languages, i.e., languages which, among other things, allow concatenation of transfinitely many (but less than ω_1) symbols to form a formula. We will not engage in this much generality. With appropriate modifications the Scott-Krauss results which are of interest to us can be specialized to finitary (i.e., ordinary) first-order or to sentential languages. Restricting attention to the sentential case, as we will now be doing, while excising the substantive results of their paper, will enable us to bring out a difference with what we present in the next chapter.

A *probability model* (for the sentential case) is an $(n + 2)$-tuple of the form

$$\langle A_1^0, \ldots, A_n^0, \mathcal{A}, m \rangle$$

where

(i) $A_1^0, \ldots, A_n^0$ are sentences (of an interpreted language),

(ii) $\mathcal{A}$ is a Boolean algebra (or, also, the set of its elements),

(iii) m is a probability (measure) on $\mathcal{A}$ which is strictly positive ($m(a) = 0$ if and only if a is the **0** of the Boolean algebra), and

(iv) there is a mapping of the sentences A_i^0 into $\mathcal{A}$, the $\mathcal{A}$-value assigned to A_i^0 being denoted by 'a_i'.

(Here $\mathcal{A}$ generalizes the role played by the two-element Boolean algebra in the case of verity logic.)

Let S be a formal sentential language with sentential variables $A_1, \ldots, A_n, \ldots$ and connectives $\neg$, $\vee$, $\wedge$; let S_n be the part of S all of whose formulas have variables contained in the set $\{A_1, \ldots, A_n\}$. Analogous to 'ϕ holds in a model $\mathfrak{A}$' for verity logic, there is the following definition of 'ϕ holds in a probability model $\mathcal{U}$ with probability α'.

First, for a probability model $\mathcal{U} = \langle A_1^0, \ldots, A_n^0, \mathcal{A}, m \rangle$ one associates a value in $\mathcal{A}$ for each $\phi \in S_n$ when the $A_1, \ldots, A_n$ are thought of, or interpreted as, $A_1^0, \ldots, A_n^0$. This is accomplished by extending the assignment (iv) via a recursively defined *valuation* function h:

$$h(A_i) = a_i, \quad \text{for } i = 1, \ldots, n$$
$$h(\neg \phi) = \mathbf{1} \sim h(\phi) \quad \text{(i.e., the complement of } h(\phi)\text{)},$$
$$h(\phi \wedge \psi) = h(\phi) \cap h(\psi),$$
$$h(\phi \vee \psi) = h(\phi) \cup h(\psi).$$

It can be shown that if ϕ is logically valid ($\vdash \phi$) then $h(\phi) = \mathbf{1}$ ($\mathbf{1}$ is the Boolean unit), and if ϕ and ψ are logically equivalent ($\vdash \phi \leftrightarrow \psi$) then $h(\phi) = h(\psi)$. Setting $\mu_{\mathcal{U}}(\phi) = m(h(\phi))$ defines what Gaifman (*1964*, 2) refers to as a probability measure (on S_n). Then

$$\mu_{\mathcal{U}}(\phi) = \alpha$$

expresses

   ϕ holds in the probability model $\mathcal{U}$ with probability α.

Thus h maps sentences into the Boolean algebra $\mathcal{A}$, whose measure m then endows sentences of $\mathcal{S}_n$ with a (probability) value. In this simplified situation the Boolean algebra $\mathcal{A}$ in the definition of probability model serves no particular function except as a carrier of the probability measure. In its place one could just as well use the Boolean algebra of sets of formulas defined for $\mathcal{S}_n$; the elements of this Boolean algebra are the equivalence classes of formulas of $\mathcal{S}_n$, modulo logical equivalence.

   Preparatory to introducing the definition of probability (logical) consequence Scott and Krauss define a probability assertion, i.e., an assertion of probability logic. The notion requires that the language include symbols for the (elementary) algebra of real numbers (= the theory of real-closed fields); these are: a binary relation symbol $\leq$, binary function symbols $+$, $\times$, and the individual constants 0, $+1$, -1. These symbols are interpreted over the real numbers in the usual manner.[8] Note that in this language the only real numbers that can be explicitly symbolized are the integers. An *algebraic formula* is a quantifier-free formula of the language. It can be shown (*Tarski 1948*, 18) that any algebraic formula is equivalent in real algebra to an alternation of conjunctions of polynomial inequations of the form $p \geq 0$ or $p > 0$, i.e., to a positive logical function of such inequations. A *probability assertion* then is an $(N+1)$-tuple $\langle \Phi, \phi_1, \ldots, \phi_N \rangle$, where Φ is an algebraic formula with N free variables and $\phi_1, \ldots \phi_N \in \mathcal{S}_n$. The definition of a probability model $\mathcal{U}$, in which for each formula ϕ of $\mathcal{S}_n$ there is an associated value $\mu_{\mathcal{U}}(\phi)$ is now to be extended to probability assertions: a probability model $\mathcal{U}$ is a *probability model of (an assertion)* $\langle \Phi, \phi_1, \ldots, \phi_N \rangle$ if the N-tuple $\langle \mu_{\mathcal{U}}(\phi_1), \ldots, \mu_{\mathcal{U}}(\phi_N) \rangle$ satisfies Φ in the reals, i.e., that Φ is true of $\langle \mu_{\mathcal{U}}(\phi_1), \ldots, \mu_{\mathcal{U}}(\phi_N) \rangle$. Now for the Scott-Krauss definition of consequence.

   If Σ is a set of probability assertions and Ψ is a probability assertion, then Ψ is a *probability consequence of* Σ iff every probability model of all assertions in Σ is also a probability model of Ψ. Ψ is a *probability law* of $\mathcal{L}$ if Ψ is a probability consequence of the empty set of assertions.

Scott and Krauss establish the following significant result for their rich (infinitary) language $\mathcal{L}$ (*1966*, 243):

---

8. We trust that the reader will not be confused by our using the same notation for the symbols of the theory of real-closed fields as for their interpretation in the reals.

Theorem 6.7. *Let* $\langle \Phi, \phi_0, \ldots, \phi_{N-1} \rangle$ *be a probability assertion of* $\mathcal{L}$ *such that the free variables of* Φ *are* $\lambda_0, \ldots, \lambda_{N-1}$*); further* $\vdash \neg(\phi_i \wedge \phi_j)$ *if* $i \neq j$*, and* $\vdash \bigvee_{i<N} \phi_i$*. Let* $I = \{i < N : \vdash \neg \phi_i\}$*. Then* $\langle \Phi, \phi_0, \ldots, \phi_{N-1} \rangle$ *is a probability law of* $\mathcal{L}$ *iff the sentence*

$$\forall \lambda_0 \ldots \forall \lambda_{N-1} \left[\bigwedge_{i \in I} \lambda_i = 0 \wedge \bigwedge_{i<N} \lambda_i \geq 0 \wedge \lambda_0 + \cdots + \Lambda_{N-1} = 1 \to \Phi \right]$$

is a theorem of real algebra.

In other words, if it is provable that $\phi_0 \vee \cdots \vee \phi_{N-1}$ and that the ϕ_i are pairwise disjoint, then the probability assertion

$$\langle \Phi, \phi_0, \ldots, \phi_{N-1} \rangle$$

is a probability law if and only if Φ is true of all sets of non-negative real numbers $\lambda_0, \ldots, \lambda_{N-1}$ which sum to 1, with the λ_i corresponding to those ϕ_i's provably false being set equal to 0.

For a sentential language $\mathcal{S}_n$ the result takes on a simpler form. Any formula ϕ of $\mathcal{S}_n$, if not logically false, is equivalent to a logical sum of the form $\bigvee K_j$, the K_j being constituents on $A_1, \ldots, A_n$. Since constituents are mutually exclusive we have, for any probability measure μ,

$$\mu(\phi) = \mu(\bigvee K_j) = \sum \mu(K_j).$$

If $\Phi(\mu(\phi_1), \ldots, \mu(\phi_N))$ is an algebraic formula with arguments $\mu(\phi_i)$ $(i = 1, \ldots, N)$ then on replacing each $\mu(\phi_i)$ by its equal sum $\sum_{\phi_i} \mu(K_j)$ (or 0, if ϕ_i is logically false) produces (after some simple algebra) an algebraic formula of the form $\Psi(\mu(K_1), \ldots, \mu(K_{2^n}))$. Note that when the ϕ_i of the Scott-Krauss Theorem 6.7 are the constituents $K_1, \ldots, K_{2^n}$ then the hypotheses are satisfied and also that the set I is empty. Thus we have for $\mathcal{S}_n$ the following corollary of Theorem 6.7:

Corollary. *Let* $\langle \Phi, K_1, \ldots, K_{2^n} \rangle$ *be a probability assertion of* $\mathcal{S}_n$ *such that the free variables of* Ψ *are* $k_1, \ldots, k_{2^n}$ *and* $K_1, \ldots, K_{2^n}$ *are the constituents on* $A_1, \ldots, A_n$*. Then* $\langle \Psi, K_1, \ldots, K_{2^n} \rangle$ *is a probability law of* $\mathcal{S}_n$ *if and only if the sentence*

$$\forall k_1 \ldots \forall k_{2^n} \left[\bigwedge_{1 \leq j \leq 2^n} (k_j \geq 0) \wedge k_1 + \cdots + k_{2^n} = 1 \to \Psi \right]$$

is a theorem of real algebra.

As a consequence, for sentential probability logic one can use a simpler definition of a probability model, namely a model is an assignment to $K_1, \ldots, K_{2^n}$ of any set of non-negative (or positive, if strict additivity is desired) real numbers $k_1, \ldots, k_{2^n}$ which sum to 1. This is the definition used in our *1984*—though with a different definition of logical consequence—and in §4.3 below. We shall be comparing the two definitions in §4.4.

Chapter 4

Formal Developments

§4.1. What is *a* logic?

Considering the conventional nature of words and that subject matter changes over the course of time, this question is a possibly impossible one to answer. A brief examination of a piece of literature lying at hand produced the following *score* of systems referred to as 'logic': alternative, Aristotelian, classical, combinatory, deontic, epistemic, erotetic, free, fuzzy, inductive, infinitary, intuitionistic, many-valued, modal, non-Aristotelian, probability, quantum, relevance, strict implication and temporal—not to mention variants and subdivisions. Rather than attempting a definitive characterization we shall describe, informally, general features which we think something should have in order to be called a logic.

There should be a formal language and an associated semantics. The internal structure of the atomic sentences of the language may or may not be specified, but the logical syntax of the language should be completely specified. This requirement involves the notion of a logical constant. In this book, aside from peripheral matters, we shall be dealing only with sentential (propositional) languages. For such languages, the internal structure of the atomic sentences playing no role, the only logical constants are the sentential connectives. The logical properties of connectives are determined by the associated semantics, which specifies 'truth'-values of compounds in terms of the 'truth'-values of their atomic sentences. (In keeping with our general viewpoint we shall now use the term *semantic value* in place of 'truth'-value.) It has generally been assumed that the semantic value of a compound sentence has to be uniquely determined by the semantic values of its atomic sentences. This feature will not be present in our probability logic.

Up until almost the middle of this century the customary form in which a logic was presented was syntactic: one has a given set of starting sentences and rules (of inference) for deriving additional ones. The current preferred approach, pioneered by Tarski, is semantic: it uses the notion of logical consequence. In turn logical consequence is framed in terms of the notion of a model. One says that a sentence κ is a logical consequence of sentences $\pi_1, \ldots, \pi_n$ if κ is true in every model in which all of $\pi_1, \ldots, \pi_n$ are true. But 'true in', while appropriate for verity logic, need not be an appropriate semantic property for other logics. In particular the probability logic we shall be presenting will define logical consequence differently. Nevertheless we will be constructing this logic so that verity logic is embedded in it, that is, when semantic values are restricted to being one 1 or 0 (corresponding to *true* and *false*) then the logic reduces to verity logic.

§4.2. Probability functions on languages.

The languages for probability logic we shall be considering in this chapter, generically denoted by '$\mathcal{L}$', are syntactically indistinguishable from those for verity logic. They include the usual sentence connectives with $\neg$, $\vee$, $\wedge$, as the basic set. So much at present for the syntax. Now for the semantics.

We assume that the verity semantics for the sentence connectives (§0.1) carries over to probability logic, i.e., that the usual 0, 1 truth tables for $\neg$, $\vee$, and $\wedge$ hold. We also assume that we have for $\mathcal{L}$ a notion of (verity) logical consequence which we shall denote by '$\vdash$'. (Customarily this symbol is used for the syntactic notion of derivability and '$\vDash$' for the semantic logical consequence. We are making the replacement to free '$\vDash$' for use with probability logic.) Thus '$\vdash \phi$' means that ϕ is logically true in $\mathcal{L}$; in particular, all truth-functional tautologies are logically true in $\mathcal{L}$. With regard to '$\vdash$' we shall need only the following properties, where the letters appearing stand for sentences.

LC1. If ϕ is a truth-functional tautology, then $\vdash \phi$.

LC2. If $\vdash \phi \wedge \psi$, then $\vdash \phi$ and $\vdash \psi$.

LC3. If $\vdash \rho \leftrightarrow \sigma$, then $\vdash \phi \leftrightarrow \phi'$, where ϕ' comes from ϕ on replacing an occurence of ρ by σ.

LC4. If $\vdash \phi$, then $\vdash \phi \leftrightarrow (\phi \vee \neg\phi)$.

References to LC1 will usually be tacit.[1]

A function $P : \mathcal{L} \to [0, 1]$, from $\mathcal{L}$ into the reals of the closed unit interval, is a *probability function on* $\mathcal{L}$ if it has the following properties:

For any ϕ and ψ of $\mathcal{L}$,

> P1. If $\vdash \phi$, then $P(\phi) = 1$.
> P2. If $\vdash \phi \to \psi$, then $P(\phi) \leq P(\psi)$.
> P3. If $\vdash \phi \to \neg\psi$, then $P(\phi \vee \psi) = P(\phi) + P(\psi)$.

In this section Greek letters ϕ, ψ, etc. will stand for elements of $\mathcal{L}$. An immediate consequence of LC2 and P2 is

> P2'. If $\vdash \phi \leftrightarrow \psi$, then $P(\phi) = P(\psi)$,

and more generally, using LC3,

> P2''. If $\vdash \rho \leftrightarrow \sigma$, then $P(\phi) = P(\phi')$,

where ϕ' comes from ϕ by replacing an occurence of ρ by σ. Since $\vdash \phi \to \phi \vee \psi$ and $\vdash \phi \wedge \psi \to \phi$, another simple consequence of P2 is

Theorem 4.20.

> (i) $\max(P(\phi), P(\psi)) \leq P(\phi \vee \psi)$
> (ii) $P(\phi \wedge \psi) \leq \min(P(\phi), P(\psi))$.

Theorem 4.21. *In* P1–P3, P1 *is replaceable by*

> P1'. $P(\phi) + P(\neg\phi) = 1$.

Proof. (a) Assume P1. Then since $\vdash \phi \vee \neg\phi$ we have $P(\phi \vee \neg\phi) = 1$ from which P1' follows by P3.

(b) Assume P1'. If $\vdash \phi$ then by LC4 $\vdash \phi \leftrightarrow (\phi \vee \neg\phi)$. Hence by P2' and P3,

$$P(\phi) = P(\phi \vee \neg\phi) = P(\phi) + P(\neg\phi) = 1.$$

---

1. In LC1–LC4 it is possible to replace '$\vdash$' by '$\pi \vdash$', i.e., by logical consequence from a premise π (so long as π is not logically false). The resulting notion of probability function would then be relative to π, i.e., be a probability function on consequences of π, with $P(\sigma) = 0$ for any σ implying $\neg\pi$. This doesn't seem as handy as having a notion which is independent of premises and then, when specifying a probability model (§4.3) include, if desired, conditions on the model so as to have $P(\sigma) = 0$ for such σ.

Corollary 4.211. *If $\vdash \neg\phi$, then $P(\phi) = 0$.*

Theorem 4.22. *A probability function whose range is the two element set $\{0, 1\}$ is a verity function.*

Proof. From P1′ of Theorem 4.21 we see that the truth table for negation is satisfied. (If $P(\phi) = 1$ then $P(\neg\phi) = 0$, and if $P(\phi) = 0$ then $P(\neg\phi) = 1$.) From Theorem 4.20 (i) one can see that three rows of the truth table for $\vee$ are satisfied. (If $P(\phi)$, $P(\psi)$, or both are 1, then so is $P(\phi \vee \psi)$.) As for the other row, by P3 and P2′,

$$P(\phi \vee \psi) = P(\phi) + P(\overline{\phi}\psi)$$
$$\leq P(\phi) + P(\psi)$$

by P2. Hence if both $P(\phi)$ and $P(\psi)$ are 0, so is $P(\phi \vee \psi)$. This completes the truth table for $\vee$. As for $\wedge$, by Theorem 4.20 (ii) three rows of the truth table are satisfied. (If $P(\phi)$, or $P(\psi)$, or both are 0 then so is $P(\phi \wedge \psi)$.) For the other row, from $\vdash \phi \leftrightarrow (\phi\psi \vee \phi\overline{\psi})$, P2′ and P3, we have

$$P(\phi\psi) = P(\phi) - P(\phi\overline{\psi})$$

which, by Theorem 4.20(ii),

$$= P(\phi) \quad \text{if } P(\psi) = 1,$$
$$= 1 \quad \text{if } P(\phi) = 1.$$

Theorem 4.23. *Every verity function on $\mathcal{L}$ is a probability function on $\mathcal{L}$.*

Proof. (a) Referring to §0.1, by V1, $V(\phi) + V(\neg\phi) = 1$. Hence property P1′ is satisfied.

(b) If $\vdash \phi \to \psi$, then by the truth table properties of $\to$ we have $V(\phi) \leq V(\psi)$. This gives us P2.

(c) By adding the corresponding sides of the equations of V2 and V3 we obtain

$$V(\phi \vee \psi) + V(\phi \wedge \psi) = V(\phi) + V(\psi).$$

Hence if $\vdash \phi \to \neg\psi$, i.e., if $\vdash \neg(\phi\psi)$, then by the truth table for $\neg$, $V(\phi\psi) = 0$. Thus $V(\phi \vee \psi) = V(\phi) + V(\psi)$, which gives P3.

§4.3. Probability logic: the finite case

Even the briefest of acquaintances with the history of probability theory suffices to make one aware of the genesic role that games of chance have played. In such games, and in similar stochastic situations, only a finite number of chance events can result. Aping this historical feature, our presentation begins with probability logic for a finite (sentential) language S_n, meaning by this a language with only a finite number of atomic sentences $A_1, \ldots, A_n$, and with the customary verity syntax.

Let $K_i(A_1, \ldots, A_n)[= K_i]$, for $i = 1, \ldots, 2^n$, be the constituents (basic conjunctions) on $A_1, \ldots, A_n$. Then a *probability model M for S_n* is an assignment of real numbers k_i to the K_i, i.e., $M(K_i) = k_i$, such that

$$\text{for each } i, \ k_i \geq 0 \quad \text{and} \quad \sum_{i=1}^{2^n} k_i = 1.$$

A probability model is a generalization of a verity model since those probability models in which exactly one k_i is 1 (and the remaining are then 0) coincide with verity models (Theorem 0.12).

Theorem 4.31. *Any probability model M for S_n is uniquely extendible to a probability function P_M on S_n.*

Proof. Let the sentences of S_n be divided into the 2^{2^n} mutually exclusive equivalence classes modulo logical equivalence. One of these classes contains $A_1 \overline{A_1}$, all the others a unique logical sum of one or more of the K_i, (generically denoted '$\bigvee K_i$'). We define a P_M value for each member of S_n by setting $P_M(A_1 \overline{A_1}) = 0$, $P_M(\bigvee K_i) = \sum k_i$ (= the sum of the values $M(K_i)$ for those K_i present in $\bigvee K_i$), and then letting sentences in a given equivalence class have as their P_M value the P_M value assigned to its $\bigvee K_i$ member. Having defined P_M we need to show that it has the properties P1–P3. By virtue of its definition, P_M has the P2' property; whence without loss of generality we may assume that our sentences are all of the $\bigvee K_i$ form. (The special case of $A_1 \overline{A_1}$ can be taken care of by considering it, and all logically false sentences, as an 'empty' logical sum.) If ϕ 'is' the sum $\bigvee^{(\phi)} K_i$, then $\neg\phi$ is the complementary sum (sum of those K_i not present in $\bigvee^{(\phi)} K_i$). Hence P1 holds for P_M. That P2 holds is clear, since if $\vdash \phi \to \psi$, then $\bigvee^{(\phi)} K_i$ is a subsum of $\bigvee^{(\psi)} K_i$. Finally, that P3 holds becomes evident when ϕ and ψ are expressed as logical sums of the K_i. The so-defined P_M is unique, for if two such functions, P_1 and P_2, agree on constituents then there can't be a ϕ with $P_1(\phi) \neq P_2(\phi)$.

By virtue of this theorem we can speak of the probability value in a model M of *any* sentence, meaning by this its P_M value.

Before going on to define logical consequence we show by a simple example that our probability models do serve to model finite stochastic situations.

Let A_i be the sentence 'The rolled die comes up with i pips showing'. Then each of the six sentences

$$\text{1.} \quad A_1\overline{A}_2\overline{A}_3\overline{A}_4\overline{A}_5\overline{A}_6$$
$$\text{2.} \quad \overline{A}_1 A_2\overline{A}_3\overline{A}_4\overline{A}_5\overline{A}_6$$
$$\vdots$$
$$\vdots$$
$$\text{6.} \quad \overline{A}_1\overline{A}_2\overline{A}_3\overline{A}_4\overline{A}_5 A_6$$

describes a (physically) possible outcome. One model ('the fair die') assigns the value $\frac{1}{6}$ to each of these constituents, and 0 to all the others. This example is an instance of what might be called a *standard probability model*. Expressed in its customary set-theoretical form there is a finite set $\Omega = \{\omega_1, \ldots, \omega_n\}$ of chance outcomes ω_i of an 'observation' or 'experiment'. One and only one of the ω_i can be an outcome, and one is sure to happen. Events are defined as subsets of Ω. Each singleton set $\{\omega_i\}$ (an elementary event) is assigned a value $P(\{\omega_i\})$ which measures the chances of this outcome; and an event E has the value $P(E)$ which is equal to the sum of the $P(\{\omega_i\})$ for those $\omega_i \in E$. For the sure event Ω one has $P(\Omega) = 1$. In our sentential version of such a stochastic situation we take (as indicated by our example) A_i to mean 'the outcome of the experiment is ω_i'. Hence constituents of the form

$$\overline{A}_1\overline{A}_2\ldots\overline{A}_{i-1}A_i\overline{A}_{i+1}\ldots\overline{A}_n$$

express the condition that ω_i, and only ω_i, occurs. Such constituents are assigned the values $k_i > 0$ (summing to 1) and all other constituents the value 0. From the sentential point of view there seems to be no point to restricting the notion of a model to just standard probability models. For example if we had a dart board as in Figure 4.31, one could be interested in the chances of striking a portion of the board describable by means of any of the 8 ultimate subdivisions and not just those labelled $\omega_1, \omega_2, \omega_3$. Of course the version using sets could also accomodate such experiments by taking the striking of appropriately chosen regions, if mutually exclusive, as the elementary outcomes. But this approach seems to us unnaturally constrained, since any experiment involving this situation can be uniformily described in terms of assignments to the 8 constituents. Note that the set-theoretic elementary events are assigned values greater that 0. In our

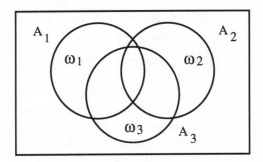

FIGURE 4.31

definition of a probability model any number, except all, of the constituents can have 0 as an assigned value.

As we have foreshadowed, the definition of 'logical consequence' for probability logic is a generalization of that for verity logic. In §0.1 we defined

$$V(\phi_1) \in \alpha_1, \ldots, V(\phi_m) \in \alpha_m \vDash V(\psi) \in \beta,$$

($\alpha_1, \ldots, \alpha_m, \beta$ non-empty subsets of $\{0, 1\}$) meaning that for all verity functions V on $\mathcal{S}$, if $V(\phi_i) \in \alpha_i$ ($i = 1, \ldots, m$) then $V(\psi) \in \beta$. Our definition of probability logical consequence is analogous. We say ψ is a *probability logical consequence of* $\phi_1, \ldots, \phi_m$ (with respect to non-empty subsets $\alpha_1, \ldots, \alpha_m, \beta$ of $[0, 1]$) if the following condition is satisfied:

For all probability *models* M :

$$\text{if } P_M(\phi_1) \in \alpha_1, \ldots, P_M(\phi_m) \in \alpha_m, \text{ then } P_M(\psi) \in \beta. \qquad (1)$$

The intuitive picture here is that of a (vertically) infinite 'truth' table entered from values assigned to the constituents $K_1, \ldots, K_{2^n}$ heading the first 2^n columns (rather than the sentences $A_1, \ldots, A_n$), which are followed by additional columns headed by $\phi_1, \ldots, \phi_m, \psi$. Thus $2^n + m + 1$ columns. The $K_1, \ldots, K_{2^n}$ are assigned all possible sets of 2^n real numbers from $[0, 1]$, the sum of such numbers in each set being 1. Each assignment (row) determines a probability model and hence (Theorem 4.31 above) values for $\phi_1, \ldots, \phi_m, \psi$. There are, by Cantorian counting, $2^{\aleph_0}$ such rows. The premise conditions in (1) select out of these rows those in which the probabilities of the ϕ_i are, respectively, in the α_i sets. The relation (1) holds if for each of these rows ψ has a value in β.

Since to each probability model M there is a unique probability function P_M, and to each probability function P a unique model M (by virtue of the $P(K_i)$ being given) we can reformulate (1) to

For all probability *functions* P :

$$\text{if } P(\phi_1) \in \alpha_1, \ldots, P(\phi_m) \in \alpha_m, \text{ then } P(\psi) \in \beta. \qquad (2)$$

Now dropping the universal quantifier and using the free variable P to indicate generality, we abbreviate (2) to

$$P(\phi_1) \in \alpha_1, \ldots, P(\phi_m) \in \alpha_m \vDash P(\psi) \in \beta. \tag{3}$$

For logical consequence relations of the form (3) we shall refer to the '$P(\phi_i) \in \alpha_i$' as its *premises*, and to '$P(\psi) \in \beta$' as its *conclusion*.

One might ask What is the rationale for using *subsets* of $[0, 1]$ rather than single values? In the case of verity logic the notion of logical consequence is easily generalizable to sets of truth values (§0.1). However there is little incentive for doing so since besides $\{1\}$ there are only two sets $\{0\}$ and $\{0, 1\}$. Neither of these latter two sets is of particular relevance to verity logic's primary business of inference, that is, of being able to assert a conclusion (as true) when the premises are verified. With probability, on the other hand, being able to assert that a sentence has probability 1 (or any other specific value) is rarely possible and not always of interest. More often it is a range of values as, for example, in statistics where assertions such as "the probability is .95 or more" is a desired conclusion. Note that in this example the set mentioned is an interval. Later we shall see that when the sets $\alpha_1, \ldots, \alpha_m$ in (3) are sub*intervals* of $[0, 1]$ then there is an effective procedure for determining the set β which provides the strongest conclusion; and it will turn out that this set is an interval. Clearly defining logical consequence so as to apply to sets, and hence to subintervals, is advantageous.

We postpone the presentation of examples of probability logic inferences until after we have extended the notion to languages with a potential infinity of atomic sentences.

§4.4. Probability logic: the potential infinite case

Our sentential languages, generically denoted by 'S', now are to have a potential infinity of atomic sentences $A_1, \ldots, A_n, \ldots$. Other than this the syntax is unchanged. For a set of sentences $\{\sigma_1, \ldots, \sigma_s\}$ of S we say $A_1, \ldots, A_N$ *supports* $\{\sigma_1, \ldots, \sigma_s\}$, and N is the *index of support*, if $A_1, \ldots, A_N$ is the shortest initial sequence of $A_1, \ldots, A_n, \ldots$ which includes any A_i present in any member of the set.

For the semantics of S we need changes in the definition of constituent and of model. A constituent of S is a sentence K which is a constituent of S_n for some n, i.e., a logical product of the form $K_i(A_1, \ldots, A_n)$ of arbitrary size n. It will be convenient to have a more compact notation for constituents of S. We write:

$$K_{111} = A_1 A_2 A_3$$

$$K_{11} = A_1 A_2 \qquad K_{110} = A_1 A_2 \overline{A_3}$$

$$K_1 = A_1 \qquad K_{10} = A_1 \overline{A_2}$$

$$K_0 = \overline{A_1} \qquad K_{01} = \overline{A_1} A_2 \qquad \vdots$$

$$K_{00} = \overline{A_1} \overline{A_2} \qquad \vdots$$

$$etc.$$

Thus any constituent of S is represented by $K_{b_1 \ldots b_n}$ where b_i $(i = 1, \ldots, n)$ is either 1 or 0, 1 if A_i is unnegated and 0 of it is negated. For any given n, the subscripts on the K's are just the binary forms for the denary numerals 0 to $2^n - 1$.

Let $S(K)$ denote the set of constituents of S. A *probability model for S* is a function $M : S(K) \to [0,1]$ which for each n $(n = 1, 2, \ldots)$ assigns a value $k_{b_1 \ldots b_n}$ to $K_{b_1 \ldots b_n}$ such that

(i) $k_{b_1 \ldots b_n} \geq 0$

(ii) $\sum k_{b_1 \ldots b_n} = 1$ (the sum over all $b_1, \ldots, b_n \in \{0, 1\}$)

(iii) $k_{b_1 \ldots b_n 1} + k_{b_1 \ldots b_n 0} = k_{b_1 \ldots b_n}$.

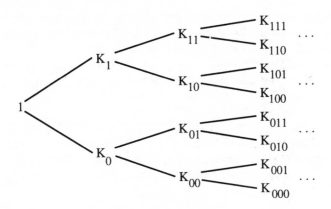

FIGURE 4.41

Pictorially, in Figure 4.41, the non-negative M values assigned to the constituents in a given column add up to 1, and the value at any node is the sum of the two values at its branch ends. For fixed n, (i) and (ii) together define a probability model for S_n. With the addition of (iii), (ii) can be replaced by the simpler (ii') $k_1 + k_0 = 1$. However for the sake of uniformity with the earlier definition we leave it as it is.

Theorem 4.41. *Any probability model M for S can be uniquely extended to a probability function P_M on S.*

Proof. Let S be divided into mutually exclusive equivalence classes modulo logical equivalence. One of these classes contains all logically false sentences and, in particular, $A_1\overline{A}_1$. All other classes, having only non-logically false sentences, must contain sums of constituents $\bigvee K_i$. Restricting our attention in any given class to those sums in which all terms of the sum have the same index of support we select the one with minimal index. Call this $\bigvee K_i$ the *representative* of the class. Then the function P_M is defined by setting $P_M(A_1\overline{A}_1) = 0$, $P_M(\bigvee K_i)=$ sum of the M values assigned to the K_i for $\bigvee K_i$ the representative of a class, and then to all other sentences in the class this same value. The proof now parallels that of Theorem 4.31, using class representatives instead of the unique $\bigvee K_i$ used in that proof.

The definition of *probability logical consequence for S*, written

$$P(\phi_1) \in \alpha_1,\ldots, P(\phi_m) \in \alpha_m \vDash P(\psi) \in \beta, \tag{1}$$

is verbally unchanged from §4.3. Since any list of sentences $\phi_1,\ldots,\phi_m$, ψ of S can be viewed as sentences of S_n, where n supports this list, we may henceforth ignore the difference between logical consequence in S_n and in S.

Our definition of probability logical consequence (1) looks quite different from that of Scott and Krauss (§3.8). Nevertheless on the sentential level the two are closely related, as we now show.

It was noted in §3.8 that as far as sentential languages are concerned one can restrict attention to probability assertions of the form

$$\langle \Psi, K_1, \ldots, K_{2^n} \rangle, \tag{2}$$

where Ψ is an algebraic formula with 2^n free variables $\lambda_1,\ldots,\lambda_{2^n}$, and the K_i are constituents on $A_1,\ldots,A_n$. By virtue of the Corollary of that section we know that (2) is a probability law if and only if

$$\forall k_1 \ldots \forall k_{2^n} \left[\bigwedge_{1 \leq i \leq 2^n} (k_i \geq 0) \wedge k_1 + \cdots + k_{2^n} = 1 \rightarrow \Psi \right] \tag{3}$$

holds for the reals. Consider now the following assertion of a probability logical consequence of the form (1):

$$P(K_2) \in \{k_2\}, \ldots, P(K_{2^n}) \in \{k_{2^n}\}$$
$$\vDash P(K_1) \in \{x \mid \Psi(x, k_2, \ldots, k_{2^n})\} \tag{4}$$

(The $\{k_i\}$ are singleton sets, so that $P(K_i) \in \{k_i\}$ is equivalent to $P(K_i) = k_i$.) Assertion (4) will be a valid probability logical consequence if and only if, for any set of non-negative reals $k_2, \ldots, k_{2^n}$ whose sum is ≤ 1, and with k_1 being the difference between 1 and $\sum_{i=2}^{2^n} k_i$, we have

$$\Psi(k_1, \ldots, k_{2^n}) \text{ holding for the reals.}$$

But this condition is equivalent to (3). Hence (4) is valid if and only if (2) is a probability law in the Scott-Krauss sense. Thus any Scott-Krauss (sentential) probability law is expressible as a valid probability consequence relation in our sense. Superficially, (1) is more general than the Scott-Krauss notion since the sets α_i, β need not be definable by algebraic formulas. However the Scott-Krauss notion assures effectivity, by virtue of Tarski's decision procedure for the theory of real-closed fields. This theory is a part of their formal syntax. For us the real number system is in the semantics.

It will be convenient to abbreviate (4) to

$$\vDash \Psi(P(K_1), \ldots, P(K_{2^n})) \tag{5}$$

To establish a result of the form (5) one need not start with (4) but simply show that $\Psi(P(K_1), \ldots, P(K_{2^n}))$ holds for all probability models (assignments of the reals k_i to $P(K_i)$). In the same way to establish a result of the form

$$\vDash \Phi(P(\phi_1), \ldots, P(\phi_N)), \tag{6}$$

defined similarly to (5), one shows that $\Phi(P(\phi_1), \ldots, P(\phi_N))$ holds for all probability functions P and there is then no need to express the probabilities in terms of the $P(K_i)$. Thus as an example,

$$\vDash P(\phi) + P(\neg\phi) = 1,$$

which in unabbreviated form is

$$P(\phi) \in \{a\} \vDash P(\neg\phi) \in \{x \mid a + x = 1\},$$

clearly holds for all probability functions. Expressions of the form (6) will be referred to as *probability logic assertions*. The following theorem states a few such assertions. They are immediate consequences of P1–P3 of §4.2 and hence hold for any $\mathcal{L}$ (S being one). Since these are well-known elementary "laws" of probability, easily derivable from P1–P3, we omit the proofs.

Theorem 4.42. *For any sentence ϕ and ψ of $\mathcal{L}$*

(a) $\vDash P(\phi \lor \neg\phi) = 1$

(b) $\vDash P(\phi) + P(\neg\phi) = 1$

(c) $\vDash P(\phi) = P(\phi\psi) + P(\phi\overline{\psi})$

(d) $\vDash P(\phi \lor \psi) + P(\phi \land \psi) = P(\phi) + P(\psi)$.

The following theorem presents necessary and sufficient conditions for some simple probability logic assertions. Proofs of the sufficiency parts depend only on P1–P3, hence hold for sentences of the $\mathcal{L}$. Since our proofs of the necessity parts assume that the sentences are expandible as logical sums of constituents on atomic letters, we need as hypothesis for the theorem that the sentences are in $\mathcal{S}$.

Theorem 4.43. *For any sentences ϕ and ψ of $\mathcal{S}$,*

(a_1) *If $\vdash \phi \rightarrow \psi$, then $\vDash P(\phi) \leq P(\psi)$.*

(a_2) *If $\vDash P(\phi) \leq P(\psi)$, then $\vdash \phi \rightarrow \psi$.*

(b) *$\vdash \phi \leftrightarrow \psi$ if and only if $\vDash P(\phi) = P(\psi)$.*

(c) *$\vdash \phi \rightarrow \neg\psi$ if and only if $\vDash P(\phi \vee \psi) = P(\phi) + P(\psi)$.*

(d) *$\vdash \phi$ if and only if $\vDash P(\phi) = 1$.*

(e) *$\vdash \neg\phi$ if and only if $\vDash P(\phi) = 0$.*

Proofs. For the first half of (a), P2 tells us that if $\vdash \phi \rightarrow \psi$, then $P(\phi) \leq P(\psi)$. Hence if $\vdash \phi \rightarrow \psi$ then for all probability functions P, $P(\phi) \leq P(\psi)$.[2] The second half of (a), its converse, may catch one by surprise. But note that '$\vDash P(\phi) \leq P(\psi)$' is a very strong condition. It says that the inequality holds in all models. So suppose this is the case and yet not $\vdash \phi \rightarrow \psi$. Consider an expansion of ϕ and ψ as logical sums of constituents on a common set of letters $A_1, \ldots, A_n$. Since not $\vdash \phi \rightarrow \psi$, then there is a constituent, say K, which is in ϕ's expansion but not in ψ's. Choose as a probability model one which assigns to K the value 1, and 0 to all others. But for this model $P(\phi) = 1$ and $P(\psi) = 0$—which contradicts $\vDash P(\phi) \leq P(\psi)$. The remaining parts (b)–(e) have similar demonstrations.

In earlier portions of this book (§§1.3, 2.3) we had occasion to refer to the notion 'A proves C'. We took this to mean '$A(A \rightarrow C)$'. The following accords with the intuition that, when $A \rightarrow C$ is necessary, the probability of A proving C is equal to the probability of A.

Theorem 4.44. *If $\vdash \phi \rightarrow \psi$, then $\vDash P(\phi(\phi \rightarrow \psi)) = P(\phi)$.*

Proof. If $\vdash \phi \rightarrow \psi$, then $\vdash \phi\psi \leftrightarrow \phi$. Also $\vdash \phi(\phi \rightarrow \psi) \leftrightarrow \phi\psi$. Hence $\vdash \phi(\phi \rightarrow \psi) \leftrightarrow \phi$ so that, by Theorem 4.43 (b), our result follows.

The invalidity of 'If $\vdash \phi_1\phi_2 \rightarrow \psi$, then $\vDash P(\phi_1)P(\phi_2) = P(\psi)$' was the subject of discussion at the end of §2.1. Bolzano's proof of its invalidity consisted in giving a physical example of random drawings from an urn containing balls which are black and/or fragrant. Now, in accordance with our

---

2. We are here using in the semantic language the quantifier rule: from $A \rightarrow B(x)$, infer $A \rightarrow \forall x B(x)$ with 'x' being the variable 'P'.

probability logic point of view, we would show invalidity by citing a suitable probability model, namely by specifying P-values for the constituents:

$$P(BF) = k_1, \quad P(B\overline{F}) = k_2 \quad P(\overline{B}F) = k_3, \quad P(\overline{B}\,\overline{F}) = k_4,$$

with the k_i appropriately chosen reals in the unit interval (summing to 1). Note that this latter way of establishing invalidity presupposes nothing whatever about the physical (or actual) world.

In the instances of probability logical consequence which we have so far presented the results were stated in terms of general sentential variables ϕ, ψ, etc. In the next section we relinquish this feature and express such results in terms of specific atomic sentence letters of S. This distinction we are making is not significant in the analogous case of verity logic where, for example,

$$A_1, \ A_1 \rightarrow A_2 \vDash A_2$$

is immediately generalizable to

$$\phi, \ \phi \rightarrow \psi \vDash \psi,$$

the specifics of the internal (sentential) structure of ϕ and ψ not being relevant. In the case of probability logical consequence we shall be interested in stating the strongest possible conclusion. But this may change on making a substitution. For example, the valid

$$P(A_1) = p, \ P(A_1 \rightarrow A_2) = 1 \vDash P(A_2) \in [p, 1]$$

remains valid if A_2 is replaced by A_1. But a much stronger conclusion, namely $P(A_1) = p$, can be given.

§4.5. Interval probability logic

In this section we treat the logical consequence relation,

$$P(\phi_1) \in \alpha_1, \ldots, P(\phi_m) \in \alpha_m \vDash P(\psi) \in \beta, \tag{1}$$

for the case in which the sets α_i $(i = 1, \ldots, m)$ are subintervals of $[0, 1]$, i.e., where each α_i is of the form $[a_i, b_i]$ with $0 \leq a_i \leq b_i \leq 1$. Note that $P(\phi) \in [a, b]$ is equivalent to a pair of inequalities

$$a \leq P(\phi) \leq b,$$

and, in particular, if $b = a$ then $P(\phi) \in [a, a]$ is just $P(\phi) = a$.

Before presenting a number of specific results of the form (1) we shall illustrate how they are obtained by going through one in some detail. For this we choose *probabilistic modus ponens*:

$$P(A_1) = p, \quad P(A_1 \to A_2) = q \vDash P(A_2) \in \beta. \tag{2}$$

Taking β in (2), or for that matter in (1), to be the entire interval $[0, 1]$ always provides a valid consequence relation, but one which is of no interest— except if it should happen to be the only conclusion which the premises warrant. Our aim is to find the β which provides for (2) the strongest conclusion, i.e., a β which is the largest set contained in every set β for which (2) is valid.

Consider an arbitrary probability model for S_2, i.e., an assignment of values k_1, k_2, k_3, k_4 to the constituents of S_2, with each $k_i \geq 0$, and $\sum_i k_i = 1$. Let P be its associated probability function. Then

$$P(A_1 A_2) = k_1, \quad P(A_1 \overline{A}_2) = k_2, \quad P(\overline{A}_1 A_2) = k_3, \quad P(\overline{A}_1 \overline{A}_2) = k_4.$$

The premise conditions in (2), together with P being a probability function, translate into the following system of linear equations and inequations:

$$
\begin{aligned}
k_1 + k_2 &\qquad\qquad\;\; = p \\
k_1 &\quad + k_3 + k_4 = q \\
k_1 + k_2 &+ k_3 + k_4 = 1 \\
k_1, \, k_2, &\, k_3, \, k_4 \geq 0.
\end{aligned}
\tag{3}
$$

Let $w = P(A_2) = k_1 + k_3$. Our objective is to find the set of values that w can take on, as the variables k_1, k_2, k_3, k_4 range over the reals, subject to the constraints expressed by (3). A first step is to determine for what values of the parameters p and q the system (3) has a solution; for if (3) has no solution then (2), though ('vacuously') valid, is useless for inference since the premises are never fulfilled.

As explained in §0.5 we eliminated from the system (3) all occurrences of k_1, k_2, k_3, k_4 so as to obtain the necessary conditions that (3) have a solution. Elimination of variables from linear inequation systems is facilitated by the presence of equations, since an equation can be solved for one of the variables and its value substituted for all of its other occurrences. Since (3) involves 3 equations we can solve them for 3 of the variables, say k_1, k_2, k_3. This produces

$$
\begin{aligned}
k_1 &= p + q - 1 \\
k_2 &= 1 - q \\
k_3 &= 1 - p - k_4.
\end{aligned}
\tag{4}
$$

Substituting these values into the last line of (3) produces

$$p + q - 1 \geq 0, \quad 1 - q \geq 0, \quad 1 - p - k_4 \geq 0, \quad k_4 \geq 0 \qquad (5)$$

so that, on eliminating k_4, we have

$$p + q - 1 \geq 0, \quad 1 - q \geq 0, \quad 1 - p \geq 0, \qquad (6)$$

as the necessary conditions for (3) to have a solution. Assuming that these consistency conditions hold, we look at the question of finding the optimum β for (2).

As explained in §0.5, the set of points (k_1, k_2, k_3, k_4) satisfying (3), i.e., the set of feasible solutions of (3), being the intersection of half-spaces, is a convex (hyper-)polyhedron. It is non-empty and bounded, i.e., is a polytope. Hence the linear function $w = k_1 + k_3$ takes on at some vertex of the polytope a minimum value $\min w$, and likewise a maximum value $\max w$, also at a vertex. These values are global extrema (§0.5); hence no feasible point of (3) can produce a value for w that is outside the interval $[\min w, \max w]$. But is there a feasible point for each value in this interval? That this is so may be seen from the following geometrical argument.

The two parallel (hyper-)planes

$$k_1 + k_3 = \min w$$
$$k_1 + k_3 = \max w$$

are in contact with, and include between them, the convex bounded polyhedron specified by (3). Hence a plane $k_1 + k_3 = w$, with variable w, moving continuously parallel to itself, first from coincidence with $k_1 + k_3 = \min w$ to coincidence with $k_1 + k_3 = \max w$, will always intersect the polyhedron. Thus at any position of this plane we have at least one feasible point (k_1, k_2, k_3, k_4) for which the value $k_1 + k_3$ is an element of $[\min w, \max w]$. This interval is the optimizing β—it provides the strongest conclusion for (2) since it contains those, and only those, values accruing to $P(A_2)$ in any model for which the premise conditions hold.

Finding the actual values of $\min w$ and $\max w$ is a linear programming problem, that is, a problem of optimizing a linear function subject to linear inequality constraints. To accomplish this one can use either of the techniques described in §0.5. For the sake of illustration we shall do both, beginning with Fourier elimination.

Here we have to adjoin the equation $k_1 + k_3 = w$ to the system (3) and eliminate all variables but w. But we have already made a good start on this in determining the consistency conditions (6). Using the results of (4) we substitute for k_1 and k_3 the values there given into the adjoined equation, so obtaining

$$w = q - k_4. \qquad (7)$$

In (5) we found that $0 \leq k_4 \leq 1 - p$. Hence replacing k_4 in (7) by the extreme values 0 and $1 - p$ results in

$$p + q - 1 \leq w \leq q.$$

Thus our optimizing β is the interval $[p + q - 1, q]$.

In the second approach we convert a linear programming problem over to its dual form. Here the primal problem (form (ii) of §0.5) is: find $\max z = k_1 + k_3$ subject to the linear inequality system (3), or, in matrix form to bring out the structural pattern,

$$\text{maximize } z = k_1 + k_3, \quad \text{subject to}$$

$$\begin{bmatrix} 1 & 1 & 0 & 0 \\ 1 & 0 & 1 & 1 \\ 1 & 1 & 1 & 1 \end{bmatrix} \begin{bmatrix} k_1 \\ k_2 \\ k_3 \\ k_4 \end{bmatrix} = \begin{bmatrix} p \\ q \\ 1 \end{bmatrix}, \quad \text{and} \quad \begin{bmatrix} k_1 \\ k_2 \\ k_3 \\ k_4 \end{bmatrix} \geq \mathbf{0}.$$

the dual of this program is to

$$\text{minimize } v = px_1 + qx_2 + x_3, \quad \text{subject to}$$

$$\begin{bmatrix} 1 & 1 & 1 \\ 1 & 0 & 1 \\ 0 & 1 & 1 \\ 0 & 1 & 1 \end{bmatrix} \begin{bmatrix} x_1 \\ x_2 \\ x_3 \end{bmatrix} \geq \begin{bmatrix} 1 \\ 0 \\ 1 \\ 0 \end{bmatrix}. \qquad (x_1, x_2, x_3 \text{ unrestricted}) \tag{8}$$

Note particularly that the parameters p, q no longer appear in the specification of the feasible set but as coefficients on variables of the objective function. The theory tells us that the minimum value of the objective function is taken on at a vertex of the feasible set (a convex polyhedron). Hence by finding all such points for (8), and taking the minimum of the values of $px_1 + qx_2 + x_3$ at these points, we have our result. To find these points we select, in all possible ways, three of the four inequations, convert them to equations and solve to find their intersection point. If the coordinates of the point satisfy the other equation (i.e., is in the feasible set), then it is a vertex. In (8) only one of the last two need be selected since, as equations, they would be inconsistent ($0x_1 + 1x_2 + 1x_3$ can't equal both 0 and 1). So there are only two possible selections of 3 out of the 4 relations in (8), and we readily find that there is only one vertex, namely $(0, 1, 0)$. Alternatively, one can drop $0x_1 + 1x_2 + 1x_3 \geq 0$ in (8) since it is implied by $0x_1 + 1x_2 + 1x_3 \geq 1$. Hence only one selection of 3 equations. Thus $p \cdot 0 + q \cdot 1 + 1 \cdot 0$, i.e., q, is the minimal value for the dual problem and, consequently, also the maximum value of the objective function for the primal, i.e., the maximum of $k_1 + k_3$. Now for the minimum of $k_1 + k_3$.

Since $P(\phi) = 1 - P(\neg\phi)$, to find the minimum of $k_1 + k_3$ we find the maximum of the complementary function and subtract from 1. In our example $\neg A_2$ is expressible as $A_1\overline{A_2} \vee \overline{A_1}\overline{A_2}$; hence we need to find the maximum of $k_2 + k_4$ subject to the same inequality constraints, namely (3). Solving this problem produces the value $-p - q + 2$. Hence a minimum value for $k_1 + k_3$ of $1 - (-p - q + 2) = p + q - 1$, once again obtaining the interval $[p + q - 1, q]$ for the optimal β in (2). To find the consistency conditions on the parameters we simply write

$$0 \leq \text{lower bound} \leq \text{upper bound} \leq 1. \tag{9}$$

this gives

$$0 \leq p + q - 1 \leq q \leq 1.$$

so that

$$1 \leq p + q, \quad p \leq 1, \quad q \leq 1.$$

(Alternatively, and more elegantly, these inequalities are obtainable from the extreme rays of the polyhedron of (8). See *Hansen-Jaumard-Poggi de Aragão 1992*, 4 where the general result is established.)

The algebraic task of finding the optimal β in (1) for sentential forms $\phi_1, \ldots, \phi_m, \psi$ involving sentences $A_1, \ldots, A_n$, a task which requires finding all vertices of a polyhedron, can be carried out by a suitably programmed computer—but only for small n. Computational complexity questions in connection with linear programs lie outside the scope we have set for ourselves in this monograph.

We now list some probabilistic inference forms of the kind (1), the first one being the one whose derivation we have been discussing in detail. We shall not bother stating the consistency conditions on the parameters which are involved. These conditions may be obtained from the conclusion interval by writing the inequalities as shown in (9).

Theorem 4.51.

$$P(A_1) = p, \; P(A_1 \to A_2) = q \vDash P(A_2) \in [p + q - 1, q].$$

This is the probabilistic form of modus ponens. Note that the probability of the conclusion sentence A_2 cannot exceed that of the premise sentence $A_1 \to A_2$; when $P(A_1) = 0$ then $q = 1$ (since $p + q - 1$ can't be negative) and the conclusion becomes $P(A_2) \in [0, 1]$. Thus, as would be expected, no probability information results in this case. In the next example the interval in the conclusion of the consequence relation is a single point, i.e., an exact value.

Theorem 4.515.

$$P(A_1 \to A_2) = p, \ P(\overline{A}_1 \to A_2) = q \vDash P(A_2) = p + q - 1.$$

Here the consistency conditions are that $0 \le p + q - 1 \le 1$, i.e., that $1 \le p + q \le 2$.

Theorem 4.52.

(a) $P(A_1) = a_1, \ P(A_2) = a_2$
$$\vDash P(A_1 A_2) \in [\max(0, \ a_1 + a_2 - 1), \ \min(a_1, \ a_2)].$$

(b) $P(A_1) = a_1, \ P(A_2) = a_2$
$$\vDash P(A_1 \lor A_2) \in [\max(a_1, \ a_2), \ \min(1, \ a_1 + a_2)].$$

These inference forms—better known in probability theory as estimates for the probability of $A_1 A_2$ (and of $A_1 \lor A_2$) when nothing is known about A_1 and A_2 except for their probabilities—were discussed in §3.5. As shown there, they are easily generalized to any number of components. The peculiar form in which the conclusion interval is written comes from the polyhedron of feasible points having two vertices. Hence, as for instance in the upper bound for (a), the value of the objective function at one vertex is a_1, at the other is a_2, so that the upper bound for the interval would be whichever value is the smaller, i.e., would be $\min(a_1, a_2)$.

In addition to modus ponens another well-known necessary inference form of verity logic, the 'hypothetical syllogism', can be given a probabilistic version:

Theorem 4.53.

$$P(A_1 \to A_2) = p, \ P(A_2 \to A_3) = q \vDash P(A_1 \to A_3) \in [p + q - 1, \ 1]$$

Note that when the sum $p + q$ is near 1 we have very little information about the probability of the conclusion sentence $A_1 \to A_3$. Only when this sum is near 2, say $2 - \epsilon$, do we get a narrowing of the interval to $[1 - \epsilon, \ 1]$.

In each of the preceding examples of inference forms the premise probabilities had just one possible value. Results extending the probabilistic modus ponens and the hypothetical syllogism to the case of intervals in the premises are quite evident, and readily obtained:

Theorem 4.54.

(a) $P(A_1) \in [p_1, p_2], \ P(A_1 \to A_2) \in [q_1, q_2]$
$$\vDash P(A_2) \in [p_1 + q_1 - 1, \ q_2]$$

(b) $P(A_1 \to A_2) \in [p_1, p_2], \ P(A_2 \to A_3) \in [q_1, q_2]$
$$\vDash P(A_1 \to A_3) \in [p_1 + q_1 - 1, \ 1].$$

The results here can be specialized to feature the nearness to 1 of the premise probabilities by using intervals $[1-\epsilon, 1]$, where ϵ is the 'uncertainty' (see §3.3 and §3.6).

Theorem 4.545.

(a) $P(A_1) \in [1 - \epsilon, 1]$, $P(A_1 \rightarrow A_2) \in [1 - \epsilon, 1]$

$$\vDash P(A_2) \in [1 - 2\epsilon, 1]$$

(b) $P(A_1 \rightarrow A_2) \in [1 - \epsilon, 1]$, $P(A_2 \rightarrow A_3) \in [1 - \epsilon, 1]$

$$\vDash P(A_1 \rightarrow A_3) \in [1 - 2\epsilon, 1].$$

For the examples so far in this section the sentence in the conclusion was a (verity) necessary consequence of those in the premise. But there is no necessity for (referring to (1)) the ψ being implied by $\phi_1, \ldots, \phi_m$. Here is a simple example where the ψ is not so implied:[3]

Theorem 4.55.

$$P(A_1 \rightarrow A_2) = p \vDash P(A_2 \rightarrow A_1) \in [1 - p, 1].$$

Another example that's not so simple: Given $P(A_i) = a_i$ $(i = 1, 2, 3)$, what is the probability of some pair of A_1, A_2, A_3 not implying the third?

Theorem 4.56. *Let ψ be $\overline{A}_1 A_2 A_3 \vee A_1 \overline{A}_2 A_3 \vee A_1 A_2 \overline{A}_3$. Then*

$$P(A_1) = a_1, \ P(A_2) = A_2, \ P(A_3) = a_3 \vDash P(\psi) \in [L, U],$$

where

$$U = \min(\frac{a_1 + a_2 + a_3}{2}, \ a_1 + a_2, \ a_1 + a_3, \ a_2 + a_3, \ 3 - (a_1 + a_2 + a_3))$$
$$L = \max(0, \ a_1 + a_2 - a_3 - 1, \ a_1 - a_2 + a_3 - 1, \ -a_1 + a_2 + a_3 - 1).$$

(The occurence of 0 in the max list for L, but not of 1 in the min list for U, may seem puzzling—but easily accounted for by noting that not all the min listings can be simultaneously greater than 1. For example if

$$\frac{a_1 + a_2 + a_3}{2} > 1 \quad \text{and} \quad 3 - (a_1 + a_2 + a_3) > 1$$

were to hold then $a_1 + a_2 + a_3 > 2$ and $2 > a_1 + a_2 + a_3$, which is impossible.)

The following theorem presents a logical consequence which, in the form of an estimate for $P(A_1 \vee A_2 \vee A_3)$ given only the probabilities of the A_i and $A_i A_j$, is of interest to statisticians. We discuss this in §6.2 below.

---

3. Logically naïve arguers sometimes confuse a necessary with a sufficient condition. Our theorem shows that the probability of the one conditional could be small (p near 0) and the other large (near 1).

Theorem 4.57.

$$P(A_1) = a_1, \; P(A_2) = a_2, \; P(A_3) = a_3,$$
$$P(A_1 A_2) = a_{12}, \; P(A_1 A_3) = a_{13}, \; P(A_2 A_3) = a_{23}$$
$$\vDash \; P(A_1 \vee A_2 \vee A_3) \in [L, U],$$

where

$$L = \max(a_1 + a_2 - a_{12},$$
$$a_1 + a_3 - a_{13},$$
$$a_2 + a_3 - a_{23},$$
$$a_1 + a_2 + a_3 - (a_{12} + a_{13} + a_{23}))$$
$$U = \min(a_1 + a_2 + a_3 - (a_{12} + a_{13}),$$
$$a_1 + a_2 + a_3 - (a_{12} + a_{23}),$$
$$a_1 + a_2 + a_3 - (a_{13} + a_{23}), \; 1).$$

In all the examples in this section our linear programs involved parameters as constants on the right-hand side of inequations or equations. Such problems can be solved algebraically by Fourier elimination, or computationally by going over, as we have done, to the equivalent dual problem. In this latter approach the parameters became coefficients of the dual objective function variables, and the dual feasible set is then defined in purely numerical terms. But now *all* vertices of the dual convex polyhedron need to be found. This can be numerically daunting if the number of possible choices needed to find all intersecting planes is large. If, however, instead of parameters we have numerical values, then there are known efficient techniques (e.g., the simplex method) for solving such linear programs which do not require finding all vertices.

§4.6. Decision procedure for interval probability logic

Our preceding section introduced the notion of interval probability logic, where the α_i in the logical consequence relation

$$P(\phi_1) \in \alpha_1, \ldots, P(\phi_m) \in \alpha_m \vDash P(\psi) \in \beta \tag{1}$$

are restricted to being subintervals of $[0, 1]$. We showed, in the specific case of probabilistic modus ponens, that the set β producing the strongest conclusion in (1) was also an interval. We now wish to prove this to be true in general. Also we shall determine the end points of this interval. By a set β providing the strongest conclusion in (1), or *being optimal for* (1), we mean first of all, that the premises of (1) (including the tacit premise that P is a probability function) are not inconsistent, and that β contains those, and only those, values of $P(\psi)$ for which the premise conditions are satisfied as P ranges over all probability functions on $\mathcal{S}_n$, n being the index of support for $\phi_1, \ldots, \phi_m, \psi$.

Theorem 4.61. *For any given probability logical consequence*

$$P(\phi_1) \in [a_1, b_1], \ldots, P(\phi_m) \in [a_m, b_m] \vDash P(\psi) \in \beta \tag{2}$$

the optimal set β is an interval whose end points are the minimum and maximum of a linear function subject to linear constraints.

Proof. Let each of $\phi_1, \ldots, \phi_m, \psi$ be expressed as logical sums of constituents K_j on $A_1, \ldots, A_n$, n being the index of support for $\phi_1, \ldots, \phi_m, \psi$. Let P be an arbitrary probability function on $\mathcal{S}_n$. Then the probabilities appearing in (2) are replaceable by (explicitly obtainable) sums of constituent probabilities, i.e., sums of $P(K_j)\,[= k_j]$, $j = 1, \ldots, 2^n$. Treating the premise conditions in (2) as two-sided inequalities and adjoining the normative conditions coming from the assumption that P is a probability function leads to the equivalent $2m + 2$ conditions

$$\begin{aligned}
\sum^{(\phi_i)} k_j &\geq a_i & (i = 1, \ldots, m) \\
-\sum^{(\phi_i)} k_j &\geq -b_i & (i = 1, \ldots, m) \\
\sum_{j=1}^{2^n} k_j &\geq 1 & \\
-\sum_{j=1}^{2^n} k_j &\geq -1 & (\text{each } k_j \geq 0)
\end{aligned} \tag{3}$$

These inequality conditions can be succinctly expressed in matrix form:

$$\mathbf{Mk} \geq \mathbf{c}, \qquad (\mathbf{k} \geq \mathbf{0}) \tag{4}$$

where $\mathbf{M}$ is a 2^n by $2m + 2$ matrix of 0's and ± 1's, where $\mathbf{k}$ is the column matrix $[k_1 \ldots k_{2^n}]^{\mathrm{T}}$ and $\mathbf{c}$ the column matrix

$$[a_1 \ldots a_m (-b_1) \ldots (-b_m) 1 (-1)]^{\mathrm{T}}.$$

Let $w = P(\psi) = \sum^{(\psi)} k_j$. As P ranges over all probability functions on $\mathcal{S}_n$, the $k_1, \ldots, k_{2^n}$ will range over all possible sets of values in $[0,1]$ summing to 1. Finding the minimum of $\sum^{(\psi)} k_j$, subject to the linear constraints (4), is a linear programming problem (of the form (iii) §0.5). Let $\mathcal{F}$ be the set of its feasible points, that is, the set of points $(k_1, \ldots, k_{2^n})$ for which (4) is satisfied.

Case 1. $\mathcal{F}$ is empty.

Here the premises are inconsistent and optimality of β is not defined.

Case 2. $\mathcal{F}$ is non-empty.

Conditions (3) imply that $\mathcal{F}$ is a bounded convex polyhedron (§0.5). Hence the minimum value of w, min w, is taken on at a vertex, say V_m, and likewise its maximum value, max w, at some vertex V_M. Moreover, these values are global extrema. Hence for no feasible point can w lie outside the interval [min w, max w]. An informal geometric argument can be used to show that each value in this interval is taken on by $\sum^{(\psi)} k_j$ at some feasible point: for the plane $w = \sum^{(\psi)} k_j$ moving continuously parallel to itself from V_m to V_M always intersects the polyhedron. Hence there is always a feasible point corresponding to any value w in the interval [min w, max w]. This interval is our optimal β since it contains precisely those values of $P(\psi)$ for which conditions (3) are satisfied as P ranges over all probability functions on $\mathcal{S}_n$.

Theorem 4.62. *Let*

$$P(\phi_1) \in [a_1, b_1], \ldots, P(\phi_m) \in [a_m, b_m] \vDash P(\psi) \in [L, U] \tag{5}$$

be a probability logical consequence in which $[L, U]$ is the optimal interval. Then the end points L and U are explicitly obtainable as the maximum, respectively minimum, of a finite number of linear (affine) forms with rational coefficients in the $2m$ arguments $a_1, \ldots, a_m, b_1, \ldots, b_m$.

Proof. Continuing with the notation used in the proof of Theorem 4.61, let $\sum^{(\psi)} k_j$ now be written $\delta^{(\psi)}\mathbf{k}$, where $\delta^{(\psi)}$ is the $2^n \times 1$ row matrix of 1's and 0's corresponding to the presence or absence of k_j in the sum $\sum^{(\psi)} k_j$, and $\mathbf{k}$ is the column matrix $[k_1 \cdots k_{2^n}]^T$. We can now express the linear program of the preceding theorem (for the minimum of w) as:

$$\text{minimize } w = \delta^{(\psi)}\mathbf{k}, \text{ subject to}$$
$$\mathbf{Mk} \geq \mathbf{c}. \qquad (\mathbf{k} \geq \mathbf{0}) \tag{6}$$

This linear program has the dual form

$$\text{maximize } u = \mathbf{c}^T\mathbf{x}, \text{ subject to}$$
$$\mathbf{M}^T\mathbf{x} \leq \delta^{(\psi)T}. \qquad (\mathbf{x} \geq \mathbf{0}) \tag{7}$$

Note that in this dual form the parameters $a_1, \ldots, a_m, b_1, \ldots, b_m$ no longer appear in the specification of the feasible set, which now depends only on the logical structure of $\phi_1, \ldots, \phi_m$ and ψ (the pattern of 1's and 0's in $\mathbf{M}$ and $\delta^{(\psi)}$ reflecting the presence or absence of k_j's). The feasible set being non-empty (optimality of a β being defined only in that circumstance) let

$$(x_1^{(v)}, \ldots, x_{2m+2}^{(v)}) \qquad (v = 1, \ldots, r)$$

be an enumeration of the vertices, r in number, of the bounded convex polyhedron defined by (7). Since $\mathbf{M}$ and $\delta^{(\psi)}$ are matrices of 1's and 0's the coordinates of these vertices are explicitly obtainable rational numbers (§0.5). Let $\mathbf{x}^{(v)}$ be the column matrix $[x_1^{(v)} \ldots x_{2m+2}^{(v)}]^{\mathrm{T}}$. Then

$$L = \min w = \max u = \max(\mathbf{c}^{\mathrm{T}}\mathbf{x}^{(1)}, \ldots, \mathbf{c}^{\mathrm{T}}\mathbf{x}^{(r)}),$$

with the items $\mathbf{c}^{\mathrm{T}}\mathbf{x}^{(v)}$ $(v = 1, \ldots, r)$ in the max listing being linear forms in $a_1, \ldots, a_m, b_1, \ldots, b_m$ with rational coefficients, gives the polyhedral function for L. To obtain that for U we apply the procedure to $\neg\psi$ and subtract from 1. The only change required is replacing (in the dual program) $\delta^{(\psi)}$ by $\delta^{(\neg\psi)}$. If the s vertices for this new polyhedron are

$$(y_1^{(v)}, \ldots, y_{2m+2}^{(v)}) \qquad (v = 1, \ldots, s)$$

then, $\mathbf{y}^{(v)}$ being a column matrix similar to $\mathbf{x}^{(v)}$,

$$U = 1 - \max(\mathbf{c}^{\mathrm{T}}\mathbf{y}^{(1)}, \ldots, \mathbf{c}^{\mathrm{T}}\mathbf{y}^{(s)})$$
$$= \min(1 - \mathbf{c}^{\mathrm{T}}\mathbf{y}^{(1)}, \ldots, 1 - \mathbf{c}^{\mathrm{T}}\mathbf{y}^{(s)}).$$

Examples of an L and U coming from a polyhedron with multiple vertices and which are functions of the parameters as described, may be seen in Theorems 4.52, 4.56 and 4.57.

In the following theorem we shall apply the phrase 'explicitly given' to a probability logical consequence. By this we mean that not only are the sentences $\phi_1, \ldots, \phi_m, \psi$ explicitly given (in terms of atomic sentences $A_1, \ldots, A_n$ and sentential connectives) but also that the end points of the intervals are given rationals, for these are the only reals that can be entered as numerical data in a computer.

Theorem 4.63. *There is, for any explicitly given probability logical consequence, an effective procedure for determining whether*

$$P(\phi_1) \in [a_1, b_1], \ldots, P(\phi_m) \in [a_m, b_m] \vDash P(\psi) \in [l, u], \qquad (8)$$

is or is not a valid consequence relation.

Proof. By Theorem 4.62 L and U are the explicitly obtainable functions of $a_1, \ldots, a_m, b_1, \ldots, b_m$ as there described. Since these arguments are given rationals, L and U are explicitly obtainable rationals. The relation (8) holds if and only if

$$l \leq L \leq U \leq u.$$

This can be effectively checked.

§4.7. Set-theoretic versus logic-theoretic probability

In the early developments relating logic and probability, as described in our chapters 1 and 2, we have seen that 'event' is rendered, or replaced, by 'proposition'. Thus Bernoulli, Lambert, Bolzano, De Morgan, Boole and MacColl all refer to the probability of *propositions* or to cognate notions such as *argument, Satz,* or *statement.* We are, of course, following in this tradition. Yet all current mathematically oriented work in probability and statistics uses the notion of *set* in place of *event*, probabilities then being assigned to sets. The practice dates from Kolmogorov's watershed monograph of 1933, which first gave a satisfactory treatment of stochastic situations in which an infinite number of possible outcomes are considered. This monograph of Kolmogorov's contains an axiomatization of probability, probability being taken to be a normalized countably additive measure on an algebra of sets. We devote this section to showing that for finite stochastic situations, whether one uses realizations of Kolmogorov's axioms or probability models in the sense of §§4.3, 4.4, equivalent results are obtained. Here we consider only the first five of Kolmogorov's axioms as that is all that is needed to model finite stochastic situations.

A (*finitely additive*) *probability space* is an ordered triple $\langle \Omega, \mathcal{A}, P \rangle$, where Ω is a non-empty set (whose elements are called elementary events), where $\mathcal{A}$ is a collection (set) of subsets of Ω, (these subsets being called chance events), where P is a set-function into the reals ($P(A)$ being called the probability of A), and the following axioms hold:

I. $\mathcal{A}$ is a field of sets.

II. Ω is a member of $\mathcal{A}$.

III. To each $A \in \mathcal{A}$ there is correlated a non-negative value $P(A)$.

IV. $P(\Omega) = 1$.

V. For any A and B in $\mathcal{A}$ which are mutually exclusive,

$$P(A \cup B) = P(A) + P(B)$$

Axiom V is the property of finite additivity. Contemporary versions replace I and II by the equivalent

II′. $\mathcal{A}$ is an algebra of subsets of Ω.

The term 'algebra' is used in the general sense of a set with finitary operations closed in the set. That $\mathcal{A}$ is an algebra of subsets of Ω means that $\mathcal{A}$ is closed under the operations of union, intersection and complementation (with respect to Ω). When Ω is finite we refer to $\langle \Omega, \mathcal{A}, P \rangle$ as a *finite probability space*. In this case it turns out that $\mathcal{A}$ is the Boolean algebra of all subsets of Ω. Thus for a finite probability space the ordered pair $\langle \Omega, P \rangle$ suffices as a designation. A finite probability space is used to represent, or model, stochastic situations of the following kind.

There is an observation, or experiment, which always results in one, and only one, of distinct outcomes $\omega_1, \ldots, \omega_n$, which one being a matter of chance. Then Ω is the set $\{\omega_1, \ldots, \omega_n\}$. The value of $P(\{\omega_i\})$ is set equal to the chances that the outcome is ω_i; and the value of $P(E)$, for E any subset of Ω, is the sum of the values $P(\{\omega_i\})$ for all $\omega_i \in E$, or is 0 if E is the empty set. The following is a precise formulation of the result, commonly taken for granted, that one can use either sets or propositions to model finite stochastic situations.

Theorem 4.71. *There is an effective pairing of probability models (in the sense of §4.4) with finite probability spaces (of sets) which is sense-preserving, i.e., is such that the corresponding members in a pairing model the same stochastic situation.*

Proof. Let $\langle \Omega, P \rangle$ be a finite probability space with $\Omega = \{\omega_1, \ldots, \omega_n\}$ (representing the set of possible outcomes of a stochastic experiment $\mathcal{X}$). We define the probability model $\langle \mathcal{L}(A_1, \ldots, A_n), P_{\mathcal{L}} \rangle$ which corresponds to it by taking as the sentential language one with atomic sentences $A_1, \ldots, A_n$ (think of A_i as saying that ω_i is the outcome of the experiment $\mathcal{X}$) and taking as the probability function $P_{\mathcal{L}}$ that defined by setting

$$P_{\mathcal{L}}(K_i) = P(\{\omega_i\}), \quad i = 1, \ldots, n$$

where the K_i are the constituents $\overline{A}_1 \overline{A}_2 \cdots \overline{A}_{i-1} A_i \overline{A}_{i+1} \cdots \overline{A}_n$, and setting $P_{\mathcal{L}}(K) = 0$ for K any other constituent on $A_1, \ldots, A_n$. In the reverse direction, suppose that $\langle \mathcal{L}(A_1, \ldots, A_n), P_{\mathcal{L}} \rangle$ is a probability model. The finite probability space which is to be its correspondent is defined as follows. The set Ω is to be $\{\omega_1, \ldots, \omega_m\}$, where the ω_i are those constituents K_i of $\mathcal{L}(A_1, \ldots, A_n)$ for which $P_{\mathcal{L}}(K_i) > 0$, $i = 1, \ldots, m$. There is at least one such K_i since the sum of $P_{\mathcal{L}}$ values taken over all constituents equals 1. (Intuitively, ω_i is the 'event' described by K_i, i.e., by the statement asserting each unnegated A_j in K_i and denying each of the negated A_j in K_i.) Then P is defined by setting

$$P(\{\omega_i\}) = P_{\mathcal{L}}(K_i), \quad i = 1, \ldots, m$$

and extending these values by additivity to all (non-empty) subsets of Ω, and setting $P(\emptyset) = 0$, for $\emptyset$ the empty set. Clearly this $\langle \Omega, P \rangle$ is a finite probability space.

Chapter 5

———

Conditional Probability

§5.1. Conjunction and multiplication of probabilities

That there is a strong connection between logical disjunction and addition of probabilities is evidenced by the basic 'law' (P3, §4.2):

$$\text{If } \vdash \phi \to \neg\psi, \text{ then } P(\phi \lor \psi) = P(\phi) + P(\psi). \tag{1}$$

But for conjunction we have only the weak inequality:

$$P(\phi \land \psi) \le \min(P(\phi),\, P(\psi)). \tag{2}$$

Although inequality (2) is weak, in that it provides no exact value for $P(\phi \land \psi)$, it is the strongest that can be said about $P(\phi \land \psi)$ if all that is known are the probabilities $P(\phi)$ and $P(\psi)$ (see §3.5). Could we perhaps obtain an exact value for $P(\phi \land \psi)$ in terms of $P(\phi)$ and $P(\psi)$ by having, as in (1), a suitable (logical) hypothesis? One such hypothesis could be that ϕ, or ψ, or both are either logically true, or logically false. Then

$$P(\phi \land \psi) = P(\phi)P(\psi) \tag{3}$$

would be trivially true, trivially in that under such circumstances the equation would be either $0 = 0$, $1 = 1$, $P(\phi) = P(\phi)$, or $P(\psi) = P(\psi)$. It seems pretty clear that no other *logical* hypothesis on ϕ and ψ could make (3) true in general. But since this assertion is capable of being proved, we shall do so.

Theorem 5.11. *Let $F(\phi, \psi)$ be a truth function of ϕ and ψ, and P a probability function. The statement*

$$\text{If } \vdash F(\phi, \psi), \text{ then } P(\phi \land \psi) = P(\phi)P(\psi) \tag{4}$$

is valid only for the trivial cases of $F(\phi, \psi)$ being one of the eight functions:
ϕ, $\neg\phi$, ψ, $\neg\psi$, $\phi\psi$, $\phi\overline{\psi}$, $\overline{\phi}\psi$, $\overline{\phi}\,\overline{\psi}$.

Proof. Suppose $F(\phi, \psi)$ is ϕ. Then $\vdash \phi$ implies that $P(\phi \wedge \psi) = P(\psi)$ and $P(\phi) = 1$. Thus the consequent in (4) holds. Suppose $F(\phi, \psi)$ is $\neg\phi$. Then $\vdash \neg\phi$ implies that $P(\phi) = P(\phi \wedge \psi) = 0$, and again the consequent holds. The remaining six cases are all just as easily established. Now, other than the eight listed functions there are, dropping the two constant functions, six other possible functions that $F(\phi, \psi)$ could be, namely

(i) $\phi \vee \psi$ (iv) $\neg\phi \vee \neg\psi$

(ii) $\neg\phi \vee \psi$ (v) $\phi \leftrightarrow \psi$

(iii) $\phi \vee \neg\psi$ (vi) $\phi \leftrightarrow \neg\psi$.

Appropriately identifying ψ with either ϕ or $\neg\phi$ in each of these six cases converts the formula to a logically true one. However, any such identification results in a false conclusion. Thus, as an example, taking ψ to be $\neg\phi$ in either (i) or (vi) converts the antecedent in (4) to a true statement, but the consequent to

$$P(\phi\overline{\phi}) = P(\phi)P(\overline{\phi}),$$

which is not true (for arbitrary ϕ).

Since there is no non-trivial formal logical condition that can tell us when $P(\phi \wedge \psi) = P(\phi)P(\psi)$ we pursue the question from the material viewpoint, i.e., by looking at examples in which probability is applied. This of course requires us to understand what is meant by probability. There are, as is well-known, a great variety of opinions as to the nature or meaning of probability. But essentially all of them lead to the elementary formal properties listed by us (in a linguistic formulation) in §4.2. Thus it will not matter which meaning we use for probability if it conforms with, i.e., satisfies, these properties. Currently a popular one is that based on relative frequency. In the early historical period, in fact at the very beginning with Huygens' *Ratiociniis in alea ludo*, properties of probability were derived via examples using the notion of expectation (of gain or loss in games of chance). This was also the case with De Moivre, in whose writings we find the first clear formulation of the multiplication rule for probabilities as well as the associated notions of independence, dependence and, though unnamed, conditional probability. We quote some relevant passages from his *The Doctrine of Chances*, the first being an example with independent events (*De Moivre 1756*, 5–6):

> 8. If the obtaining of any Sum requires the happening of several Events that are independent on each other, then the Value of the

Expectation of that Sum is found by multiplying together the several Probabilities of happening, and again multiplying the product by the Value of the Sum expected.

Thus supposing that in order to obtain $90£$ two Events must happen; the first whereof has 3 Chances to happen, and 2 to fail, the second has 4 Chances to happen, and 5 to fail, and I would know the value of that Expectation; I say,

The Probability of the first's happening is $\frac{3}{5}$, the Probability of the second's happening is $\frac{4}{9}$; now multipling these two Probabilities together, the product will be $\frac{12}{45}$ or $\frac{4}{15}$; and this product being again multiplied by 90, the new product will be $\frac{360}{15}$ or 24, therefore that Expectation is worth $24£$.

The Demonstration of this will be very easy, if it be consider'd, that supposing the first Event had happened, then that Expectation depending now intirely upon the second, would, before the determination of the second, be found to be exactly worth $\frac{4}{9} \times 90£$, or $40£$. (by Art. 5th)[1] We may therefore look upon the happening of the first, as a condition of obtaining an Expectation worth $40£$ but the Probability of the first's happening has been supposed $\frac{3}{5}$, wherefore the Expectation sought for is to be estimated by $\frac{3}{5} \times 40$, or by $\frac{3}{5} \times \frac{4}{9} \times 90$; that is, by the product of the two Probabilities of happening multiplied by the Sum expected.

Note that De Moivre computes the expectation of obtaining the $90£$ on the happening of both events, by first *supposing that the first event has happened*, giving then an expectation of obtaining $\frac{4}{9} \times 90£$ (the probability of the second happening times the "Sum"), and then considering this expectation as the "Sum" to be obtained if the first happens. He states the general rule for several [independent] events (*1756*, 6):

COROLLARY.

If we make abstraction of the Value of the Sum to be obtained, the bare Probability of obtaining it, will be the product of the several Probabilities of happening, which evidently appears from this 8th Art. and from the Corollary to the 5th.

(Note the evanescent role of the "Sum".) Continuing:

Hitherto, I have confined myself to the consideration of Events independent; but for fear that, in what is to be said afterwards, the

---

1. De Moivre's Article is: 5. In all cases, the Expectation of obtaining any Sum is estimated by multiplying the value of the Sum expected by the Fraction which represents the Probability of obtaining it.

terms independent or dependent might occasion some obscurity, it will be necessary, before I proceed any farther, to settle intirely the notion of those terms.

Two Events are independent, when they have no connexion one with the other, and that the happening of one neither forwards nor obstructs the happening of the other.

Two Events are dependent, when they are so connected together as that the Probability of either's happening is altered by the happening of the other.

He illustrates the notions by two "Problems". These, or similar ones, are now stock examples:

1°. Suppose there is a heap of 13 Cards of one colour, and another heap of 13 Cards of another colour, what is the Probability that taking a Card at a venture out of each heap, I shall take the two Aces?

The Probability of taking the Ace out of the first heap is $\frac{1}{13}$: now it being very plain that taking or not taking the Ace out the the first heap has no influence in the taking or not taking the Ace out of the second; it follows, that supposing that Ace taken out, the Probability of taking the Ace out of the second will also be $\frac{1}{13}$; and therefore, those two Events being independent, the probability of their both happening will be $\frac{1}{13} \times \frac{1}{13} = \frac{1}{169}$.

2°. Suppose that out of one single heap of 13 Cards of one colour, it should be undertaken to take out the Ace in the first place, and then the Deux, and that it were required to assign the Probability of doing it; we are to consider that altho' the Probability of the Ace's being in the first place be $\frac{1}{13}$, and the Probability of the Deux's being in the second place, would also be $\frac{1}{13}$, if that second Event were considered in itself without any relation to the first; yet that the Ace being supposed as taken out at first, there will remain but 12 Cards in the heap, and therefore that upon the supposition of the Ace being taken out at first, the Probability of the Deux's being next taken will be alter'd, and become $\frac{1}{12}$; and therefore, we may conclude that those two Events are dependent, and the Probability of their both happening will be $\frac{1}{13} \times \frac{1}{12} = \frac{1}{156}$.

The results are summed up in a Rule (*1756*, 7):

From whence it may be inferred, that the Probability of happening of two Events dependent, is the product of the Probability of the happening of one of them, by the Probability which the other will have of happening, when the first is considered as having happened;

and the same Rule will extend to the happening of as many Events as may be assigned.

This rule that De Moivre states, for "two Events dependent", applies equally well for independent events since the "Probability which the other will have of happening, when the first is considered as having happened" will, in the case of independent events, be the original probability of the second—hence yielding the proper product for the independent events. It is to be noted that the supposition that the first has happened results, in general, in a revision of the probability of the second. But what were originally equally likely outcomes for the second are still considered as equally likely after the supposition.

To have the multiplication rule for a conjunction in probability logic will require replacing De Moivre's physicalistic criterion by a logico-semantic one. But the logical semantics we have been using so far has no means of expressing the notion of an event happening when another is considered as having happened. Our logic needs to be extended if we are to have such a notion. This will be accomplished in §5.7. In the meantime much of conditional probability theory can be developed on the basis of adjoining to (unconditional) probability logic a special function on ordered pairs of sentences of S. Later we shall be able to replace this ordered pair by a sentential connective in the so extended logic. Anticipating success in this regard we shall write '$\psi \mid \phi$' in place of the pair '(ψ, ϕ)', and refer to it as a *probability conditional* (since it has resemblances to the verity conditional).

Let M be a probability model for S and P_M the associated probability function. We define a 'semantic function' P_M^* on probability conditionals by setting

K_1	K_2	$\cdot$	$\cdot$	$\cdot$	K_{2^n}	ϕ	$\phi \wedge \psi$	$\psi \mid \phi$
k_1	k_2	$\cdot$	$\cdot$	$\cdot$	k_{2^n}	$\sum_\phi k_i$	$\sum_{\phi\psi} k_i$	$\begin{cases} \sum_{\phi\psi} k_i / \sum_\phi k_i \\ c, \quad c \in [0, 1] \end{cases}$

where $\sum_\phi k_i$ is the sum of the k_i's corresponding to those K_i which imply (in the verity sense) ϕ, where $\sum_{\phi\psi} k_i$ similarly is the sum of those k_i for K_i which imply $\phi\psi$ (n being the index of support), and the value of $P_M^*(\psi \mid \phi)$ is the quotient $\sum_{\phi\psi} k_i / \sum_\phi k_i$ if $\sum_\phi k_i \neq 0$, and otherwise is an arbitrary (unspecified or indefinite) value c in the unit interval. (This c is analogous to the constant of integration in calculus, which is an arbitrary value $C \in (-\infty, +\infty)$). Thus $P_M^*(\psi \mid \phi)$ is essentially the usual notion of conditional probability defined as a quotient of ordinary probabilities except that, because of our special meaning when $P(\phi) = 0$, we can write

$$P_M(\phi \wedge \psi) = P_M(\phi) P_M^*(\psi \mid \phi),$$

assuming, as we are, that the 'constant' c has the property $0 = 0 \cdot c$.

Although we have as yet no logical concept associated with '$\psi \mid \phi$' the function $P_M^*(\psi \mid \phi)$ accords with the intended probability meaning for it. For if ϕ is considered as having happened then the only possibilities are those represented by constituents implying ϕ. Hence what was $\sum_\phi k_i$ now has to equal 1. The tacit probability assumption (see De Moivre's examples) is that each possibility is changed in the same proportion, that is, that each k_i becomes ak_i, so that $\sum_\phi ak_i = 1$. This gives $a = 1/\sum_\phi k_i$. Then the chances of getting ψ, when ϕ is considered as having happened, is the same as the chances of getting $\phi \wedge \psi$ when ϕ is considered as having happened. Thus

$$\sum_{\phi\psi} ak_i = a \sum_{\phi\psi} k_i = \frac{\sum_{\phi\psi} k_i}{\sum_\phi k_i}.$$

It is evident that the probability conditional is not expressible in terms of the probability connectives $\neg$, $\wedge$, $\vee$. For any syntactic combination of these latter three with atomic sentences $A_1, \ldots, A_n$, i.e., any element of $\mathcal{S}_n$, has a probability value which is a linear combination, with coefficients either 0 or 1, of the $k_1, \ldots, k_{2^n}$ (probabilities of the constituents on $A_1, \ldots, A_n$). On the other hand, for ϕ and ψ in $\mathcal{S}_n$, the value of $\psi \mid \phi$ is a linear fractional form in the k_i whose denominator is, in general, different from 1.

It should be noted that although $P_M^*(\psi \mid \phi)$ is referred to as a (conditional) probability, it is not to be assumed that P_M^* is a probability function; indeed, since as yet no meaning has been attached to sentential combinations whose components are probability conditionals, it makes no sense to ask if P_M^* satisfies the properties P1–P3 of §4.2.

We interrupt our exposition of conditional probability with a section whose import is mainly historical.

§5.2. Independence: logical and stochastic

In the preceding section we used De Moivre's two 'Problems', illustrating the difference between independent and dependent events, as a discussion point for arriving at the multiplication rule for the probability of a conjunction of two events. We noted that the rule applied irrespective of whether the two events were independent or dependent. The concept expressed by $P(\phi \wedge \psi) = P(\phi)P(\psi)$, i.e., *stochastic independence*, is of far-reaching importance in probability and statistics. But as it is a property relative to a given probability model it is not *per se* a topic for probability logic where the interest is in properties holding for all models. Accordingly, when appearing in this book stochastic independence will usually be incidental to other matters. However in this section we shall feature a relationship between stochastic and logical independence. These two notions played an important role in the way Boole viewed probability, though were never clearly distinguished by him. But before describing these views of Boole's we present some results involving the two kinds of independence.

We consider the sentences in a language $S_n = S(A_1, \ldots, A_n)$ with atomic sentence letters $A_1, \ldots, A_n$. It will be convenient for us in what follows to assume that S_n has an additional (atomic) sentence letter '1'. This letter is always to be assigned the value 1 (*true*) in any model, and hence represents a logically true sentence.[2] We shall use '0' as an abbreviation for '¬1'.

A letter A_i ($i > 0$) is said to occur *inessentially* in a sentence ϕ if $\phi[1/A_i]$ and $\phi[0/A_i]$, the sentences obtained from ϕ by replacing A_i by 1 and 0, are logically equivalent; otherwise it occurs *essentially*. Two sentences ϕ and ψ are *logically independent* if they are equivalent, respectively, to sentences ϕ' and ψ' which have no essential A_i in common. *A fortiori* two sentences with no A_i ($i > 0$) in common, as well as any two atomic sentences, are logically independent. Also sentences having only occurrences of 1 (recall that '0' is an abbreviation for '¬1'), and hence no essentially occurring letter, are logically independent. Such sentences are, of course, either logically true or logically false.

Theorem 5.21. *Any sentence ϕ of S_n is logically equivalent to a sentence with no inessential letters.*

Proof. If A_i were inessential then, by developing ϕ with respect to the A_i and using the definition of inessential, we have

$$\phi \leftrightarrow \phi[1/A_i]A_i \vee \phi[0/A_i]\overline{A_i}$$
$$\leftrightarrow \phi[1/A_i](A_i \vee \neg A_i)$$
$$\leftrightarrow \phi[1/A_i].$$

Thus inessential letters can be removed until none are left.

We shall say a sentence ϕ is *logically contingent* if it is neither logically true nor logically false. As an immediate consequence of the preceding theorem we have

---

2. We trust that there will be no confusion in our using the same symbol for a formal sentence as for a semantic value which is its constant value—as is the practice in standard mathematics, which uses the same symbol for a numerical constant as for the function having this one value for any argument value.

Theorem 5.22. *Any ϕ in S_n is logically contingent if and only if it has at least one essential letter.*

Our introduced notion of logical independence between two sentences is connected with the idea of these sentences not having any logical relationship, that is, other than trivial ones enjoyed by any two sentences. As a simple example of such a trivial relation, consider the two-place $\Phi(A_1, A_2) = (A_1 \vee \overline{A}_1)(A_2 \vee \overline{A}_2)$. Clearly $\Phi(\phi, \psi)$ holds (is logically true) for any two sentences ϕ and ψ.

Theorem 5.23. *Let* (i) *ϕ and ψ be logically contingent sentences of S_n which are logically independent, and* (ii) *$\Phi(A_1, A_2)$ be a sentence of S_n having only A_1 and A_2 as atomic sentences. Then $\Phi(\phi, \psi)$ is logically true if and only if $\Phi(A_1, A_2)$ is.*

Proof.

(a) If $\Phi(A_1; A_2)$ is logically true then so also is $\Phi(\phi, \psi)$; for any truth values which ϕ and ψ take on will be among those which A_1 and A_2 take on.

(b) Suppose that $\Phi(\phi, \psi)$ is logically true. Then its expansion

$$\Phi(1,1)\phi\psi \vee \Phi(1,0)\phi\overline{\psi} \vee \Phi(0,1)\overline{\phi}\psi \vee \Phi(0,0)\overline{\phi}\,\overline{\psi}, \tag{1}$$

is also logically true. If we can prove that each of the coefficients (i.e., $\Phi(1,1)$, etc.) in (1) is (logically) true then so also is $\Phi(A_1, A_2)$ since its expansion is

$$\Phi(1,1)A_1 A_2 \vee \Phi(1,0)A_1\overline{A}_2 \vee \Phi(0,1)\overline{A}_1 A_2 \vee \Phi(0,0)\overline{A}_1\overline{A}_2. \tag{2}$$

Now since $\phi(A_{i_1}, \ldots, A_{i_k})$ and $\psi(A_{j_1}, \ldots, A_{j_l})$, with essential letters as shown, are logically contingent and independent, there are assignments of values, 0 or 1, to $A_{i_1}, \ldots, A_{i_k}$ which make ϕ have the value 1 (otherwise it would be logically false), and values which make it take on the value 0 (otherwise it would be logically true). The same is true for ψ and $A_{j_1}, \ldots, A_{j_l}$. Since the sets $\{A_{i_1}, \ldots, A_{i_k}\}$ and $\{A_{j_1}, \ldots, A_{j_l}\}$ are disjoint there are assignments which can make ϕ and ψ take on the values $(1,1)$ or $(1,0)$ or $(0,1)$ or $(0,0)$. Hence by considering each of these four assignments in turn we see that each of the coefficients in (1) is logically true, and then so also is $\Phi(A_1, A_2)$.

To put the result of this theorem succinctly: A logical relation $\Phi(\phi, \psi)$ holds necessarily between logically contingent and independent ϕ and ψ, only if Φ is *per se* necessary, i.e., holds for any two sentences as arguments.

We turn now to stochastic independence. When referring to a *set* of sentences being independent, rather than pairwise independence, a stronger notion is found more useful and is generally the one in use: A set of sentences $\{\phi_1, \ldots, \phi_m\}$ $(m > 0)$ is said to be *stochastically independent* (with respect to a given probability function) if the probability of the conjunction of any two or more elements of the set is equal to the product of the probabilities of the respective elements of the set. Since there are $2^m - \binom{m}{1} - \binom{m}{0}$ $(= 2^m - m - 1)$ selections of such subsets the definition requires that that many equations hold. There is an equivalent condition in terms of the 2^m constituents constructible on $\phi_1, \ldots, \phi_m$ (see *Feller 1957*, 117 or *Rényi 1970*, 110):

Theorem 5.24. *The system of 2^m equations which states, for each constituent on $\phi_1, \ldots, \phi_m$, equality of the probability of the constituent with the product of the probabilities of the respective factors (conjuncts) of the constituent, is equivalent to the system of $2^m - m - 1$ equations in the definition of stochastic independence for a set of sentences.*

Theorem 5.25. *If $\{\phi_1, \ldots, \phi_m\}$ is a stochastically independent set then so is each of the $2^m - 1$ other sets of the form $\{(\neg)\phi_1, \ldots, (\neg)\phi_m\}$, where '$(\neg)\phi_i$' indicates the element is either ϕ_i or $\neg\phi_i$, ad libitum.*

Proof. On removal of any double negations, the set of constituents based on any one of the sets $\{(\neg)\phi_1, \ldots, (\neg)\phi_m\}$ is the same as the set of constituents on $\{\phi_1, \ldots, \phi_m\}$.

By virtue of this theorem the probability of a sentence, whose set of atomic sentences represent stochastically independent events, can be given a particularly simple form, a form which was extensively employed by Boole:

Theorem 5.26. *Let V be a sentence of S_n expressed in disjunctive normal form (on letters $A_1, \ldots, A_n$). If $A_1, \ldots, A_n$ are stochastically independent with respect to P then $P(V) = [V]$, where '$[V]$' stands for the result obtained from V on replacing each occurrence of A_i by $P(A_i)$, each $\overline{A}_i$ by $P(\overline{A}_i)$ $(i = 1, \ldots, n)$ and taking logical sum and product as, respectively, numerical sum and product.*

Proof. Immediate by virtue of the preceding theorem, and that the disjuncts in the disjunctive normal form are mutually exclusive constituents on the stochastically independent factors making up a constituent term.

Our next theorem shows that logical independence is a sufficient condition for stochastic independence, *provided that* the atomic sentences involved are stochastically independent.

Theorem 5.27. *Let the atomic sentences $A_1, \ldots, A_n$ of S_n be stochastically independent. If sentences ϕ and ψ of S_n are logically independent then they are stochastically independent.*

Proof. Without loss of generality we may assume that ϕ and ψ have no inessential letters present. Also that both are logically contingent. For if either were logically true, or false, then they are both logically and stochastically independent. So suppose ϕ and ψ are logically independent. Then, not being logically false and being contingent (so that there are essential letters present) we may write disjunctive normal forms as shown:

$$\phi \leftrightarrow \bigvee^{(\phi)} K_r(A_{i_1}, \ldots, A_{i_k})$$

$$\psi \leftrightarrow \bigvee^{(\psi)} L_s(A_{j_1}, \ldots, A_{j_l}),$$

where the two sets of essential letters, $\{A_{i_1}, \ldots, A_{i_k}\}$ and $\{A_{j_1}, \ldots, A_{j_l}\}$, are non-empty and disjoint. Then no logical product of the form $K_r L_s$ can vanish (i.e., be logically false), and any two distinct ones of this form are mutually exclusive (by having either distinct K_r's or distinct L_s's). Making use of the stochastic independence of $A_1, \ldots, A_n$ so that (by Theorem 5.25) for each $K_r L_s$, $P(K_r L_s) = P(K_r)P(L_s)$, we have

$$
\begin{aligned}
P(\phi \wedge \psi) &= P(\bigvee^{(\phi)} K_r \bigvee^{(\psi)} L_s) \\
&= P(\bigvee^{(\phi)} \bigvee^{(\psi)} K_r L_s) \\
&= \sum^{(\phi)} \sum^{(\psi)} P(K_r L_s) \\
&= \sum^{(\phi)} \sum^{(\psi)} P(K_r)P(L_s) \\
&= \sum^{(\phi)} P(K_r) \sum^{(\psi)} P(L_s) \\
&= P(\phi)P(\psi).
\end{aligned}
$$

Thus, under the hypothesis that $A_1, \ldots, A_n$ are stochastically independent, if ϕ and ψ are logically independent they are also stochastically independent. That the converse implication doesn't hold is readily seen:

Example. Let A_1 and A_2 be stochastically independent and each of probability $\frac{1}{2}$. Then it is easily verified that $\phi = A_1 A_2 \vee \overline{A_1} \overline{A_2}$ and $\psi = \overline{A_1} \overline{A_2} \vee \overline{A_1} A_2$ are stochastically independent but not logically since they both have A_1 as an essential atomic letter.

Boole's probability theory has been discussed in detail by us in our *1976* (revised and enlarged second edition *1986*). A summary of key aspects of the theory were critically described above in §§2.4–2.5. Here we examine Boole's notion of independence. We find that it fails to sharply distinguish between stochastic and logical independence.

In his enumeration of the principles of probability he states the product rule in the form (*Boole 1854*, 253):

[Principle] II. The probability of the occurrence of any two events is the product of the probability of either of those events by the probability that if that event occur, the other will occur also.

Then, as an immediate consequence, conditional probability (though unnamed) is expressed as a quotient of probabilities:

[Principle] III. The probability that if an event x occur, the event y will occur, is a fraction whose numerator is the probability of their joint occurrence, and the denominator the probability of the occurrence of the event x.

Independence is then introduced (p. 255):

2. DEFINITION.—Two events are said to be independent when the probability of the happening of either of them is unaffected by our expectation [?] of the occurrence or failure of the other.

From this definition, combined with Principle II., we have the following conclusion:

[Principle] V. The probability of the concurrence of two independent events is equal to the product of the separate probabilities of those events.

For if p be the probability of an event x, q that of an event y regarded as [?] quite independent of x, then is q also the probability that if x occur y will occur. Hence, by Principle II., pq is the probability of the concurrence of x and y.

Aside from the dubious psychologistic features (indicated by our insertion of question marks) this is essentially a standard account, resulting in the condition $P(xy) = P(x)P(y)$ as an equivalent to Boole's x and y being independent events. Here, then, his independence is stochastic independence. But the term acquires additional import when we read, after some discussion (p. 256): "When the probabilities of events are given, but all information respecting their dependence withheld, the mind regards them as independent." The examples and discussion on the page shows that Boole is taking 'independent' to mean stochastic independence, but that 'dependence' refers to a logical connection. The ideas discussed are codified in the statement of a principle which he wishes to add to the previous (generally accepted) ones:

[Principle] VI. The events whose probabilities are given are to be regarded as independent of any connexion but such as is either expressed, or necessarily implied, in the data; and the mode in which our knowledge of that connexion is to be employed is independent of the nature of the source from which such knowledge has been derived.

Superficially this principle seems to be a truism of practical logic—that one shouldn't assume a connection exists unless it is expressed (explicitly or implied) by the data. But, since Boole is taking absence of a connection to imply (stochastic) independence, as he uses it the content is that

absence of logical connection (i.e., logical independence)

implies

stochastic independence.

We contrast this principle of Boole's with our Theorem 5.27. Our proof that logical independence implied stochastic independence required the hypothesis that the atomic letters, out of which ϕ and ψ were compounded, were themselves stochastically independent.

Interestingly enough, Boole has a concept resembling our atomic sentences, namely that of 'simple unconditioned' events. Events out of which an event is compounded (logically) are *simple* (in relation to it). Simple events $x, y, z, \ldots$ are *conditioned* "when they are not free to occur in every possible combination; in other words, when some compound event depending upon them is precluded from happening." Boole asserts: "Simple unconditioned events are by definition independent." Like Boole's simple unconditioned events, our atomic sentences are uncompounded and every possible combination (constituent) is free to occur. But they are not "by definition" independent, *stochastically*. For that to happen we need to be in a specific probability model with appropriate values for the $P(K_i)$.

§5.3. Logical consequence with probability conditionals

This section presents results, mostly well-known, on conditional probability. They are stated in the framework of our hybrid combination of probability logic with the special function P_M^*, which has the property that for any probability model M,

$$P_M(\phi \wedge \psi) = P_M(\phi)P_M^*(\psi \mid \phi).$$

In addition to this product (or multiplication) rule we have

$$0 \leq P_M^*(\psi \mid \phi) \leq 1$$

(assuming, as we are, that $0 \leq c \leq 1$ holds), and

$$\text{If } \vdash \phi \leftrightarrow \phi', \text{ then } P_M^*(\psi \mid \phi) = P_M^*(\psi \mid \phi')$$
$$\text{and } \quad P_M^*(\phi \mid \psi) = P_M^*(\phi' \mid \psi).$$

In other words, one can made replacements of a sentence of S by a logically equivalent one without altering its P_M^* value. Moreover, since the results involving P_M and P_M^* hold for all models M, we shall take over from probability logic use of the symbol '$\vDash$', in conjunction with dropping the subscript M, to indicate the notions of (conditional) probability consequence and (conditional) probability assertion. This usage will be justified in §5.7. And, since P_M^* will not occur except with an argument which is a probability conditional, we can unclutter the notation by dropping the asterisk, and remembering that only if the argument of a 'P' is an element of S can P then be considered to be a probability function (in an arbitrary M). Thus we will write the product rule as

$$\vDash P(\phi \wedge \psi) = P(\phi)P(\psi \mid \phi).$$

The following theorem lists a number of consequences of this rule (restated as item (a) in the list):

Theorem 5.30.

(a) $\vDash P(\phi \wedge \psi) = P(\phi)P(\psi \mid \phi)$.

(b) $\vDash P(\phi\psi\chi) = P(\phi \mid \psi\chi)P(\psi \mid \chi)P(\chi)$.

(c) $\vDash P(\phi)P(\psi \mid \phi) = P(\psi)P(\phi \mid \psi)$.

(d) $\vDash P(\sigma)P(\phi \mid \sigma) = P(\sigma(\sigma \rightarrow \phi))$.

(e) $\vDash P(\psi \mid \phi) = P(\psi\phi \mid \phi)$.

None of these assertions need premises. However in general we shall be seeing premises to the effect that the probability of a conditioning sentence be different from 0—understandably, since the algebra of $P(\phi \mid \sigma)$ is peculiar when $P(\sigma) = 0$. (Note that $P(\sigma) \neq 0$ is equivalent to $P(\sigma) \in \alpha$, where the set α is the half-open interval $(0, 1]$.)

The following results, reflecting the properties P1–P3 of the definition of a probability function (§4.2), are readily established:

Theorem 5.31.

(a) $P(\sigma \rightarrow \phi) = 1$, $P(\sigma) \neq 0 \vDash P(\phi \mid \sigma) = 1$.

(b) $P(\sigma \rightarrow (\phi \rightarrow \psi)) = 1$, $P(\sigma) \neq 0 \vDash P(\phi \mid \sigma) \leq P(\psi \mid \sigma)$.

(c) $P(\sigma \rightarrow (\phi \rightarrow \neg\psi)) = 1$, $P(\sigma) \neq 0$
$$\vDash P(\phi \vee \psi \mid \sigma) = P(\phi \mid \sigma) + P(\psi \mid \sigma).$$

Proof. We sketch that of (a) as a sample.

If $P(\sigma \to \psi) = 1$ then $\sum_{\sigma\phi} k_i = \sum_\sigma k_i$. Hence if $\sum_\sigma k_i \neq 0$ then

$$\frac{\sum_{\sigma\phi} k_i}{\sum_\sigma k_i} = \frac{\sum_\sigma k_i}{\sum_\sigma k_i} = 1.$$

Theorem 5.31 then leads naturally to results corresponding to those stemming from P1–P3:

Theorem 5.32.

(a) $P(\phi \leftrightarrow \psi) = 1$, $P(\sigma) \neq 0 \vDash P(\phi \mid \sigma) = P(\psi \mid \sigma)$.

(b) $P(\phi \leftrightarrow \psi) = 1$, $P(\phi)P(\psi) \neq 0 \vDash P(\sigma \mid \phi) = P(\sigma \mid \psi)$.

(c) $P(\sigma) \neq 0 \vDash P(\phi \mid \sigma) + P(\neg\phi \mid \sigma) = 1$.

((d) $P(\sigma) \neq 0 \vDash P((\phi \vee \psi) \mid \sigma) = P(\phi \mid \sigma) + P(\psi \mid \sigma) - P(\phi\psi \mid \sigma)$.

Parts (a) and (b) of Theorem 5.32 readily generalize to permit, under allowable circumstances, replacement of a sentence(-part) in antecedent or consequent of a probability conditional by a sentence implied by σ to be equivalent to it.

In Theorem 5.30 item (d), a variant of (a), has been referred to earlier in §2.4 as providing a connection between the probability conditional $\phi \mid \sigma$, and the verity conditional, $\sigma \to \phi$. When written in the form

$$P(\phi \mid \sigma) = \frac{P(\sigma(\sigma \to \phi))}{P(\sigma)} \qquad (P(\sigma) \neq 0)$$

it tells us that the probability of $\phi \mid \sigma$ is that of $\sigma \to \phi$ when the latter conditional is confined to the cases for which σ holds, and with that probability renormalized so that $P(\sigma)$ is the unit. Equality occurs only at the extreme cases of $P(\sigma) = 1$ or $P(\sigma \to \phi)) = 1$:

Theorem 5.33. *If $P(\sigma) \neq 0$, then*

$$P(\sigma \to \phi) = P(\phi \mid \sigma)$$

if and only if

$$P(\sigma) = 1 \text{ or } P(\sigma \to \phi) = 1.$$

Proof. Suppose $P(\sigma) \neq 0$. Then (using '$\Leftrightarrow$' for 'if and only if')

$$P(\sigma \rightarrow \phi) = P(\phi \,|\, \sigma) \quad \Leftrightarrow \quad 1 - P(\sigma\overline{\phi}) = \frac{P(\sigma\phi)}{P(\sigma)}$$

$$\Leftrightarrow \qquad\qquad = \frac{P(\sigma) - P(\sigma\overline{\phi})}{P(\sigma)}$$

$$\Leftrightarrow \qquad\qquad = 1 - \frac{P(\sigma\overline{\phi})}{P(\sigma)}$$

$$\Leftrightarrow \quad P(\sigma) = 1 \text{ or } P(\sigma\overline{\phi}) = 0$$

$$\Leftrightarrow \quad P(\sigma) = 1 \text{ or } P(\sigma \rightarrow \phi) = 1.$$

It is interesting to note that this equality between the probabilities of the two conditionals holds if σ and $\sigma \rightarrow \phi$ are stochastically independent (and $P(\sigma) \neq 0$). For if, by stochastic independence,

$$P(\sigma(\sigma \rightarrow \phi)) = P(\sigma)P(\sigma \rightarrow \phi),$$

then by Theorem 5.30 (d)

$$P(\sigma)P(\phi \,|\, \sigma) = P(\sigma)P(\sigma \rightarrow \phi).$$

so that, with $P(\sigma) \neq 0$,

$$P(\phi \,|\, \sigma) = P(\sigma \rightarrow \phi).$$

But this only indicates how rare the condition of stochastic independence of σ and $\sigma \rightarrow \phi$ is, for it occurs if and only if $P(\sigma) = 1$ or $P(\sigma \rightarrow \phi) = 1$.[3]

The following generalization is a simple consequence of the multiplication rule (Theorem 5.30 (a)).

Theorem 5.34.

$$P(\phi\sigma) \neq 0 \models P(\phi\psi \,|\, \sigma) = P(\phi \,|\, \sigma)P(\psi \,|\, \phi\sigma).$$

In §1.7 we had noted that Boole had shown the invalidity of the inference form which we now can describe as contraposing a probability conditional. His example was of an A and B such that $P(\overline{B} \,|\, A) \neq P(\overline{A} \,|\, B)$; additionally he remarked that equality held if either probability "rises to 1". Here is a formal proof of this particular case, stated as a probability logical consequence:

---

3. *Dubois and Prade 1990*, 24, gives an incorrect necessary and sufficient condition for the equality of the two probabilities of Theorem 5.33. Their statement is "$P(a \,|\, b)$ and $P(\neg b \vee a)$ coincide if and only if $P(b) > 0$ and $P(\neg b \vee a) = 1$." However, if $P(b) = 1$ then $P(a \,|\, b) = P(a), P(\neg b \vee a) = P(a)$, and their statement reduces to

$$P(a) = P(a) \quad \text{if and only if} \quad P(a) = 1.$$

Theorem 5.35.

$$P(\overline{\phi}) \neq 0, \ P(\phi \mid \psi) = 1 \ \vDash \ P(\overline{\psi} \mid \overline{\phi}) = 1.$$

Proof. Consider an arbitrary probability function P on $\mathcal{S}^v$ with $P(\overline{\phi}) \neq 0$. Then (using '$\Rightarrow$' as short for 'implies')

$$P(\phi \mid \psi) = 1 \Rightarrow P(\phi\psi) = P(\psi)[= P(\phi\psi) + P(\overline{\phi}\psi)]$$
$$\Rightarrow P(\overline{\phi}\psi) = 0$$
$$\Rightarrow P(\overline{\phi}\,\overline{\psi}) = P(\overline{\phi}\,\overline{\psi}) + P(\overline{\phi}\psi)[= P(\overline{\phi})]$$
$$\Rightarrow \frac{P(\overline{\phi}\,\overline{\psi})}{P(\overline{\phi})} = 1.$$

Boole did not explore the question of what could be said about $P(\overline{\psi} \mid \overline{\phi})$ if $P(\phi \mid \psi)$ were to have, not the value 1, but an arbitrary value p. Could there be a significant probability logical consequence of the form

$$P(\overline{\phi}) \neq 0, \ P(\phi \mid \psi) = p \ \vDash \ P(\overline{\psi} \mid \overline{\phi}) \in \beta$$

with β depending only on p? We shall be responding to this question in our next section.

The following theorem conforms with the view that 'high' conditional probability goes with 'high' probability of the corresponding verity conditional.[4]

Theorem 5.36. $P(\phi \mid \sigma) = 1$ *if and only if* $P(\sigma) \neq 0$ *and* $P(\sigma \to \phi) = 1$.

Proof. (a) Suppose $P(\phi \mid \sigma) = 1$. Then (Theorem 5.30 (a)) $P(\phi\sigma) = P(\sigma)$ so that (Theorem 4.42 (c)) $P(\overline{\phi}\sigma) = 0$, and thus $P(\sigma \to \phi) = 1$.

(b) Suppose $P(\sigma) \neq 0$ and $P(\sigma \to \phi) = 1$. Then by Theorem 5.33, $P(\phi \mid \sigma) = 1$.

---

4. *Adams 1965, 1966.*

§5.4. Logical consequence featuring intervals

In this section we shall be presenting a number of probability logical consequences. Although proofs can be given in a uniform general manner we shall use a variety of approaches, the most general not always being the simplest or most direct.

Given the conditional probability of a sentence ψ with respect to a sentence ϕ, and also its conditional probability with respect to the denial of ϕ, then its (unconditional) probability lies between the two given values (note the open interval in the conclusion):

Theorem 5.41.

$$P(\psi\,|\,\phi) = p,\; P(\psi\,|\,\overline{\phi}) = q \;\vDash\; P(\psi) \in (\min(p,\,q),\; \max(p,\,q)).$$

Proof. From the premises we have

$$P(\psi\phi) = pP(\phi)$$
$$P(\psi\overline{\phi}) = qP(\overline{\phi}) = q - qP(\phi).$$

Hence, by addition,
$$P(\psi) = (p - q)P(\phi) + q. \tag{1}$$

Thus $P(\psi)$ is a linear function of $P(\phi)$. There is a tacit restriction on $P(\phi)$: since p and q are actual real numbers neither $P(\phi)$ nor $P(\overline{\phi})$ can be 0. Thus $0 < P(\phi) < 1$ but otherwise $P(\phi)$ is unrestricted. If (case 1) $p \geq q$, then from (1) the least possible value for $P(\psi)$ is q (but which is not attained since $P(\phi) > 0$), and the greatest is p (but which is not attained since $P(\phi) < 1$). Thus $q < P(\psi) < p$. If (case 2) $p \leq q$, then in a similar manner $p < P(\psi) < q$. Hence the conclusion as stated.

A stock situation, found in many elementary probability texts, illustrates this Theorem 5.41. A room contains two urns of balls, of which one urn has a ratio of p white balls to the total number in it; the other a ratio of q. A person is allowed in, required to choose an urn and from it blindly draw a ball. What is the probability of its being white? Setting $P(W\,|\,U_1) = p$ and $P(W\,|\,U_2) = P(W\,|\,\overline{U}_1) = q$, and using (1) gives

$$P(W) = (p - q)P(U_1) + q.$$

If the choice of urn is 'random' then $P(U_1) = \frac{1}{2}$ and $P(W) = \frac{p+q}{2}$. But if nothing is known about how the urn is selected then the best one can say is that $P(W)$ is some value between p and q.

Up to now our results on probability logical consequence have been presented in terms of general sentence variables ϕ, ψ, $\sigma, \ldots$ ranging over elements of $\mathcal{S}$. In such cases general systematic techniques are not always available, although (as our example proof of Theorem 5.31(a) showed) it can be helpful to abstractly consider an expansion into disjunctive normal form, even if the structure of the sentences is not given. However, to take full advantage of the logical interrelations of sentences we need to express them in terms of the atomic sentence components. We now shall be doing this.

Our general method to be used is much like that presented in §4.5, namely, to express all sentences in terms of the probabilities of the constituents in an arbitrary probability model. Then, using linear programming methods, determine the minimum and maximum of the probability in the conclusion, subject to the conditions in the premises. But now when conditional probabilities are involved the objective function to be optimized may not be a linear function but a linear fractional function of the constituent probabilities. Nevertheless the program can be converted to an equivalent one in which the objective function is a linear function, and one whose optimal values are the same as that for the original program. All this will be clearer after going through some examples.[5]

Our first example responds to the question asked just after Theorem 5.35 of the preceding section. Atomic letters are used to emphasize absence of (known) logical connection.

Theorem 5.42. *For the probability logical consequence*

$$P(A_1 \mid A_2) = p \vDash P(\overline{A}_2 \mid \overline{A}_1) \in \beta,$$

and for any p, $0 \le p < 1$, *the optimal* β *is* $[0, 1]$. *That is, subject to the given condition in the premise, 0 and 1 are the minimum and maximum values of* $P(\overline{A}_2 \mid \overline{A}_1)$.

Proof. In an arbitrary probability model for $\mathcal{S}^v$ let $A_1 A_2$, $A_1 \overline{A}_2$, $\overline{A}_1 A_2$, $\overline{A}_1 \overline{A}_2$ have respectively the probabilities k_1, k_2, k_3, k_4. Our problem is to

$$\text{optimize } w = P(\overline{A}_2 \mid \overline{A}_1) = \frac{k_4}{k_3 + k_4}$$

subject to

$$\frac{k_1}{k_1 + k_3} = p, \text{ i.e. } (1 - p)k_1 - pk_3 = 0,$$
$$k_1 + k_2 + k_3 + k_4 = 1, \quad k_1, k_2, k_3, k_4 \ge 0.$$

---

5. In some of these which include parameters we shall be using Fourier elimination. While the topic of Fourier elimination is on the level of school algebra, it is not usually taught. Accordingly we think that many readers of this book might welcome seeing how it can be tactically done—at least for the small size problems we will be looking at. If handled by rote the process can be tediously long and one is then prone to calculating errors.

Reexpressed in matrix form by setting

$$\mathbf{c} = [0 \quad 0 \quad 0 \quad 1], \; \mathbf{a} = [0 \quad 0 \quad 1 \quad 1], \; \mathbf{k} = [k_1 \quad k_2 \quad k_3 \quad k_4]^{\mathrm{T}},$$

$$\mathbf{A} = \begin{bmatrix} 1-p & 0 & -p & 0 \\ 1 & 1 & 1 & 1 \end{bmatrix}, \; \mathbf{b} = \begin{bmatrix} 0 \\ 1 \end{bmatrix},$$

the problem is:

$$\text{optimize} \quad \frac{\mathbf{ck}}{\mathbf{ak}}$$

subject to

$$\mathbf{Ak} = \mathbf{b}$$
$$\mathbf{k} \geq \mathbf{0}.$$

Case 1. $p = 0$.
Here the problem is reduced to the simple

$$\text{optimize } w = \frac{k_4}{k_3 + k_4}$$

subject to

$$k_1 = 0$$
$$k_1 + k_2 + k_3 + k_4 = 1. \qquad (k_i \geq 0)$$

The solution is readily obtained without elaborate mathematical machinery. For, starting with any valued of $w = k_4/(k_3 + k_4)$, $0 < w < 1$, the value of k_4 can be continuously decreased to 0 while maintaining $k_2 + k_3 + k_4 = 1$ (by increasing $k_2 + k_3$ by an equal amount). Similarly starting with the same w one can continuously decrease k_3 to 0 while satisfying $k_2 + k_3 + k_4 = 1$. Thus the constrained w can take on any value in $[0, 1]$, hence showing that $\beta = [0, 1]$ is optimal for this case.

Case 2. $0 < p < 1$.
Since the objective function is linear fractional in form, we employ Theorem 0.53. According to it an equivalent problem is

$$\text{optimize } y_4$$

subject to

$$(1 - p)y_1 - py_3 = 0$$
$$y_1 + y_2 + y_3 + y_4 = t \qquad\qquad (1)$$
$$y_3 + y_4 = 1$$
$$t, y_1, y_2, y_3, y_4 \geq 0.$$

We shall eliminate the variables one by one until only y_4 remains. Starting with t, we use $t \geq 0$ and the second item in (1) to obtain $y_1 + y_2 + y_3 + y_4 \geq 0$. But this can be deleted since it is implied by $y_3 + y_4 = 1$. Then we have

$$(1 - p)y_1 - py_3 = 0$$
$$y_3 + y_4 = 1 \tag{2}$$
$$y_1, y_3, y_4 \geq 0.$$

Solving for y_3 in the second equation and substituting the value into $y_3 \geq 0$ and also into the first equation produces

$$1 - y_4 \geq 0, \quad \text{i.e., } 1 \geq y_4$$
$$(1 - p)y_1 - p(1 - y_4) = 0, \quad \text{i.e.,}$$
$$y_1 = \frac{p}{1 - p}(1 - y_4).$$

Substituting the value obtained for y_1 into $y_1 \geq 0$ yields $1 - y_4 \geq 0$. Combined with $y_4 \geq 0$ produces $1 \geq y_4 \geq 0$. Hence the optimal values are 0 and 1 and again our optimal β is $[0, 1]$.[6]

The result of the preceding theorem, that $[0, 1]$ is the best that one can do for the probability, isn't very exciting. Our next example provides something more interesting. It is a form of probabilistic modus ponens but having a probability conditional in place of the verity conditional of Theorem 4.51.

Theorem 5.43.

$$P(A_1 \mid A_2) = p, \ P(A_2) = q \ \vDash \ P(A_1) \in [pq, \ 1 - \overline{p}q],$$

($\overline{p}$ abbreviating $1 - p$) with the interval in the conclusion being optimal.

Proof. Since the value for $P(A_1 \mid A_2)$ is a fixed numerical value and not an arbitrary value in $[0, 1]$ we are tacitly assuming that $P(A_2) \neq 0$. Using k_1, k_2, k_3, k_4 for the probabilities of $A_1 A_2, A_1 \overline{A}_2, \overline{A}_1 A_2, \overline{A}_1 \overline{A}_2$, the premises are expressible as

$$\frac{k_1}{k_1 + k_3} = p, \qquad k_1 + k_3 = q.$$

Combining this with the usual standard probability conditions gives the linear problem

$$\text{optimize } w = k_1 + k_2$$

---

6. Note the discontinuity in β as a function of p. For any $p < 1$, β is $[0, 1]$. But when $p = 1$ (and $P(\overline{A}_1) \neq 0$), β is $[1, 1]$, by Theorem 5.36.

subject to

$$\bar{p}k_1 - pk_3 = 0$$
$$k_1 + k_3 = q \tag{3}$$
$$k_1 + k_2 + k_3 + k_4 = 1, \qquad k_1, k_2, k_3, k_4 \geq 0.$$

This is a linear programming problem with the objective function to be optimized a linear function. But by virtue of the presence of parameters in the specification of the feasible region, standard numerical techniques are not readily usable. We resort to Fourier elimination. The equation $w = k_1 + k_2$ is then adjoined to the system (3) and all variables but w are eliminated:

(i) Eliminating k_1 by using $k_1 \geq 0$ and $k_1 = q - k_3$ results in

$$q - k_3 \geq 0, \text{ i.e., } q \geq k_3$$
$$k_3 = \bar{p}q$$
$$k_2 + k_3 = \bar{q}$$
$$q - k_3 + k_4 = w, \quad k_2, k_3, k_4 \geq 0$$

(ii) Eliminating k_2 by using $k_2 \geq 0$ and $k_2 = \bar{q} - k_4$ produces

$$\bar{q} - k_4 \geq 0, \text{ i.e., } \bar{q} \geq k_4$$
$$k_3 = \bar{p}q$$
$$1 - k_3 - k_4 = w, \quad k_3, k_4 \geq 0.$$

(iii) Eliminating k_3 produces

$$\bar{q} \geq k_4$$
$$1 - \bar{p}q - k_4 = w, \quad k_4 \geq 0$$

as well as $q \geq 0$, $\bar{p}q \geq 0$.

(iv) Eliminating k_4 yields

$$\bar{q} \geq 1 - \bar{p}q - w$$
$$\bar{p}q \geq 0$$
$$1 - \bar{p}q - w \geq 0.$$

Then, from the first and third of these inequalities, we obtain the optimal bounds

$$pq \leq w \leq 1 - \bar{p}q.$$

Alternatively to this solution one can solve the three equations in (3) for k_1, k_2, k_3 and obtain

$$k_1 = q - \overline{p}q$$
$$k_2 = \overline{q} - k_4 \qquad (4)$$
$$k_3 = \overline{p}q$$

so that $w = k_1 + k_2 = 1 - \overline{p}q - k_4$. The smallest k_4 can be is 0, and the largest is $\overline{q}$ (see (iii)). Hence, once again, $pq \leq w \leq 1 - \overline{p}q$.

Part of the result, the lower bound value, was given in *Suppes 1966* (see (5) in §3.3 above). In connection with this modus ponens type inference Suppes wished to feature nearness to 1 of the probabilities and cited

$$\frac{P(A \mid B) \geq 1 - \epsilon}{P(B) \geq 1 - \epsilon}$$
$$\therefore P(A) \geq (1 - \epsilon)^2$$

The following result shows that Suppes' bound is the best possible if nothing is known about the component events:

Theorem 5.44. *For* $0 < \epsilon < 1$

$$P(A_1 \mid A_2) \in [1 - \epsilon, 1], \; P(A_2) \in [1 - \epsilon, 1] \; \vDash \; P(A_1) \in [(1 - \epsilon)^2, 1],$$

with the interval in the conclusion optimal.

Proof. As in the preceding theorem we express the problem in terms of the probabilities of the constituents $A_1 A_2$, $A_1 \overline{A}_2$, $\overline{A}_1 A_2$, $\overline{A}_1 \overline{A}_2$ obtaining:

$$\text{optimize } w = k_1 + k_2$$

subject to (setting $\overline{\epsilon} = 1 - \epsilon$)

$$\overline{\epsilon} \leq \frac{k_1}{k_1 + k_3} \leq 1, \quad \overline{\epsilon} \leq k_1 + k_3 \leq 1, \quad \sum_{i=1}^{4} k_i = 1, \quad k_i \geq 0.$$

The condition $k_1 + k_2 + k_3 + k_4 = 1$ can be replaced by $k_1 + k_2 + k_3 \leq 1$; for if we obtain an optimal feasible value (k_1^0, k_2^0, k_3^0) with the inequality, then we also have one for the equality case by setting $k_4^0 = 1 - (k_1^0 + k_2^0 + k_3^0)$.

The feasible region is then defined by the inequalities

$$
\begin{aligned}
k_1 + k_2 + k_3 &\leq 1 \\
-\epsilon k_1 \phantom{{}+k_2} + \overline{\epsilon} k_3 &\leq 0 \\
k_1 \phantom{{}+k_2} + k_3 &\leq 1 \\
-k_1 \phantom{{}+k_2} - k_3 &\leq -\overline{\epsilon}
\end{aligned}
\qquad (5)
$$

together with $k_i \geq 0$. Optimal values for the linear objective function $k_1 + k_2$ are obtained at vertices of the polyhedron specified by these inequalities using the method described in §0.5. The third inequality in (5), being implied by the first, can be deleted. This results in six inequalities specifying the polytope of feasible points—the three remaining in (5) together with $k_i \geq 0$ ($i = 1, 2, 3$). The following displays the five vertices of this polytope and the corresponding values of $k_1 + k_2$:

vertices	$k_1 + k_2$
$(\bar{\epsilon}^2, \epsilon, \epsilon\bar{\epsilon})$	$\bar{\epsilon}^2 + \epsilon$
$(\bar{\epsilon}, 0, \epsilon)$	$\bar{\epsilon}$
$(1, 0, 0)$	1
$(\bar{\epsilon}, \epsilon, 0)$	1
$(\bar{\epsilon}^2, 0, \epsilon\bar{\epsilon})$	$\bar{\epsilon}^2$

Thus the optimal bounds are $\bar{\epsilon}^2$ (which is less than $\bar{\epsilon}$) and 1.

In §2.6, after describing various attempts on how to combine two items of 'evidence' for a conclusion, we mentioned Boole's conception. It was formulated by him as one of solving the problem

$$\text{Given:} \quad P(C \mid A) = p, \quad P(C \mid B) = q$$
$$\text{find:} \quad P(C \mid AB).$$

Boole's solution (*1952*, 355–56) involved two additional parameters besides p and q, namely the probabilities of the two items of evidence, A and B. But if no additional information, probabilistic or logical, is given the probability of the conclusion on the conjoint evidence can be any value in the unit interval:

Theorem 5.45. For $0 < p \leq 1$, $0 < q \leq 1$,

$$P(A_3 \mid A_1) = p, \ P(A_3 \mid A_2) = q \ \vDash \ P(A_3 \mid A_1 A_2) \in [0, 1],$$

with the interval in the conclusion being optimal.

Proof. In an arbitrary probability model for S let $k_1, \ldots, k_8$ be, respectively, the probabilities of the constituents K_{111}, K_{110}, K_{101}, K_{100}, K_{011}, K_{010}, K_{001}, K_{000} (see §4.4). Our probability logical consequence becomes a linear programming problem, namely to

$$\text{optimize} \quad \frac{k_1}{k_1 + k_2}$$

subject to

$$k_1 + k_3 = p(k_1 + k_2 + k_3 + k_4)$$
$$k_1 + k_5 = q(k_1 + k_2 + k_5 + k_6),$$

or, introducing $a = (1-p)/p$, $b = (1-q)/q$, to

$$ak_1 - k_2 + ak_3 - k_4 = 0$$
$$bk_1 - k_2 + bk_5 - k_6 = 0.$$

To these two equations we need to add $\sum_{i=1}^{8} k_i = 1$ and $k_i \geq 0$, $1 \leq i \leq 8$. Since the objective function which is to be optimized is a linear fractional function we employ Theorem 0.53 (as for Case 2 of Theorem 5.42) to convert it to an equivalent program, namely, to

$$\text{optimize} \quad y_1$$

subject to

$$\sum_{i=1}^{8} y_i = t$$
$$ay_1 - y_2 + ay_3 - y_4 = 0$$
$$by_1 - y_2 + by_5 - y_6 = 0 \qquad\qquad (6)$$
$$y_1 + y_2 = 1 \qquad\qquad t, y_i \geq 0$$

We rewrite these equations in matrix form:

$$\begin{bmatrix} 1 & 1 & 1 & 1 & 1 & 1 & 1 & 1 & -1 \\ a & -1 & a & -1 & 0 & 0 & 0 & 0 & 0 \\ b & -1 & 0 & 0 & b & -1 & 0 & 0 & 0 \\ 1 & 1 & 0 & 0 & 0 & 0 & 0 & 0 & 0 \end{bmatrix} \begin{bmatrix} y_1 \\ y_2 \\ y_3 \\ y_4 \\ y_5 \\ y_6 \\ y_7 \\ y_8 \\ t \end{bmatrix} = \begin{bmatrix} 0 \\ 0 \\ 0 \\ 1 \end{bmatrix}.$$

Going over to the dual program, we need to optimize

$$0x_1 + 0x_2 + 0x_3 + 1x_4 \qquad\qquad (7)$$

subject to

$$\begin{bmatrix} 1 & a & b & 1 \\ 1 & -1 & -1 & 1 \\ 1 & a & 0 & 0 \\ 1 & -1 & 0 & 0 \\ 1 & 0 & b & 0 \\ 1 & 0 & -1 & 0 \\ 1 & 0 & 0 & 0 \\ 1 & 0 & 0 & 0 \\ -1 & 0 & 0 & 0 \end{bmatrix} \begin{bmatrix} x_1 \\ x_2 \\ x_3 \\ x_4 \end{bmatrix} \leq \text{ or } \geq \begin{bmatrix} 1 \\ 0 \\ 0 \\ 0 \\ 0 \\ 0 \\ 0 \\ 0 \\ 0 \end{bmatrix}$$

(x_i unrestricted) where '$\leq$' is used when minimizing, and '$\geq$' for maximizing, the objective function. It isn't necessary to find (algebraically) all the vertices of these polytopes. By inspection we see that $(0,0,0,0)$ is a vertex for the $\leq$-case and $(0,0,0,1)$ for the $\geq$-case. On substituting these coordinates in for the x's in (7) we obtain the values 0 and 1. Thus, irrespective of the values of a and b, the optimal bounds on $P(A_3 \mid A_1 A_2)$ are 0 and 1.[7]

A similar calculation shows also an 'empty' conclusion for the 'hypothetical syllogism' with probability conditionals in place of verity conditionals:

Theorem 5.455. *For* $0 < p \leq 1,\ 0 < q \leq 1$,

$$P(A_3 \mid A_2) = p,\ P(A_2 \mid A_1) = q \ \vDash \ P(A_3 \mid A_1) \in [0, 1],$$

with the interval in the conclusion being optimal.

The following is a simple example whose optimal bounds are more interesting than just 0 and 1. Given the probabilities of two events, what is the probability of their conjunction if it is known that one or the other (or both) has happened? We have

Theorem 5.46. *For* $p_1 p_2 \neq 0$,

$$P(A_1) = p_1,\ P(A_2) = p_2 \ \vDash \ P(A_1 A_2 \mid (A_1 \vee A_2)) \in [L, U],$$

where

$$L = \max(0,\ p_1 + p_2 - 1), \quad U = \min\left(\frac{p_1}{p_2},\ \frac{p_2}{p_1}\right),$$

and these bounds are optimal. Consistency condition on the parameters is that $0 \leq L \leq U \leq 1$.

Proof. Expressed in terms of the constituent probabilities the problem is to

$$\text{optimize}\ \ \frac{k_1}{k_1 + k_2 + k_3}$$

subject to

$$\begin{bmatrix} 1 & 1 & 0 & 0 \\ 1 & 0 & 1 & 0 \\ 1 & 1 & 1 & 1 \end{bmatrix} \begin{bmatrix} k_1 \\ k_2 \\ k_3 \\ k_4 \end{bmatrix} = \begin{bmatrix} p_1 \\ p_2 \\ 1 \end{bmatrix}, \qquad k_i \geq 0$$

By Theorem 0.53 an equivalent linear program is to

$$\text{optimize}\ y_1$$

---

7. The statement of this theorem in *Hailperin 1986*, 405, has an erroneous lower bound for $P(A_3 \mid A_1 A_2)$.

subject to

$$
\begin{bmatrix} 1 & 1 & 0 & 0 \\ 1 & 0 & 1 & 0 \\ 1 & 1 & 1 & 1 \\ 1 & 1 & 1 & 0 \end{bmatrix} \begin{bmatrix} y_1 \\ y_2 \\ y_3 \\ y_4 \end{bmatrix} = \begin{bmatrix} tp_1 \\ tp_2 \\ t \\ 1 \end{bmatrix}, \qquad t, y_i \geq 0
$$

Here the problem is small enough that Fourier elimination (§0.5) is practical—expecially since there are 4 equations to lighten the task of elimination. On eliminating all variables but y_1, one finds

$$
0 \leq y_1, \quad p_1 + p_2 - 1 \leq y_1
$$
$$
y_1 \leq \frac{p_1}{p_2}, \quad y_1 \leq \frac{p_2}{p_1},
$$

and hence the bounds L, U as stated.

Our final example of this section has the same premises as Theorem 5.45 but in place of $A_1 A_2$, the common part of the two items of 'evidence', it has $A_1 \vee A_2$. This represents a kind of pooling of the evidence—perhaps 'information' would be a better word here.

Theorem 5.47. *Let $0 < p, q < 1$. Then*

$$
P(A_3 \mid A_1) = p, \; P(A_3 \mid A_2) = q \; \vDash \; P(A_3 \mid (A_1 \vee A_2)) \in [L, U],
$$

where

$$
L = \frac{pq}{pq + \bar{p}q + \bar{q}p}, \quad U = \frac{\bar{p}q + \bar{q}p}{\bar{p}q + \bar{q}p + \bar{p}\bar{q}},
$$

and the interval $[L, U]$ is optimal.

Proof. The premises here are the same as for Theorem 5.45, but instead of $k_1/(k_1 + k_2)$ we have to optimize

$$
\frac{P(A_3(A_1 \vee A_2))}{P(A_1 \vee A_2)} = \frac{k_1 + k_3 + k_5}{k_1 + k_2 + k_3 + k_4 + k_5 + k_6}.
$$

Using Theorem 0.53 gives us a set of equations like (6) but with the first equation being $y_1 + y_2 + y_3 + y_4 + y_5 + y_6 = 1$, and with $y_1 + y_3 + y_5$ to be optimized. In matrix form the feasible region is (setting $a = (1-p)/p$, $b = (1-q)/q$)

$$
\begin{bmatrix} 1 & 1 & 1 & 1 & 1 & 1 & 1 & 1 & -1 \\ 1 & 1 & 1 & 1 & 1 & 1 & 0 & 0 & 0 \\ a & -1 & a & -1 & 0 & 0 & 0 & 0 & 0 \\ b & -1 & 0 & 0 & b & -1 & 0 & 0 & 0 \end{bmatrix} \begin{bmatrix} y_1 \\ y_2 \\ y_3 \\ y_4 \\ y_5 \\ y_6 \\ y_7 \\ y_8 \\ t \end{bmatrix} = \begin{bmatrix} 0 \\ 1 \\ 0 \\ 0 \end{bmatrix}. \quad t, y_i \geq 0
$$

For the corresponding dual program we need to optimize

$$0x_1 + 1x_2 + 0x_3 + 0x_4$$

subject to

$$
\begin{bmatrix}
1 & 1 & a & b \\
1 & 1 & -1 & -1 \\
1 & 1 & a & 0 \\
1 & 1 & -1 & 0 \\
1 & 1 & 0 & b \\
1 & 1 & 0 & -1 \\
1 & 0 & 0 & 0 \\
1 & 0 & 0 & 0 \\
-1 & 0 & 0 & 0
\end{bmatrix}
\begin{bmatrix}
x_1 \\
x_2 \\
x_3 \\
x_4
\end{bmatrix}
\leq \text{ or } \geq
\begin{bmatrix}
1 \\
0 \\
1 \\
0 \\
1 \\
0 \\
0 \\
0 \\
0
\end{bmatrix}
$$

(x_i unrestricted). A computer was programmed to do the algebraic task of finding the vertices of the respective two polytopes. For the minimizing case the polytope vertices are

$$(0, 0, 0, 0)$$
$$\left(0, 0, 0, \frac{1}{b}\right)$$
$$\left(0, 0, \frac{1}{a}, 0\right)$$
$$\left(0, \frac{1}{1+a+b}, \frac{1}{1+a+b}, \frac{1}{1+a+b}\right),$$

while for the maximizing case the vertices are

$$(0, 1, 0, 0)$$
$$(0, 1, 0, 1)$$
$$\left(0, \frac{a+b}{a+b+ab}, \frac{b}{a+b+ab}, \frac{a}{a+b+ab}\right).$$

Reintroducing p and q by replacing a and b ($a = (1-p)/p$, $b = (1-q)/q$) yields

$$L = \max\left(0, \frac{pq}{pq + \overline{p}q + \overline{q}p}\right) = \frac{pq}{pq + \overline{p}q + \overline{q}p}$$
$$U = \min\left(1, \frac{\overline{p}q + \overline{q}p}{\overline{p}q + \overline{q}p + \overline{p}\,\overline{q}}\right) = \frac{\overline{p}q + \overline{q}p}{\overline{p}q + \overline{q}p + \overline{p}\,\overline{q}}.$$

By algebra we find $L \leq U$ for any p and q in $(0,1)$.

The following corollary, which features nearness to 1 of the premise probabilities, will be referred to in §5.8.

Theorem 5.475. *Let* $0 < 1 - \delta < 1$. *Then*

$$P(A_3 \mid A_1) = 1 - \delta, \ P(A_3 \mid A_2) = 1 - \delta \ \vDash \ P(A_3 \mid A_1 \vee A_2) \in [L, U],$$

where

$$L = 1 - \frac{2\delta}{1 + \delta}, \qquad U = 1 - \frac{\delta}{2 - \delta}.$$

Thus $P(A_3 \mid A_1 \vee A_2)$ can be made arbitarily close to 1 by having δ sufficiently close to 0. This result is obtained by replacing both p and q in Theorem 5.47 with $1 - \delta$ and doing some simple algebra.

The following logical consequence result (Theorem 5.48) was suggested by an educational article in *Mathematics Magazine* (*Ortel and Rossi 1981*). By way of illustrating how students may be led into error the authors present an example (dressed up in a story) in which the probability of an event is obtained by using a directed graph with edges corresponding to transition probabilities. If it is assumed that the graph is a tree then only a single value is computed, whereas there are multiple paths in the graph leading to the desired event. Shorn of its colorful story and reduced to essentials, the example amounts to asking for $P(A_1 A_2 A_3)$ if given $P(A_1) = \frac{1}{2}$, $P(A_2 \mid A_1) = \frac{2}{3}$, $P(A_3 \mid A_2) = \frac{1}{2}$. It is shown that with the given data the probability is only determined to being less than $\frac{1}{3}$. (The question asked is considered to be "ill-posed".) Generalizing the given probabilities to p, q, and r we find by our methods

Theorem 5.48.

$$P(A_1) = p, \ P(A_2 \mid A_1) = q, \ P(A_3 \mid A_2) = r$$
$$\vDash \ P(A_1 A_2 A_3) \in [L, U],$$

where

$$L = \max(0, \ -\overline{p} + qr + \overline{p}\,\overline{q}r)$$
$$U = \min(qr, \ p - p\overline{q}r),$$

with consistency conditions on the parameters $0 \leq L \leq U \leq 1$.

§5.5. Boole's challenge problem

This is a problem proposed by Boole in his *1851c* as a test of the "received" theory versus his own (not yet then published) theory of probability. The problem and its subsequent history has been described in detail in our *1986*, §§6.2, 6.3. Here we will present our solution as a probability logical consequence and briefly remark on its relevance to risk assessment.

The following statement of the problem in taken from Boole's *Laws of Thought* (*1854*, 321).

> PROBLEM I—The probabilities of two causes A_1 and A_2 are c_1 and c_2 respectively. The probability that if the cause A_1 present itself, an event E will accompany it (whether as a consequence of the cause A_1 or not) is p_1, and the probability that if the cause A_2 present itself, that event E will accompany it, whether as a consequence of it or not, is p_2. Moreover, the event E cannot appear in the absence of both the causes A_1 and A_2.* [Boole's footnote will be quoted later.] Required the probability of the event E.

Stated in modern notation the problem is

$$\text{Given:} \quad P(A_1) = c_1, \; P(A_2) = c_2$$
$$P(E \mid A_1) = p_1, \; P(E \mid A_2) = p_2$$
$$E \to (A_1 \vee A_2) \quad (\text{or, } P(E\overline{A_1}\overline{A_2}) = 0)$$
$$\text{find:} \quad P(E).$$

Using his theory and the associated peculiar logical method Boole obtains (*1854*, 308–309, 322–25) 'the' value for $P(E)$. However, as explained in §2.5, this might be, and here actually is, but one value in a range of possible values for $P(E)$. A treatment using standard probability methods would run like this.

From $E \to (A_1 \vee A_2)$ we have $E \leftrightarrow (EA_1 \vee EA_2)$. Hence

$$P(E) = P(EA_1 \vee EA_2)$$
$$= P(EA_1) + P(EA_2) - P(EA_1 A_2)$$
$$= c_1 p_1 + c_2 p_2 - P(EA_1 A_2). \tag{1}$$

From (1), inasmuch as $P(EA_i) \geq P(EA_1 A_2)$, we see that $P(E)$ can't be less that either $c_1 p_1$ or $c_2 p_2$. Hence these are lower bounds. Since $P(EA_1 A_2) \geq 0$, $P(E)$ can't be more than $c_1 p_1 + c_2 p_2$. Hence this is an upper bound. But this upper bound can be improved upon with two additional possible bounds, as the following theorem shows.

Theorem 5.51.

$$P(A_1) = a_1, \ P(A_2) = a_2, \ P(A_3 \,|\, A_1) = p_1, \ P(A_3 \,|\, A_2) = p_2,$$
$$P(A_3\overline{A}_1\overline{A}_2) = 0 \vDash P(A_3) \in [L, U],$$

where ($\overline{a}$ abbreviating $1 - a$)

$$L = \max(a_1 p_1, a_2 p_2),$$
$$U = \min(a_1 p_1 + a_2 p_2, \ a_1 p_1 + \overline{a}_1, \ a_2 p_2 + \overline{a}_2).$$

The interval $[L, U]$ is optimal, and the consistency condition on the parameters is $0 \le L \le U \le 1$.

Proof. Although the premises involve conditional probabilities they can be replaced by equivalent ones in which there are only unconditional probabilities. Thus the pair of equations $\{P(A_1) = a_1, \ P(A_3 \,|\, A_1) = p_1\}$ is equivalent to the pair $\{P(A_1) = a_1, \ P(A_3 A_1) = a_1 p_1\}$. Similarly for $P(A_2)$ and $P(A_3 \,|\, A_2)$. Expressing the premise conditions in terms of the constituent probabilities (the k_i) and adjoining the standard probability condition $\sum_{i=1}^{8} k_i = 1$, leads to the matrix equation

$$
\begin{bmatrix}
1 & 1 & 1 & 1 & 0 & 0 & 0 & 0 \\
1 & 1 & 0 & 0 & 1 & 1 & 0 & 0 \\
1 & 0 & 1 & 0 & 0 & 0 & 0 & 0 \\
1 & 0 & 0 & 0 & 1 & 0 & 0 & 0 \\
0 & 0 & 0 & 0 & 0 & 0 & 1 & 0 \\
1 & 1 & 1 & 1 & 1 & 1 & 1 & 1
\end{bmatrix}
\begin{bmatrix}
k_1 \\ k_2 \\ k_3 \\ k_4 \\ k_5 \\ k_6 \\ k_7 \\ k_8
\end{bmatrix}
=
\begin{bmatrix}
a_1 \\ a_2 \\ a_1 p_1 \\ a_2 p_2 \\ 0 \\ 1
\end{bmatrix}. \tag{2}
$$

(The 1's and 0's in the first five rows of the leftmost matrix correspond, as usual, to the presence or absence of a constituent in the expansions of A_1, A_2, $A_1 A_3$, $A_2 A_3$, and $\overline{A}_1 \overline{A}_2 A_3$.) Equation (2) together with $k_i \ge 0$, $1 \le i \le 8$, prescribes the feasible region within which we wish to optimize

$$P(A_3) = k_1 + k_3 + k_5 + k_7$$
$$= \begin{bmatrix} 1 & 0 & 1 & 0 & 1 & 0 & 1 & 0 \end{bmatrix} \mathbf{k},$$

where $\mathbf{k} = \begin{bmatrix} k_1 & k_2 & k_3 & k_4 & k_5 & k_6 & k_7 & k_8 \end{bmatrix}^T$.

To solve this linear programming problem we go over to the equivalent dual problem. Then the parameters in the primal form only appear in the objective function (as coefficients) and the feasible region would be expressed in purely numerical terms (§0.5).

For the dual form we wish to find the minimum of

$$a_1 x_1 + a_2 x_2 + a_1 p_1 x_3 + a_2 p_2 x_4 + 0 x_5 + x_6 \tag{3}$$

subject to

$$\begin{bmatrix} 1 & 1 & 1 & 1 & 0 & 1 \\ 1 & 1 & 0 & 0 & 0 & 1 \\ 1 & 0 & 1 & 0 & 0 & 1 \\ 1 & 0 & 0 & 0 & 0 & 1 \\ 0 & 1 & 0 & 1 & 0 & 1 \\ 0 & 1 & 0 & 0 & 0 & 1 \\ 0 & 0 & 0 & 0 & 1 & 1 \\ 0 & 0 & 0 & 0 & 0 & 1 \end{bmatrix} \begin{bmatrix} x_1 \\ x_2 \\ x_3 \\ x_4 \\ x_5 \\ x_6 \end{bmatrix} \leq \begin{bmatrix} 1 \\ 0 \\ 1 \\ 0 \\ 1 \\ 0 \\ 1 \\ 0 \end{bmatrix} \tag{4}$$

(x_i unrestricted), and the maximum of (3) subject to the condition (4) with the inequality reversed. For the minimum case one finds the vertices of (4) to be

$$(0,\ 0,\ 1,\ 0,\ 0,\ 0) \quad \text{and} \quad (0,\ 0,\ 0,\ 1,\ 0,\ 0),$$

while for the maximum case the vertices are

$$(0,\ 0,\ 1,\ 1,\ 1,\ 0),\ (-1,\ 0,\ 1,\ 0,\ 0,\ 1),\ \text{and}\ (0,\ -1,\ 0,\ 1,\ 0,\ 1).$$

(That these are vertices is easily checked. Obtaining them, and being sure that one has them all, is a chore best left to a computer.) Substituting these values in for the $x_1, \ldots, x_6$ of (3) produces the L and U of the theorem.[8]

Boole's problem readily generalizes—and this is the way he proposed it in his *1851c*—to any number $A_1, \ldots, A_n$ of 'causes' with $E \rightarrow (A_1 \vee \cdots \vee A_n)$.

Boole's footnote to the above stated form of the problem describes in picturesque language a situation which could give rise to the question (*1854*, 321):

> The mode in which such data as the above might be furnished by experience is easily conceivable. Opposite the window of the room in which I write is a field, liable to be overflowed from two causes, distinct, but capable of being combined, viz., floods from the upper sources of the River Lee, and tides from the ocean. Suppose that observations made on N separate occasions have yielded the following results: On A occasions the river was swollen by freshets, and on P of those occasions it was inundated, whether from this cause or not. On B occasions the river was swollen by the tide, and on Q

---

8. These bounds were obtained by Boole using Fourier elimination. See *Boole 1854a*, 284.

of those occasions it was inundated, whether from this cause or not. Supposing, then, that the field cannot be inundated in the absence of *both* causes above mentioned, let it be required to determine the total probability of its inundation.

Here the elements a, b, p, q [c_1, c_2, p_1, p_2] of the general problem represent the ratios

$$\frac{A}{N}, \frac{P}{A}, \frac{B}{N}, \frac{Q}{B},$$

or rather the values to which those ratios approach, as the value of N is indefinitely increased.

Flooding of a field is usually not a matter of great concern, but the event whose probability is sought might equally well be an event whose occurrence would have serious consequences. Estimating the risk is then important. Note that the event E doesn't necessarily happen if A_1 or A_2 does but only conditionally, with given (or estimated) probability. Boole's story shows, moreover, that he is indeed using conditional probabilities for the probability of E being 'caused by', say, A_1. For he uses as an estimate the ratio P/A, where P is the number of occasions there was inundation and swollen freshets, and A is the number of occasions of swollen freshets.

There is a problem closely related to the challenge problem with which it can be compared. Boole includes the hypothesis that the event in question cannot happen in the absence of either cause. In the conclusion, then

$$P(A_3) = P(A_3(A_1 \vee A_2)).$$

Hence Boole's problem (in terms of the story) is asking for the chances of flooding and either swollen freshets or tides. On the other hand one might be interested in knowing the chances of flooding, given that there were either swollen freshets or tides, i.e., instead of Boole's $P(A_3(A_1 \vee A_2))$ with the premise $P(A_3\overline{A}_1\overline{A}_2) = 0$, the conditional probability

$$P(A_3 \,|\, (A_1 \vee A_2)).$$

The following theorem gives bounds for this conditional probability. To remove uninteresting special cases we shall assume that the parameters are strictly between 0 and 1.

Theorem 5.52. *Let* $0 < a_1, a_2, p_1, p_2 < 1$. *Then*

$$P(A_1) = a_1, \ P(A_2) = a_2, \ P(A_3 \,|\, A_1) = p_1, \ P(A_3 \,|\, A_2) = p_2$$
$$\vDash P(A_3 \,|\, (A_1 \vee A_2)) \in [L, U],$$

where

$$L = \max(\frac{a_2 p_2}{a_2 + a_1 \bar{p}_1}, \frac{a_1 p_1}{a_1 + a_2 \bar{p}_2}, a_1 p_1, a_2 p_2)$$

$$U = \min(\frac{a_1 p_1 + a_2 p_2}{a_1 p_1 + a_2}, \frac{a_2 p_2 + a_1 p_1}{a_2 p_2 + a_1}, \bar{a}_1 + a_1 p_1, \bar{a}_2 + a_2 p_2).$$

The interval $[L, U]$ is optimal, and the consistency condition on the parameters is $(0 \le)L \le U(\le 1)$.

Proof. As in Theorem 5.51, corresponding to (2), the premise conditions lead to the matrix equation

$$\begin{bmatrix} 1 & 1 & 1 & 1 & 1 & 1 & 1 & 1 \\ 1 & 1 & 1 & 1 & 0 & 0 & 0 & 0 \\ 1 & 1 & 0 & 0 & 1 & 1 & 0 & 0 \\ 1 & 0 & 1 & 0 & 0 & 0 & 0 & 0 \\ 1 & 0 & 0 & 0 & 1 & 0 & 0 & 0 \end{bmatrix} \begin{bmatrix} k_1 \\ k_2 \\ k_3 \\ k_4 \\ k_5 \\ k_6 \\ k_7 \\ k_8 \end{bmatrix} = \begin{bmatrix} 1 \\ a_1 \\ a_2 \\ a_1 p_1 \\ a_2 p_2 \end{bmatrix}. \tag{5}$$

The objective function to be optimized is the linear fractional function

$$P(A_3 \mid (A_1 \vee A_2)) = \frac{k_1 + k_3 + k_5}{\sum_{i=1}^{6} k_i}.$$

By Theorem 0.53 an equivalent linear programming problem is to

$$\text{optimize} \quad y_1 + y_3 + y_5$$

subject to

$$\begin{bmatrix} 1 & 1 & 1 & 1 & 1 & 1 & 1 & 1 & -1 \\ 1 & 1 & 1 & 1 & 0 & 0 & 0 & 0 & -a_1 \\ 1 & 1 & 0 & 0 & 1 & 1 & 0 & 0 & -a_2 \\ 1 & 0 & 1 & 0 & 0 & 0 & 0 & 0 & -a_1 p_1 \\ 1 & 0 & 0 & 0 & 1 & 0 & 0 & 0 & -a_2 p_2 \\ 1 & 1 & 1 & 1 & 1 & 1 & 0 & 0 & 0 \end{bmatrix} \begin{bmatrix} y_1 \\ y_2 \\ y_3 \\ y_4 \\ y_5 \\ y_6 \\ y_7 \\ y_8 \\ t \end{bmatrix} = \begin{bmatrix} 0 \\ 0 \\ 0 \\ 0 \\ 0 \\ 1 \end{bmatrix}. \tag{6}$$

Going over to the dual form requires optimizing

$$0x_1 + 0x_2 + 0x_3 + 0x_4 + 0x_5 + 1x_6$$

subject to

$$
\begin{bmatrix}
1 & 1 & 1 & 1 & 1 & 1 \\
1 & 1 & 1 & 0 & 0 & 1 \\
1 & 1 & 0 & 1 & 0 & 1 \\
1 & 1 & 0 & 0 & 0 & 1 \\
1 & 0 & 1 & 0 & 1 & 1 \\
1 & 0 & 1 & 0 & 0 & 1 \\
1 & 0 & 0 & 0 & 0 & 0 \\
-1 & -a_1 & -a_2 & -a_1 p_1 & -a_2 p_2 & 0
\end{bmatrix}
\begin{bmatrix}
x_1 \\ x_2 \\ x_3 \\ x_4 \\ x_5 \\ x_6
\end{bmatrix}
\le \text{ or } \ge
\begin{bmatrix}
1 \\ 0 \\ 1 \\ 0 \\ 1 \\ 0 \\ 0 \\ 0
\end{bmatrix}
\tag{7}
$$

(x_i unrestricted). With the aid of a computer all vertices for the two polytopes (for $\le$ and for $\ge$) subject to the conditions of the hypothesis that $0 < a_1, a_2, p_1, p_2 < 1$ were found resulting in the L and U as stated in the theorem.

We note that these bounds just obtained, the L and U of Theorem 5.52, are included between those of Theorem 5.51, since

$$\max(a_1 p_1, a_2 p_2) \le L, \text{ and}$$
$$U \le \min(a_1 p_1 + a_2 p_2, a_1 p_1 + \bar{a}_1, a_2 p_2 + \bar{a}_2).$$

§5.6. Suppositional logic

The literature on the nature of the conditional in logic is an unusually extensive one. Beginning in ancient times we find contention as to its meaning: Cicero in his *Academica Priora* mockingly reports on the great dispute among the Stoic philosophers on how to judge the truth or falsity of a conditional. In the fourteenth century there were numerous tracts on *consequentiae*, a topic closely related to the conditional. Leibniz's baccalaureate thesis of 1665 was entitled *De conditionibus*. In this century, C. I. Lewis (*1918*, and earlier papers) introduced a notion of 'strict implication' as a version of the conditional which would not be subject to such 'paradoxical' properties as that of a false proposition implying any proposition. A recently appearing collection of essays, *Conditionals* (*Jackson 1991*), lists in its selected bibliography nearly a dozen books, with dates in the 1970's and 1980's, on the topic. There is no one generally accepted meaning for the 'if, then' as used in ordinary discourse, though many have been proposed. The probability conditional notion introduced by us, when reduced to the true-false level, suggests another.

Earlier, in §4.3, we noted that probability models in which one of the k_i is 1 and the remaining are 0 coincide with verity models. In such models

$\neg\phi$, $\phi \wedge \psi$, $\phi \vee \psi$ would have (probability) semantic tables identical with the usual truth tables for negation, conjunction and alternation of verity logic. And, since in such models the only probability values are 0 and 1, the unit interval reduces to just the two-element set $\{0, 1\}$. How does the conditional probability function P_M^* fare in such models? If ψ and ϕ can only take on the values 0 and 1, then the associated value of $\psi \,|\, \phi$ (i.e. $P_M^*(\psi \,|\, \phi)$) can only be 0, 1, or an indefinite (unspecified) value in $\{0, 1\}$. Calling this indefinite value now u, what results in such models can be displayed in a table:

ϕ	ψ	$\psi \,|\, \phi$
1	1	1
1	0	0
0	1	u
0	0	u

(1)

This suggests basing conditional probability logic on a 3-valued logic, parallel to the way that (unconditional) probability logic is based on 2-valued verity logic. For this logic we shall use the don't-care logic of §0.3.

We shall refer to the statement connective defined by (1) in this logic as the *suppositional*. This term recommends itself to us on historical grounds since 'supposition' (for the antecedent) has been used with the related notion of conditional probability as least as early as Bayes. (See his Prop. 3 in §1.6 above.) In its non-probabilistic sense it is a new notion and we shall take the opportunity to replace the ' $|$ ' which we have been using by the symbol '$\dashv$', already introduced above by us in §0.3 for this notion.[9] There $\dashv$ was defined in terms of the unary don't-care connective $\triangle$. Reciprocally, $\triangle$ can be defined in terms of $\dashv$ when given by the semantic table (1) extended with five more lines each being assigned the value u. Thus, as can be readily verified, we have ('=' standing for 'has the same semantic table as')

$$\Psi \dashv \Phi \;=\; \triangle\overline{\Phi} \vee \Phi\Psi \tag{2}$$

$$\triangle\Phi = \Phi \dashv \overline{\Phi}. \tag{3}$$

We believe the suppositional (with components that are in $\mathcal{S}$) adequately represents an 'if, then' construction when there is no interest in its truth value for those cases when its antecedent is false. As an example, consider a poll in which respondents are to express their opinion on the proposition 'Taxes will go up if the Democrats are elected'. Only a wiseacre logician

---

9. The vertical portion of '$\dashv$' is to remind us of its connection with conditional probability. In general it is desirable to represent a non-commutative notion by an asymmetrical (left-to-right) symbol. This applies as well to conditional probability; but since ' $|$ ' is so ingrained in the probability literature—having been in use since the 1930's—it would not be sensible to advocate a change.

would respond "Since the Democrats won't be elected the proposition, having a false antecedent, is automatically true without my having to consider the matter any further." Clearly it is not the verity (material, Philonean) conditional which the formulator of the poll had in mind but one in which one is to *suppose* that *The Democrats are elected* is the case. In general, if one wishes to rule out vacuous implications—the kind that our wiseacre logician used—then $\psi \dashv \phi$ rather than $\phi \to \psi$ should be chosen. In which case it is then u-validity, introduced in §0.3, and not validity, which justifies correct inferences.

Since our interest is logic, not combinational circuits, and since the suppositional $\dashv$ plays a central role in distinguishing this logic from others, we shall now drop the name 'don't-care' and refer to it as *suppositional logic*.

Here is an assortment of some easily established results.[10]

Theorem 5.60.

$$\psi \dashv \phi = \psi\phi \dashv \phi$$

where '$=$' means 'has the same semantic table as'.

Theorem 5.61. *If $\psi \dashv \phi$ is a u-valid, then $\phi \to \psi$ is valid.*

Proof. Any assignment of 0 or 1 to atomic sentence letters in $\psi \dashv \phi$ which results in $\psi \dashv \phi$ having a value u or 1, will produce the value 1 for $\phi \to \psi$.

Theorem 5.62. *A formula Φ of S^u which is a sentence of S (and hence has no occurrence of '$\dashv$') is u-valid if and only if it is valid.*

Theorem 5.63. *The following are all correct suppositional inference forms, i.e., if the fomulas above the line are u-valid, then so is the formula below the line:*

$$\frac{\psi \dashv \phi \quad \phi}{\psi} \quad \text{(modus ponens)} \tag{4a}$$

$$\frac{\psi \dashv \phi \quad \phi \dashv \chi}{\psi \dashv \chi} \quad \text{(hypothetical syllogism)} \tag{4b}$$

$$\frac{\psi \dashv \phi \quad \psi \dashv \neg\phi}{\psi} \quad \text{(proof by cases)} \tag{4c}$$

$$\frac{\psi \dashv \phi \quad \neg\psi \dashv \phi}{\neg\phi} \quad \text{(reductio ad absurdum)} \tag{4d}$$

Proof. Consider first (4a). If $\psi \dashv \phi$ and ϕ are u-valid then by Theorem 5.61 $\phi \to \psi$ is valid, and by Theorem 5.62 ϕ is valid. Hence by modus

---

10. Recall that ϕ, ψ, χ, ... range over elements of S and that in any (0,1)-model (the only kind in suppositional logic) their values can only be 0 or 1.

ponens of verity logic ψ is valid; being an element of S it is then u-valid. Proofs of (4c) and (4d) are similar. As for (4b), suppose $\psi \dashv \phi$ and $\phi \dashv \chi$ are u-valid. The conclusion formula $\psi \dashv \chi$ can fail to be u-valid only if either

   (i) in some model $U(\psi) = 0$ and $U(\chi) = 1$ (so that $\psi \dashv \chi$ is 0) or else
   (ii) $U(\chi) = 0$ in all models (so that $U(\psi \dashv \chi)$ has no 1 value).

If it is (i) that is the case, then in that model $\psi \dashv \phi$ and $\phi \dashv \chi$ would, respectively, reduce to $0 \dashv \phi^*$ and $\phi^* \dashv 1$, with ϕ^* being 0 or 1. But whatever the value of ϕ^* one of the two has the value 0. If (ii) is the case then $\phi \dashv \chi$ could not be u-valid as its value would have to be u in all models.

Contraposition, in the form

$$\frac{\phi \dashv \psi}{\neg\psi \dashv \neg\phi}, \tag{5}$$

fails to be a correct inference as the example $\phi = \psi = A_1 \vee \overline{A}_1$ shows. Such suppositionals as[11]

   (i) $A_1 \dashv A_1$ (ii) $A_1 \dashv A_1 A_2$ (iii) $A_1 \vee A_2 \dashv A_1$ (6)

are clearly u-valid, but not

$$\phi \dashv A_1 \overline{A}_1 \tag{7}$$

for any ϕ whatever. Likewise, if

$$\psi \dashv \phi_1 \phi_2 \ldots \phi_n \tag{8}$$

is u-valid then $\phi_1 \phi_2 \ldots \phi_n$ can't be inconsistent. Moreover, when (8) is u-valid its value in a model is 1 if and only if both ψ and $\phi_1 \phi_2 \ldots \phi_n$ have the value 1.

Unlike 2-valued sentential validity, u-validity is not preserved under substitution of a formula for an atomic sentence letter. Thus if the A_1 of (i) in (6) is replaced by $A_1 \overline{A}_1$ the result, namely $A_1 \overline{A}_1 \dashv A_1 \overline{A}_1$, is not u-valid since all of its values are u. Clearly replacement of a formula by a semantically equivalent one does preserve u-validity.

As is the case with validity in sentential verity logic, *there is a decision procedure for u-validity*. Any Φ of S^u is constructed from a finite subset of the atomic sentence letters $A_1, \ldots, A_n, \ldots$. If this set has n members then there are only 2^n different $(0,1)$-models which have to be inspected

---

11. Conventions as to omission of parentheses for $\dashv$ follow those for $\rightarrow$. Thus '$A_1 \dashv A_1 A_2$' means '$A_1 \dashv (A_1 A_2)$' and '$A_1 \vee A_2 \dashv A_1$' means '$(A_1 \vee A_2) \dashv A_1$'.

to determine whether or not Φ has the value 0 in some model, and if not, whether it has at least one model in which it has the value 1.

Also, as is the case for verity logic, there is a normal form for a formula Φ of $\mathcal{S}^u$ in terms of constituents $K_1, \ldots, K_{2^n}$ (assuming Φ to be constructed from $A_1, \ldots, A_n$). As already seen in §0.3, this suppositional normal form is

$$K_{i_1} \vee \cdots \vee K_{i_r} \dashv \overline{K}_{j_1}\overline{K}_{j_2}\ldots\overline{K}_{j_s}, \tag{9}$$

where $i_1, \ldots, i_r$ correspond to the rows in Φ's semantic table for which the value is 1, and $j_1, \ldots j_s$ the rows for which the value is u. If $r = 0$ then (9) is to be interpreted as

$$\triangle K_{j_1} \vee \cdots \vee \triangle K_{j_s}.$$

If $s = 0$ then (9) is to be interpreted as

$$K_{i_1} \vee \cdots \vee K_{i_r}.$$

And if both $r = 0$ and $s = 0$ then (9) is to be interpreted as $A_1\overline{A}_1$.

Theorem 5.635. *For any ϕ_1, ϕ_2, ψ_1, ψ_2 in $\mathcal{S}$,*

$$\psi_1 \dashv \phi_1 = \psi_2 \dashv \phi_2 \text{ if and only if } \psi_1\phi_1 = \psi_2\phi_2 \text{ and } \phi_1 = \phi_2.$$

Proof. From (1), the semantic table for $\psi \dashv \phi$, it is seen that its value is 1 if and only if $\psi\phi$ is 1 and its value is u if and only if ϕ is 0. Hence in the normal form (9) $K_{i_1} \vee \cdots \vee K_{i_r}$ is the normal form for $\psi\phi$ and the $K_{j_1} \vee \cdots \vee K_{j_s}$ is that for ϕ. The result then readily follows on considering the normal forms for $\psi_1 \dashv \phi_1$ and $\psi_2 \dashv \phi_2$.

Compounds constructed from suppositionals and connectives $\neg$, $\wedge$, $\vee$ have, by (4) of §0.3, a uniquely associated semantic table with 2^n rows (n being the number of atomic sentence letters) and values which are 1, 0, or u. Thus any such compound has the same semantic table as a suppositional, namely that obtained by selecting the appropriate K's to construct one of the form (9). But of more interest and significance is the following result. It shows that for such a suppositional its antecedent and consequent are expressible as ($\neg$, $\wedge$, $\vee$)-compounds of the antecedents and consequents of the component suppositionals. (As before, '=' means 'has the same semantic table as'.) The alternative forms for (12) and (13) are obtained by simple logical transformations and use of Theorem 5.60.

Theorem 5.64. *For ϕ, ψ, ϕ_1, ϕ_2, ψ_1, ψ_2 elements of $\mathcal{S}$,*

$$\phi = \phi \dashv A_1 \vee \overline{A_1} \tag{10}$$

$$\neg(\psi \dashv \phi) = \overline{\psi} \dashv \phi \tag{11}$$

$$(\psi_1 \dashv \phi_1) \wedge (\psi_2 \dashv \phi_2) = \psi_1 \psi_2 \dashv (\phi_1 \vee \phi_2 \overline{\psi_2})(\phi_2 \vee \phi_1 \overline{\psi_1}) \tag{12}$$

$$= \psi_1 \psi_2 \dashv \phi_1 \phi_2 \vee \phi_1 \overline{\psi_1} \vee \phi_2 \overline{\psi_2}$$

$$= \phi_1 \phi_2 \psi_1 \psi_2 \dashv \phi_1 \phi_2 \vee \phi_1 \overline{\psi_1} \vee \phi_2 \overline{\psi_2}$$

$$(\psi_1 \dashv \phi_1) \vee (\psi_2 \dashv \phi_2) = \psi_1 \vee \psi_2 \dashv (\phi_1 \vee \phi_2 \psi_2)(\phi_2 \vee \phi_1 \psi_1)$$

$$\tag{13}$$

$$= \psi_1 \vee \psi_2 \dashv \phi_1 \phi_2 \vee \phi_1 \psi_1 \vee \phi_2 \psi_2$$

$$= \phi_1 \psi_1 \vee \phi_2 \psi_2 \dashv \phi_1 \phi_2 \vee \phi_1 \psi_1 \vee \phi_2 \psi_2.$$

These semantic equivalences can be verified by constructing semantic tables—those for (12) and (13) would be somewhat of a chore since there would be $2^4 = 16$ rows in the tables. However there is a rather pretty 'algebraic' way which we would like to present. This replaces a suppositional $\psi \dashv \phi$ by $\triangle\overline{\phi} \vee \phi\psi$ ((2) above) and uses replacement of formulas by semantically equivalent ones. (These equivalences, i.e., De Morgan, distributivity, absorption, etc., are readily justified when they also include occurrences of $\triangle$.)

Proof of (11).

$$\neg(\psi \dashv \phi) = \neg(\triangle\overline{\phi} \vee \phi\psi) \qquad \text{(by (2))}$$

$$= \neg(\triangle\overline{\phi}) \wedge (\overline{\phi} \vee \overline{\psi})$$

$$\text{(by De Morgan)}$$

$$= (\triangle \vee \phi) \wedge (\overline{\phi} \vee \overline{\psi})$$

$$\text{(by De Morgan)}$$

$$= (\triangle\overline{\phi} \vee \phi)(\overline{\phi} \vee \overline{\psi})$$

$$= \triangle\overline{\phi}(\overline{\phi} \vee \overline{\psi}) \vee \phi(\overline{\phi} \vee \overline{\psi})$$

$$\text{(distributivity)}$$

$$= \triangle\overline{\phi} \vee \phi\overline{\psi} \qquad \text{(absorption)}$$

$$= \overline{\psi} \dashv \phi. \qquad \text{(by (2))}$$

Proof of (12).

$$(\psi_1 \dashv \phi_1) \wedge (\psi_2 \dashv \phi_2) = (\triangle\overline{\phi}_1 \vee \phi_1\psi_1)(\triangle\overline{\phi}_2 \vee \phi_2\psi_2)$$

$$= \triangle(\overline{\phi}_1\phi_2\psi_2 \vee \phi_1\overline{\phi}_2\psi_1 \vee \overline{\phi}_1\overline{\phi}_2) \vee \phi_1\phi_2\psi_1\psi_2,$$

by distributivity and (5) of §0.3. Letting

$$\overline{A} = \overline{\phi}_1 \phi_2 \psi_2 \vee \phi_1 \overline{\phi}_2 \psi_1 \vee \overline{\phi}_1 \overline{\phi}_2$$

we find that

$$A = (\phi_1 \vee \phi_2 \overline{\psi}_2)(\phi_2 \vee \phi_1 \overline{\psi}_1)$$

and

$$\psi_1 \psi_2 A = \phi_1 \phi_2 \psi_1 \psi_2,$$

so that

$$(\psi_1 \dashv \phi_\mathrm{I}) \wedge (\psi_2 \dashv \phi_2) = \triangle \overline{A} \vee \psi_1 \psi_2 A$$
$$= \psi_1 \psi_2 \ \dashv \ (\phi_1 \vee \phi_2 \overline{\psi}_2)(\phi_2 \vee \phi_1 \overline{\psi}_1).$$

Proof of (13).

$$(\psi_1 \dashv \phi_1) \vee (\psi_2 \dashv \phi_2) = \neg(\overline{\psi_1 \dashv \phi_1} \wedge \overline{\psi_2 \dashv \phi_2})$$
$$= \neg((\overline{\psi}_1 \dashv \phi_1) \wedge (\overline{\psi}_2 \dashv \phi_2)) \qquad \text{(by (11)}$$
$$= \neg(\overline{\psi}_1 \overline{\psi}_2 \ \dashv \ (\phi_1 \vee \phi_2 \psi_2)(\phi_2 \vee \phi_1 \psi_1))$$
$$\text{(by (12)}$$
$$= \psi_1 \vee \psi_2 \ \dashv \ (\phi_1 \vee \phi_2 \psi_2)(\phi_2 \vee \phi_1 \psi_1)$$
$$\text{(by (11))}$$

Until now all $\mathcal{S}^u$ formulas in theorems were such that only elements of $\mathcal{S}$ occurred in antecedent and consequent of suppositionals. In the following results this is no longer the case.

Theorem 5.65. For $\phi_1, \phi_2, \psi_1, \psi_2 \in \mathcal{S}$,

$$(\psi_1 \dashv \phi_1) \dashv (\psi_2 \dashv \phi_2) \ = \ \psi_1 \dashv \phi_1 \phi_2 \psi_2.$$

Proof. Again, rather than appealing to semantic tables we use algebraic techniques.

$$(\psi_1 \dashv \phi_1) \dashv (\psi_2 \dashv \phi_2)$$
$$= \triangle(\triangle \overline{\phi}_2 \vee \phi_2 \overline{\psi}_2) \vee (\triangle \overline{\phi}_1 \vee \phi_1 \psi_1)(\triangle \overline{\phi}_2 \vee \phi_2 \psi_2)$$
$$= \triangle(\overline{\phi}_2 \vee \phi_2 \overline{\psi}_2) \vee \triangle(\overline{\phi}_1 \phi_2 \psi_2 \vee \phi_1 \overline{\phi}_2 \psi_1 \vee \overline{\phi}_1 \overline{\phi}_2) \vee \phi_1 \phi_2 \psi_1 \psi_2$$
$$= \phi_1 \phi_2 \psi_1 \psi_2 \vee \triangle \overline{\phi_1 \phi_2 \psi_2}$$
$$= \psi_1 \dashv \phi_1 \phi_2 \psi_2.$$

Corollary 5.66.

(a) $(\psi \dashv \phi) \dashv \sigma = \psi \dashv \phi\sigma$

(b) $\psi \dashv (\phi \dashv \sigma) = \psi \dashv \phi\sigma$

(c) $(\psi \dashv \phi) \dashv (\psi \dashv \phi) = \psi\phi \dashv \psi\phi.$

Proof. For (a) use (10) of Theorem 5.64 to replace σ by $\sigma \dashv A_1 \vee \overline{A}_1$ and then apply Theorem 5.65. Proof of (b) is similar. For (c), applying Theorem 5.65 to the left-hand side we have $\psi \dashv \phi\psi$ which, by Theorem 5.60, is equivalent to $\phi\psi \dashv \phi\psi$.

In the light of Theorems 5.64 and 5.65 we may assume, without loss of generality, that the set of formulas of suppositional logic—for which set we still will use $\mathcal{S}^u$—to be all sentences of $\mathcal{S}$ together with all formulas of the form $\phi \dashv \psi$ for $\phi, \psi \in \mathcal{S}$—since any constructed using $\neg, \wedge, \vee,$ and $\dashv$ is semantically equivalent to one of this form.

§5.7. Probability on conditional events

An offhand attempt at introducing a notion of conditional event dates back to *Peirce 1867a*. We have made a few remarks about this in §2.7. Though the notion does not occur in Boole's work, it can be read into his treatment of conditional probability involved in his method of solving the 'general' probability problem. We shall describe this in our next section. Bolzano, Keynes, and Reichenbach viewed probability in terms of two propositional components (§§2.1, 3.1, 3.2). However neither Bolzano nor Keynes conceived of a fusion of the two propositions into a propositional concept; and, while Reichenbach did (in terms of frequency), we have deemed his theory to be inadequate (§3.2 above). The contemporary axiomatic set-theoretic approach in general use by probabilists does not have a notion of conditional event, and simply works with quotients of probabilities of unconditional events. However in the latter half of this century, principally among workers in the fields of approximate reasoning, artificial intelligence and the like, the topic has become active. In this milieu there have been proposals to provide an entity, i.e., an 'event', associated with conditional probability, such entities being freely combinable with connectives, just as unconditional events are. We shall be describing these proposals—all of which use an axiomatic algebraic approach—in our next section. Here we present our contribution to the topic in the form of a conditional probability *logic*.

The construction of this logic, which we will base on suppositional logic, parallels that of our construction of (unconditional) probability logic from verity logic. Since this latter construction was carried out piecemeal over a number of sections, it will be helpful to have it all together in a brief summary.

On our view (stated in §4.1) a logic is specified by a triple consisting of a syntax, a semantics and a notion of (logical) consequence. In the case of (unconditional) probability logic the syntax is the same as that of verity logic—the set of sentences denoted by S being common to both. The semantics of probability logic is, however, an extension of that of verity logic. Models for verity logic can be obtained either

(i) by assignment of a semantic value (0 or 1) to atomic sentences, or
(ii) by assignment of values (only one of which can be 1) to constituents (Theorem 0.42).

For probability logic the latter approach is the one we chose to generalize: a model M is an assignment of values (from $[0,1]$, summing to 1) to constituents $K_1, \ldots, K_{2^n}$ ($n = 1, 2, \ldots$), the assignment being consistent from one n to the next (§4.4). The model M is then extended to a probability function P_M over all elements of S. Meanings for the connectives $\neg, \wedge, \vee$ are thereby determined for probability logic; however not in the simple way as in verity logic: the P_M value of a compound sentence is not (necessarily) a function of the P_M values of its immediate sub-components but of the compound's constituents. For any $\phi \in S$, $P_M(\phi) = \sum_\phi P_M(K_i)$, the indicated sum taken over those K_i present in ϕ's disjunctive normal form. Finally, the third component of the triple, logical consequence, is obtained by generalizing

$$V(\phi_1) \in \alpha_1, \ldots, V(\phi_m) \in \alpha_m \ \vDash \ V(\psi) \in \beta,$$

where V is an arbitrary verity function on S, and $\alpha_1, \ldots, \alpha_m, \beta$ are non-empty subsets of $\{0,1\}$, to

$$P(\phi_1) \in \alpha_1, \ldots, P(\phi_m) \in \alpha_m \ \vDash \ P(\psi) \in \beta,$$

i.e., by replacing V by P, an arbitrary probability function on S, and taking $\alpha_1, \ldots, \alpha_m, \beta$ to be subsets of $[0,1]$.

The parallel construction of *conditional probability logic*, on the basis of suppositional logic, is as follows. Its syntax is to be the same as that of suppositional logic (§5.6). Elements of S^u will represent *conditional events*.[12]

---

12. We think that the term 'suppositional event' would be more fitting for the concept.

As for the semantics of conditional probability logic, suppositional logic's semantic values, consisting of $\{0,1\}$ together with u, is extended to $[0,1]$ together with c. (If $\{0,1\}$ is replaced by $[0,1]$ it is 'natural' to replace $u = \nu x(x \in \{0,1\})$ by $c = \nu x(x \in [0,1])$.) Models for conditional probability logic will be the same as models for unconditional probability logic (§4.4). Accordingly, for any such model M there is a probability function $P_M : S \to [0,1]$ having a value $P_M(\phi)$ for each $\phi \in S$. Then P_M is extended to a conditional probability function P_M^* defined on formulas of S^u:

DEFINITION OF A CONDITIONAL PROBABILITY FUNCTION

Let Φ be an element of S^u and let

$$K_{i_1} \vee \cdots \vee K_{i_r} \dashv \overline{K}_{j_1} \overline{K}_{j_2} \ldots \overline{K}_{j_s}$$

be its suppositional normal form. Set

$$A = \overline{\overline{K}_{j_1} \overline{K}_{j_2} \ldots \overline{K}_{j_s}} = K_{j_1} \vee K_{j_2} \cdots \vee K_{j_s}$$
$$C = K_{i_1} \vee K_{i_2} \vee \cdots \vee K_{i_r},$$

and let $\sum_A k_i$ be the sum of $P_M(K_i)$ for those K_i present in A and $\sum_{AC} k_i$ be the sum of $P_M(K_i)$ for those present in AC. Then

$$P_M^*(\Phi) = \begin{cases} \sum_{AC} k_i / \sum_A k_i, & \text{if } \sum_A k_i \neq 0 \\ c, & \text{otherwise,} \end{cases}$$

where $c = \nu x(x \in [0,1])$.

It can be readily shown that

(i) $P_M^*(\Phi) = P_M^*(\Psi)$ if Φ and Ψ are semantically equivalent in suppositional logic; that

(ii) $P_M^*(\phi) = P_M(\phi)$, for any $\phi \in S \subseteq S^u$; and that

(iii) $P_M^*(\psi \dashv \phi)$ is unchanged in value on replacement of ϕ, or ψ, by a logically equivalent sentence (of S).

As for (i), that the equality holds is clear since semantically equivalent formulas have the same suppositional normal form. (Note that in case the P_M^* value is c, the '$=$' has the idiomatic meaning used in '$\nu x(x \in [0,1]) = \nu x(x \in [0,1])$'.) Item (ii) follows from the fact that the A in the definition

However, since a suppositional can be viewed as an instance of the general notion of a conditional, and since 'conditional event' is already in use in the literature, it seems best not to disturb a somewhat settled usage for the sake of a small gain in aptness of terminology.

of P_M^* would be logically true, while (iii) is evident in that the respective A and C formulas would be logically equivalent.

Completion of our specification of what constitutes conditional probability logic requires defining logical consequence.

We say Ψ is a *logical consequence* (in conditional probability logic) of $\Phi_1, \ldots, \Phi_m$ with respect to $\alpha_1, \ldots, \alpha_m, \beta$ (these being non-empty subsets of $[0, 1]$) if for each model M,

$$\text{when } P_M^*(\Phi) \in \alpha_1, \ldots, P_M^*(\Phi_m) \in \alpha_m, \text{ then } P_M^*(\Psi) \in \beta. \tag{1}$$

In keeping with previous notation for logical consequence in (unconditional) probability logic we write this relation as

$$P^*(\Phi_1) \in \alpha_1, \ldots, P^*(\Phi_m) \in \alpha_m \vDash P^*(\Psi) \in \beta, \tag{2}$$

except that now there is no claim that P^* is a probability function (P1–P3 of §4.2), though it is one on $\mathcal{S}$ (by (ii) above). Since a value of P^* can be c, the indefinite constant, the meaning of $c \in \alpha$ needs to be fixed: we take $c \in \alpha$ to be false except when α is $[0, 1]$.

We now have succeeded, as anticipated in §5.1, in introducing into logic—more exactly defining a logic—so as to have a means of expressing that an event happens when another event is considered as having happened. The value of $P_M^*(\psi \dashv \phi)$ is the same as $P(\psi \mid \phi)$ of §5.1. Hence the properties of $P(\psi \mid \phi)$ referred to in §5.3 may now be considered as applying to $P^*(\psi \dashv \phi)$ of conditional probability logic. But, in addition, we can state results referring to logical compounds of conditional events, something we couldn't do earlier since no meaning was then attached to $\psi \mid \phi$ outside of the specific contexts of the form $P(\psi \mid \phi)$.

NOTE.— In the interests of not proliferating notation we shall now drop the asterisk from P^* and also replace '$\dashv$' by '$\mid$' wherever it occurs within the scope of a P.

Adapting to our present circumstances the conventions of §4.4 regarding the use of '$\vDash$', we have the following immediate consequences of (i)–(iii) and Theorem 5.64:

Theorem 5.71.

(a) $\models P(\phi) = P(\phi \,|\, A_1 \vee \overline{A_1})$

(b) $\models P(\neg(\psi \,|\, \phi)) = P(\overline{\psi} \,|\, \phi)$

(c) $\models P((\psi_1 \,|\, \phi) \wedge (\psi_2 \,|\, \phi)) = P(\psi_1 \psi_2 \,|\, \phi)$

(d) $\models P((\psi_1 \,|\, \phi) \vee (\psi_2 \,|\, \phi)) = P(\psi_1 \vee \psi_2 \,|\, \phi)$

(e) $\models P((\psi \,|\, \phi_1) \wedge (\psi \,|\, \phi_2)) = P(\psi \,|\, \phi_1 \phi_2 \vee \phi_1 \overline{\psi} \vee \phi_2 \overline{\psi})$

(f) $\models P((\psi \,|\, \phi_1) \vee (\psi \,|\, \phi_2)) = P(\psi \,|\, \phi_1 \phi_2 \vee \phi_1 \psi \vee \phi_2 \psi)$

(g) $\models P(((\psi_1 \,|\, \phi_1) \wedge (\psi_2 \,|\, \phi_2)) = P(\psi_1 \psi_2 \,|\, \phi_1 \phi_2 \vee \phi_1 \overline{\psi}_1 \vee \phi_2 \overline{\psi}_2)$

(h) $\models P((\psi_1 \,|\, \phi_1) \vee (\psi_2 \,|\, \phi_2)) = P(\psi_1 \vee \psi_2 \,|\, \phi_1 \phi_2 \vee \phi_1 \psi_1 \vee \phi_2 \psi_2)$

As immediate consequences of Theorem 5.65 and Corollary 5.66:

Theorem 5.72.

(a) $\models P((\psi \,|\, \phi) \,|\, \sigma) = P(\psi \,|\, \phi\sigma)$

(b) $\models P(\psi \,|\, (\phi \,|\, \sigma)) = P(\psi \,|\, \phi\sigma)$

(c) $\models P((\psi \,|\, \phi) \,|\, (\psi \,|\, \phi)) = P(\psi \,|\, \phi\psi) = P(\phi\psi \,|\, \phi\psi)$

(d) $\models P((\psi_1 \,|\, \phi_1) \,|\, (\psi_2 \,|\, \phi_2)) = P(\psi_1 \,|\, \phi_1 \phi_2 \psi_2)$

The linear programming methods used in §5.4 to obtain optimal interval bounds apply equally well to (2). For, by Theorem 5.64, any $\Phi \in \mathcal{S}^u$ can be expressed as a semantically equivalent suppositional of the form $\psi \dashv \phi$, with ψ and ϕ being determined explicitly from the given Φ, whose probability can then be expressed as a quotient of unconditional probabilities.

The following theorem illustrates a kind of 'independence' of ϕ and $\psi \,|\, \phi$ (where $\phi, \psi \in \mathcal{S}$):

Theorem 5.73. $\models P(\phi \wedge (\psi \,|\, \phi)) = P(\phi)P(\psi \,|\, \phi)$

Proof.

$$\begin{aligned}
P(\phi \wedge (\psi \,|\, \phi)) &= P((\phi \,|\, A_1 \vee \overline{A_1}) \wedge (\psi \,|\, \phi)) &&\text{by Thm 5.64(10)} \\
&= P(\phi\psi \,|\, \phi \vee \overline{\phi}) &&\text{by Thm 5.71(g)} \\
&= P(\phi\psi) &&\text{by Thm 5.71(a)} \\
&= P(\phi)P(\psi \,|\, \phi) &&\text{by Thm 5.30(a).}
\end{aligned}$$

We conclude this section with two brief remarks.

1. Resuming the discussion of §3.3 on the problematic inference scheme

$$\frac{\begin{array}{c} P(A \,|\, B) = r \\ B \end{array}}{P(A) = r}$$

we now see that involved are two different logics—the first premise refers to conditional probability logic, the second to verity logic. A combination of the two makes no sense to us.

2. The so-called "trivialization" results of *Lewis 1976*, to the effect that a notion of probability conditional (i.e. $\psi \mid \phi$) cannot be sensibly introduced into probability theory, become, in the present context, proofs that additivity and the product rule,

$$P(\Phi\Psi \vee \Phi\overline{\Psi}) = P(\Phi\Psi) + P(\Phi\overline{\Psi})$$
$$P(\Phi \wedge \Psi) = P(\Phi)P(\Psi \mid \Phi),$$

are not valid for $\Phi, \Psi \in \mathcal{S}^u$, that is, do not hold in all probability models.

§5.8. Other writers on conditional events

There is an extensive coverage of the history of the notion of a conditional event in chapter 1 of *Goodman-Nguyen-Walker 1991*, a treatise almost entirely devoted to "measure-free conditioning". Here in this section we shall be restricting attention to items directly related to our approach to the topic.

Earlier (§§2.4, 2.5) in presenting an account of Boole's theory of probability there was an aspect of it pertaining to the notion of a conditional event which we did not mention. Involved in Boole's solution of his general probability problem is a (peculiar) logical technique which obtains the event w, whose probability is sought, as a quotient of Boolean polynomials, say $w = E/F$. The method then expands E/F (by assigning all possible 1's and 0's to its variables $x, y, z, \dots$) into the form

$$1A + 0B + \frac{0}{0}C + \frac{1}{0}D. \tag{1}$$

Here A, B, C, D are, respectively, mutually exclusive sums of constituents on the (event) variables $x, y, z \dots$, these sums being grouped in accordance with the value obtained for E/F. Boole asserts (*1854a =1952*, 280–88):

1. A represents those combinations of events $x, y, z, \dots$ which must happen if w happen.
2. B represents those combinations which cannot happen if w happen, but may otherwise happen

3. C those combinations which may or may not happen if w happen, and

4. D those combinations which cannot happen at all.

If C and D are empty (this happening of F were the Boolean 1) then $1A + 0B$ is, essentially, the disjunctive normal form of E/F, A being the usual sum of constituents and B its complement. To the contemporary logician Boole's explanation suggests an extension of two-valued logic in which sentences—specifically, constituents—are relegated to one of four categories marked by $1, 0, \frac{0}{0}, \frac{1}{0}$. The occurrence of $\frac{0}{0}$ terms in his expansions is a concomitant of his not having a symbol for class inclusion and forcing all logical relations into equational form. As we have explained in §2.5, his interpretation for

$$w = 1A + 0B + \frac{0}{0}C + \frac{1}{0}D \tag{2}$$

is that

$$\begin{cases} w = A + \nu C, & \nu \text{ arbitrary} \\ D = 0 \end{cases}$$

or, if we did use inclusion,

$$\begin{cases} A \subseteq w \subseteq A + C \\ D = 0. \end{cases}$$

To avoid having to use $\frac{0}{0}$—thereby simplifying matters—and yet adhere to Boole's *modus operandi*, we will replace (1) by

$$(A + \nu C) + \frac{1}{0}D. \tag{3}$$

Boole contended that (2) shows that, as a solution based on the logical relations of the data of the problem, w is not an event free to happen in all circumstances but is *conditioned* by $D = 0$, i.e., by $V = 1$, where $V = \overline{D} = A + B + C$. Hence (according to his theory) its probability is given by

$$\frac{P((A + \nu C)V)}{P(V)},$$

i.e., by the conditional probability of $A + \nu C$, given V. If (3) is rewritten in the form

$$(A + \nu C)V + \frac{1}{0}\overline{V},$$

it evidently mirrors our

$$\psi\phi \vee \Delta\overline{\phi},$$

i.e., the suppositional $\psi \dashv \phi$. Had Boole thought of $(A+\nu C)+\frac{1}{0}D$ as being subject to algebraic operations, and had he endowed his $\frac{1}{0}$ with suitable algebraic properties (e.g., $1 \cdot \frac{1}{0} = \frac{1}{0}$, $\frac{1}{0} \cdot \frac{1}{0} = \frac{1}{0}$) he could conceivably have had an algebra of conditional events.[13]

Apparently the first conscious attempt at such an algebra—more that a 100 years after Boole—is that of Schay who posed the question (*1968*, 334):

> ... could $A \mid B$ be defined in such a manner consistent with general usage in probability theory, that is, so that $P_B(A) [= P(AB)/P(B)]$ may be interpreted as the probability of $A \mid B$

Though not explicitly saying so, Schay's intention was to define a mathematical structure, and a notion of probability on the structure, so that an $A \mid B$ of the structure has probability equal to $P_B(A)$. Noting that, by virtue of $P_B(A) = P_B(A \cap B)$, the only part of A contributing to $P_B(A)$ is the part in B, he proposes (p. 334):

> DEFINITION 1. Consider a set Ω and a Boolean algebra $\mathcal{B}$ of its subsets, with $\Omega \in \mathcal{B}$. The basic operations on $\mathcal{B}$ will be denoted by $\cap$, $\cup$ and $\sim$. For any two elements A and B of $\mathcal{B}$ we define $A \mid B$ as the indicator function of A restricted to B. In other words $A \mid B$ is a function given by
>
> $$(A \mid B)(\omega) = \begin{cases} 1 & \text{if} \quad \omega \in A \cap B \\ 0 & \text{if} \quad \omega \in \tilde{A} \cap B \\ \text{undefined} & \text{if} \quad \omega \in \tilde{B}. \end{cases}$$

We define $\mathcal{B} \mid \mathcal{B}$ as the set of all such functions, and call $A \mid B$ a conditional event "A given B."

On rephrasing Schay's Boolean-algebra-based definition[14] into the form

---

13. Thus, as an example, by this 'algebra'

$$(\phi_1 V_1 + \frac{1}{0}\overline{V}_1)(\phi_2 V_2 + \frac{1}{0}\overline{V}_2) = \phi_1\phi_2 V_1 V_2 + \frac{1}{0}(\phi_1 V_1\overline{V}_2 + \phi_2 V_2\overline{V}_1 + \overline{V}_1\overline{V}_2),$$

which agrees with our Theorem 5.64 (12), (See the second line of the proof of (12).)

14. According to Goodman, Nguyen and Walker (*1991*, 31) the idea of having such an indicator function representing a conditional event occurred earlier to B. DeFinetti who, however, did not develop the idea.

	$B(\omega)$	$A(\omega)$	$(A\,|\,B)(\omega)$
$(\omega \in A \cap B)$	1	1	1
$(\omega \in A \cap \tilde{B})$	1	0	0
$(\omega \in \tilde{A} \cap B)$	0	1	undefined
$(\omega \in \tilde{A} \cap \tilde{B})$	0	1	undefined

it appears that Schay's definition is essentially the same as our suppositional $A \dashv B$ (§§0.4, 5.6) except that he has the word 'undefined' where we have u, u being $\nu x(x \in \{0, 1\})$. Seemingly slight, the difference nevertheless produces, as we shall see, a major difference in the consequent properties of the notion.

The elements of Schay's $\mathcal{B}\,|\,\mathcal{B}$ are partial functions on Ω, i.e., functions not necessarily defined for all elements of Ω. An expression, e.g., '$(A\,|\,B)(\omega)$', may be non-designating for some ω's. Operations on, or definitions of, partial functions need to be accompanied by specification of the domain.

Schay provides an algebraic structure for $\mathcal{B}\,|\,\mathcal{B}$ by introducing five operations. The first is a unary operation $\sim (A\,|\,B)(\omega)$ given by the definition

$$\sim (A\,|\,B)(\omega) = (\tilde{A}\,|\,B)(\omega) \quad \text{with domain } B.$$

This corresponds to our $\neg(\psi \dashv \phi) = \overline{\psi} \dashv \phi$. Then there are two pairs of binary operations $(\cap, \cup)$ and $(\wedge, \vee)$ given by (p. 335):

$$(A\,|\,B \cap C\,|\,D)(\omega) = \min\{(A\,|\,B)(\omega), (C\,|\,D)(\omega)\}, \text{ with domain } B \cap D$$

$$(A\,|\,B \cup C\,|\,D)(\omega) = \max\{(A\,|\,B)(\omega), (C\,|\,D)(\omega)\}, \text{ with domain } B \cup D$$

$$(A\,|\,B \wedge C\,|\,D)(\omega) = \min\{(A\,|\,B)(\omega), (C\,|\,D)(\omega)\}, \text{ with domain } B \cup D$$

$$(A\,|\,B \vee C\,|\,D)(\omega) = \max\{(A\,|\,B)(\omega), (C\,|\,D)(\omega)\}, \text{ with domain } B \cap D$$

As rationale for these definitions there is only the brief remark "As for the intuitive meaning of the operations, it seems that $\cup$, $\wedge$, and $\sim$ correspond to the usual meanings of the words 'or', 'and' and 'not'. The other two operations are introduced mainly for algebraic reasons."

Schay's definitions for the operations on conditional events seem reasonable, and resemble those for the suppositional as given in our §0.4. But the results he derives from them are quite discordant with what we have. In his Theorem 1 he gives equations equivalent to

$$A\,|\,B \cap C\,|\,D = A \cap C\,|\,B \cap D$$

$$A\,|\,B \cup C\,|\,D = (A \cap B) \cup (C \cap D)\,|\,B \cup D$$

$$A\,|\,B \wedge C\,|\,D = (A \cup \tilde{B}) \cap (C \cup \tilde{D})\,|\,B \cup D$$

$$A\,|\,B \vee C\,|\,D = A \cup C\,|\,B \cap D,$$

whereas Theorem 5.64 gives for our $\wedge$ and $\vee$ on suppositionals

$$(A \dashv B) \wedge (C \dashv D) = AC \dashv (B \vee \overline{C}D)(D \vee \overline{A}B)$$
$$(A \dashv B) \vee (C \dashv D) = A \vee C \dashv (B \vee CD)(D \vee AB).$$

It is easily checked that neither of these agrees with any of Schay's.

The source of the discordance is in the different meaning he attributes to max and min in his definition of the binary operations when one of the arguments is other than 0 or 1. For Schay "... if a is defined and b is undefined, then we put $\max\{a, b\} = \min\{a, b\} = a$". It is difficult to conceive of a rationale for defining max and min in this manner. For us, on the other hand, that $\max(0, u)$ should be u, and $\min(1, u)$ should be u, is a natural outcome of u being a value which is ambiguous, or not determined, as to whether it is 0 or 1.

In his book *The Logic of Conditionals (1975)* E.W. Adams argues that the "indicative conditional" (If ϕ, then ψ) between factual propositions accords best with rational deductive inference if taken to be the probability conditional $\psi \mid \phi$. (Adams' symbolization is $\phi \Rightarrow \psi$.) His views are supported by illustrations of actual simple inference schemes, tested informally by Venn-diagram-like pictures in which areas are proportional to probabilities. The appropriateness of compounding (probability) conditionals with logical connectives and with $\Rightarrow$—which is our particular interest here—occasions difficulties for Adams (*1975*, 31):

> It is a drawback to our probabilistic approach in comparison to the usual truth-functional one that it does not provide us with a theory of inferences involving compound propositions with conditional constituents, whereas truth-functional logic does provide such a theory. We shall shortly see that there are serious difficulties to confront in extending probabilistic analysis to these problematic propositions and inferences involving them, ... The application is a very poor one, and it might be much better to frankly admit that at present we don't understand such constructions rather than to delude ourselves with a very inadequate theory.

A formal language is introduced consisting of a "factual" sentential language (essentially the same as our S) extended by the addition of formulas of the form '$\phi \Rightarrow \psi$', for any ϕ, ψ which are factual sentences *provided that ϕ be not logically false*. (Thus one can't tell whether '$\phi \Rightarrow \psi$' is a well-formed expression of the language until it has been determined that ϕ is not logically false.) Although Adams' language does not allow for the conjoining of conditionals with $\wedge$ (and also $\vee$) he does introduce a notion of the *quasi-conjunction* of a set of conditionals $\{\phi_1 \Rightarrow \psi_1, \ldots, \phi_n \Rightarrow \psi_n\}$, namely, as the conditional

$$(\phi_1 \vee \cdots \vee \phi_n) \Rightarrow ((\phi_1 \rightarrow \psi_1) \wedge \cdots \wedge (\phi_n \rightarrow \psi_n)).$$

This is axactly Schay's conjunction of conditional events (the one denoted by his '∧') extended from 2 to n conditionals. Adams also defines the quasi-disjunction of a set of conditionals, which is the same as Schay's disjunction of conditional events. Unlike Schay, who gives essentially no reasons, Adams notes that quasi-conjunction has many of the properties associated with conjunction in formal deduction, though he notes that "In the general case quasi-conjunction does not entail all formulas of the set."

Adams' definition of entailment is quite different from our notion of logical consequence for conditional probability and merits a brief discussion. The definition is of the ϵ-δ kind, like that of a limit in analysis (*1975*, 57):

> DEFINITION 1. Let $\mathcal{L}$ be a factual language, let $\mathcal{A}$ be a formula of its conditional extension [i.e., be of the form $\phi \Rightarrow \psi$, ϕ not logically false], and let X be a set of such formulas. Then X *probabilistically entails* $\mathcal{A}$ (abbr. 'X *p*-entails $\mathcal{A}$') if for all $\epsilon > 0$ there exists a $\delta > 0$ such that for all probability-assignments p for $\mathcal{L}$ which are proper for X and $\mathcal{A}$ [i.e., which do not result in an antecedent having a p-value 0], if $p(\mathcal{B}) \geq 1 - \delta$ for all $\mathcal{B}$ in X, then $p(\mathcal{A}) \geq 1 - \epsilon$.

Using this definition, and general results derived from it, Adams shows, for example, that (p. 61, R4)

$$(\phi \vee \psi) \Rightarrow \eta \quad \text{is entailed by} \quad \phi \Rightarrow \eta \text{ and } \psi \Rightarrow \eta.$$

We compare this inference form with our Corollary 5.475. We have, for $P(\eta \mid \phi) = 1 - \delta$ and $P(\eta \mid \psi) = 1 - \delta$,

$$P(\eta \mid \phi \vee \psi)) \in [1 - \frac{2\delta}{1 + \delta}, 1 - \frac{\delta}{2 - \delta}].$$

Thus not only can we conclude that this probability can be made arbitrarily close to 1 by having δ sufficiently close to 0, but we also provide exact values for the end points of the interval which bounds it.

Independent of Schay and Adams, *Calabrese 1987* presents a type of algebraic structure L/L based on a Boolean logic L, whose elements are ordered pairs of the logic, written '$q \mid p$', provided that p is not impossible. These elements are referred to as conditional propositions. Definitions and axioms are provided which result in disjunction and conjunction of conditional propositions that are the same as Schay's $\cup$ and $\wedge$. (We omit any mention of negation as this operation is uniformly the same for all investigators.) In a subsequent paper, *Calabrese 1991*, the topic is revisited, this time via the DeFinetti-Schay indicator function approach. Of interest here is Calabrese's discussion of an example involving the probability of a disjunction of two conditionals. The discussion contrasts the result which the Schay-Adams-Calabrese formula obtains with that given by Goodman-Nguyen [to be presently discussed], this latter being equivalent to our Theorem 5.64 (13),

$$(\psi_1 \dashv \phi_1) \vee (\psi_2 \dashv \phi_2) = \psi_1 \vee \psi_2 \dashv (\phi_1 \vee \phi_2\psi_2)(\phi_2 \vee \phi_1\psi_1).$$

Calabrese's example (*1991*, 95–96) is:

> For instance, consider the experiment of rolling a single die once. The compound proposition 'if the roll is even then it will be a six or if the roll is odd it will be a five' reduces by the Goodman/Nguyen operations to 'if the roll is five or six then it will be five or six', which of course, is certain and so has conditional probability 1. In contrast, according to the operations of Theorem 2, the compound conditional reduces to 'the roll will be five or six' and has probability 2/6 or 1/3, which corresponds nicely with intuition.

To us the matter is not so clean-cut. As the following table shows, the different probabilities result because of differing conceptions of how 'or' between conditionals is to be understood:

roll	5 | O	6 | E	GN-or	SC-or
1	0	u	u	0
2	u	0	u	0
3	0	u	u	0
4	u	0	u	0
5	1	u	1	1
6	u	1	1	1

Calabrese seems to think that the *SC*-or "corresponds nicely with intuition". But before betting with anybody on this proposition it would be prudent to first ask *them* to say what their understanding of 'or' is when occurring between conditional statements.

We also disagree with Calabrese's concluding comment (*1991*, 97–98):

> The combining of logic with probability is fraught with the danger of contaminating absolute (certain) information with partially true information, with absolute nonsense being the result. For instance, it is known with certainty (by definition) that "if an animal is a penguin

then it is a bird" and it is also known with very high probability that "if a randomly chosen animal is a bird then it will fly". Therefore one might conclude with high probability that "if an animal is a penguin then it will fly". Such examples should give pause to those who could cavalierly fuse data with suboptimal techniques.

The error here is not in a mixing of certain with partially true information but in using an invalid inference form. By our Theorem 5.445, for $0 < p \leq 1$ and $0 < q \leq 1$,

$$P(A_3 \,|\, A_2) = p, \; P(A_2 \,|\, A_1) = q \vDash P(A_3 \,|\, A_1) \in [0, 1]$$

with the interval in the conclusion being best possible. Hence even if p were 1 (in which case, by Theorem 5.33, $P(A_3 \,|\, A_2)$ is replaceable by $P(A_2 \to A_3)$) and q had a high probability there is no warrant for concluding that $P(A_3 \,|\, A_1)$ has high probability. The best that one can do is to say that its probability is between 0 and 1.

The conditional events theory to be next described is that of I. R. Goodman and N. T. Nguyen. We shall use as our reference *Goodman-Nguyen-Walker 1991*. As with Schay and Calabrese, the approach is from an algebraic rather than a logical standpoint, though the Goodman-Nguyen analysis is much deeper.

The idea of representing a conditional event $a \,|\, b$ by a coset (or residue class modulo a prime ideal) $a + Rb' = \{\, a + rb' : r \in R \,\}$ (R a Boolean ring, b' the complement of b) can be teased out of Boole's theory of probability— though it takes some doing on account of his peculiar algebraic techniques and eccentric probability ideas. (See our *1986* chapter 5, and the summary in *Goodman-Nguyen-Walker 1991*, 25–29.) Without going through a circuitous and murky course of this kind one can nevertheless see that the idea is a natural one from the following consideration.

The intention is to have a notion $a \,|\, b$ to which one can associate a value $P(ab)/P(b)$, when a and b are elements of a Boolean ring R and P is an arbitrary probability function on R. Since any part of a outside of b (any part of a in b') contributes nothing to this value we ought to have

$$(a + r_1 b' \,|\, b) = (a + r_2 b' \,|\, b), \quad \text{any } r_1, r_2 \in R \tag{1}$$

i.e., that the notion should treat $a + r_1 b'$ and $a + r_2 b'$ as being the 'same'. Now what $a + r_1 b'$ and $a + r_2 b'$ have in common is the property

$$\textit{being equal to } a + rb', \textit{ for some } r \in R. \tag{2}$$

Sets, being 'objects', are preferred to predicates in mathematics. Hence we replace (2) by the set

$$\{\, a + rb' : r \in R \,\}. \tag{3}$$

Then defining $a \mid b$ to be (3) produces the result (1) since for $i = 1$ or 2,

$$
\begin{aligned}
(a + r_i b') \mid b &= \{\, a + r_i b' + rb' \ : \ r \in R \,\} \\
&= \{\, a + (r_i + r)b' \ : \ r \in R \,\} \\
&= \{\, a + rb' \ : \ r \in R \,\} \\
&= a \mid b.
\end{aligned}
$$

Goodman-Nguyen-Walker establish the interesting result that $f(a, b) = a + Rb'$ is essentially the only function on $R \times R$ which satisfies a few simple and evident properties (such as (1)) which they list (*1991*, 50, Theorem 2). Taking $R \mid R$ then to be the space (set) of all cosets $a \mid b = a + Rb'$ for $a, b \in R$, their next objective then is to provide an algebraic structure on $R \mid R$ whose operations extend, in some meaningful fashion, those of R. Considering the possibility of using as these operations set operations on cosets, i.e., defining

$$
(a + Rb')' = \{\, x' \ : \ x \in a + Rb' \,\}
$$
$$
(a + Rb') \, \rho \, (c + Rd') = \{\, x \rho y \ : \ x \in a + Rb', \ y \in c + Rd' \,\},
$$

ρ being either $+$, $\cdot$, or $\vee$, it is shown that these operations are closed in $R \mid R$. For example,

$$
x' = a + rb' \rightarrow x = 1 + x' = 1 + a + rb' = a' + rb'
$$

so that

$$
(a + Rb')' = a' + Rb'.
$$

Most surprising is it that closure holds for multiplication, as it isn't necessarily the case for a general ring R but does hold, as the authors prove, for Boolean rings. They comment: "This unusual fact suggests that perhaps set addition and multiplication are appropriate operations on any pair of elements of $R \mid R$. Similar remarks hold for the set operations ' and $\vee$ on $R \mid R$."

Using these definitions produces (items (1), (3) and (4) of their Theorem 2, p. 63):

$$
(a \mid b)' = a' \mid b
$$
$$
(a \mid b)(c \mid d) = (ac \mid a'b \vee c'd \vee bd)
$$
$$
(a \mid b) \vee (c \mid d) = (a \vee c \mid ab \vee cd \vee bd),
$$

which exactly correspond to our results in Theorem 5.64, taking conditional events as elements of suppositional logic. It is not surprising then that when they determine three-valued tables for connectives associated with their operations ', $\vee, \wedge$, these turn out to be the Łukasiewicz-Kleene kind. (They also show that those for the Schay-Calabrese operations turn out to be of the Sobocinski kind.) The algebra defined on $R|R$ is extensively investigated in their chapter 4.

Chapter 6

Technical Applications

§6.1. A theorem of Rényi's and extensions

This theorem, stated earlier by us in §3.5, suffices to establish all the usual formulas encountered in elementary probability, save those involving independence (or conditional probability). We rephrase it here in terms of the notions of probability logic (§4.3).

Theorem 6.11. *Let $\phi_1, \ldots, \phi_m$ be sentences of $S_n = S(A_1, \ldots, A_n)$ and K_j $(j = 1, \ldots, 2^n)$ the constituents on $A_1, \ldots, A_n$. Let $c_1, \ldots, c_m$ be real numbers. A necessary and sufficient condition for*

$$\vDash \sum_{i=1}^{m} c_i P(\phi_i) = 0 \quad (or \ \leq 0, \ or \ \geq 0)$$

that is, that the equality (or inequality) be a law (identity) of probability, is that it be true in each of the 2^n models obtained by assigning 1 to one of the $P(K_j)$ and 0 to all the others.

Proof. Considering first the equality case, assume that

$$\sum_{i=1}^{m} c_i P(\phi_i) = 0 \tag{1}$$

is true in each of the 2^n models described in the statement of the theorem. Replacing each ϕ_i by its disjunctive normal form $\bigvee^{(\phi_i)} K_j$, and distributing

267

P over the mutually exclusive disjuncts, enables us to reexpress the left-hand side of (1) as a linear form in the $k_j = P(K_j)$:

$$\sum_{i=1}^{m} c_i P(\phi_i) = \sum_{i=1}^{m} c_i P(\bigvee^{(\phi_i)} K_j)$$

$$= \sum_{j=1}^{2^n} d_j P(K_j)$$

$$= \sum_{j=1}^{2^n} d_j k_j,$$

where each d_j is some linear combination of $c_1, \ldots, c_m$ (determined by the logical structures of $\phi_1, \ldots, \phi_m$ in terms of their constituents). We then examine the linear form

$$z(k_1, \ldots, k_{2^n}) = \sum_{j=1}^{2^n} d_j k_j,$$

subject to the linear constraints

$$\sum_{j=1}^{2^n} k_j = 1,$$

$$k_j \geq 0, \qquad j = 1, \ldots, 2^n.$$

Geometrically the constraints represent a polytope in 2^n dimensions with vertices the 2^n points

$$(1, 0, \ldots, 0), (0, 1, 0, \ldots, 0), \ldots (0, \ldots, 0, 1).$$

As described in §0.5, a linear function such as $z(k_1, \ldots, k_{2^n})$, subject to such constraints, has to take on its maximum $\max z$, respectively its minimum $\min z$, at some vertex of the polytope. The current hypothesis implies that $z(k_1, \ldots, k_{2^n})$ is 0 at each vertex. Then $\min z = \max z = 0$ and $z(k_1, \ldots, k_{2^n}) = 0$ for all values. Similar arguments yield the results for the other two relations (i.e., ≤ 0, ≥ 0). Thus if $z(k_1, \ldots, k_{2^n}) \leq 0$ at all vertex points then $\max z \leq 0$, and hence $z(k_1, \ldots, k_{2^n}) \leq 0$ for all values. This concludes the sufficiency part of the theorem. The necessary part clearly holds.

GENERAL REMARKS

1. The elementary probability formula

$$P(A_1 \vee A_2) = P(A_1) + P(A_2) - P(A_1 A_2),$$

or in the homogeneous form of Theorem 6.11,

$$P(A_1 \vee A_2) - P(A_1) - P(A_2) + P(A_1 A_2) = 0,$$

is a special case ($n = 2$) of the well-known Poincaré formula

$$P(\textstyle\bigvee_{i=1}^n A_i) = \sum_i P(A_i) - \sum_{i<j} P(A_i A_j) + \ldots (-1)^{n+1} P(A_1 A_2 \ldots A_n). \quad (2)$$

As Rényi shows (*1970*, 66), (2) is readily established by his theorem. This formula provides a way of obtaining the probability $P(A_1 \vee \cdots \vee A_n)$ that at least one of $A_1, \ldots, A_n$ occur. However to obtain this value one needs to have the probabilities of the A_i and also of all the logical products of any two or more of them. When this full information is lacking the linear programming approach presented in §4.5 above enables one to obtain upper and lower bounds on $P(A_1 \vee \cdots \vee A_n)$ given any subset of these probabilities. Moreover the bounds so obtained are best possible if nothing else is known of the events. The Fréchet example in §3.5 is one for which just the $P(A_i)$ are given. In the next section we describe Kounias and Marin's treatment of the case when both the $P(A_i)$ and $P(A_i A_j)$ are given.

2. The linear programming ideas used in the proof of Theorem 6.11 do not readily extend to probability assertions which are not linear functions of the $P(\phi_i)$, i.e., which do not lead to a $z(k_1, \ldots, k_{2^n})$ which is a linear function of the k_i. However *Ferenczi 1977*, using the Scott-Krauss approach (§3.8), develops results applicable to probability assertions leading to polynomial functions of degree d, where $1 \leq d < 2^n$. (See therein Theorem 2.3, p. 107.)

3. Ideas for extending results such as in Theorem 6.11 to non-linear probability assertions are obtainable from a geometric visualization of the problem. The interest is in knowing whether a probability assertion

$$\Phi(P(\phi_1), \ldots, P(\phi_m)) = \Psi(P(K_1), \ldots, P(K_{2^n})) = Z(k_1, \ldots, k_{2^n}),$$

where Z is a general algebraic relation (and not simply $z(k_1, \ldots, k_{2^n}) \geq 0$ with z linear), holds in all probability models if it holds in the 2^n special cases of assigning 1 to one of the $P(K_j) = k_j$ and 0 to all the others. These 2^n models correspond to the 2^n points lying on the coordinate axes of a 2^n-dimensional Euclidean space, each at a unit distance from the origin. The condition $\sum_{i=1}^{2^n} k_i = 1$ represents a hyperplane passing through these unit coordinate points, while the set of all models corresponds to the set of points which span these coordinate points, i.e., to the polytope referred to above. To the algebraic relation $Z(k_1, \ldots, k_{2^n})$ corresponds the 'space' of points which satisfy this relation; for example, in the Rényi theorem $Z(k_1, \ldots, k_{2^n})$, when taken as $z(k_1, \ldots, k_{2^n}) \geq 0$, is the half-space of points lying on one side of the hyperplane $z(k_1, \ldots, k_{2^n}) = 0$, which passes through the origin (points *on* the plane included in the half-space). It is geometrically evident that if all the unit coordinate points are in this half-space they are then all on one side of the plane, and hence so is the polytope of points which spans these unit coordinate points. Equally evident is it that, in addition to the half-spaces determined by planes, a large variety of regions (relations $Z(k_1, \ldots, k_{2^n})$) lend themselves to this type of argument, i.e., being able to conclude that a probability assertion holds in all models if it holds in the 2^n simple models.

§6.2. Linear programming in probability/statistics

In the 1970's, in addition to the Adams-Levine paper discussed in §3.6, two papers appeared which also applied linear programming methods to logico-probability problems. Here also we find no reference to earlier use of this technique. Written by statisticians, the interest in these papers was not in probabilistic inference, i.e., in the logical aspects, but in obtaining useful bounds on the probability of an event (logically specified) when the data are insufficient to obtain an exact value and nothing concerning dependence or independence is assumed.

Strictly speaking, the first of these two papers, *Kwerel 1975*, is not part of probability logic as we are conceiving it. In Kwerel's paper the bounds are determined not in terms of probabilities of logical functions of events but in terms of certain *sums* of such probabilities. In particular he uses the sums $S_1^{(n)}$ and $S_2^{(n)}$ where

$$S_1^{(n)} = \sum_{i=1}^{n} P(A_i) \quad \text{and} \quad S_2^{(n)} = \sum_{i<j}^{n} P(A_i A_j) \tag{1}$$

(the latter sum having $\binom{n}{2} = n(n-1)/2$ terms). Nevertheless the method is so close to our topic that a brief description is warranted, if only to appreciate the difference.

The probabilities which Kwerel wishes to find bounds for are the 'aggregated' probabilities $P_{[r]}^{(n)}$ $(r = 0, 1, \ldots, n)$ which, for a given r, is the probability that exactly r out of the n chance events $A_1, \ldots, A_n$ occur. These probabilities are expressible as probabilities of logical functions of the n events; for example, for $n = 3$,

$$P_{[0]}^{(3)} = P(\overline{A}_1 \overline{A}_2 \overline{A}_3)$$

$$P_{[1]}^{(3)} = P(A_1 \overline{A}_2 \overline{A}_3 \vee \overline{A}_1 A_2 \overline{A}_3 \vee \overline{A}_1 \overline{A}_2 A_3)$$

$$P_{[2]}^{(3)} = P(A_1 A_2 \overline{A}_3 \vee A_1 \overline{A}_2 A_3 \vee \overline{A}_1 A_2 A_3)$$

$$P_{[3]}^{(3)} = P(A_1 A_2 A_3).$$

However this is not the case for the associated quantities $S_r^{(n)}$, defined to be the sums of probabilities of all possible logical products of r out of n chance events—for example the $S_1^{(n)}$ and $S_2^{(n)}$ of (1). In general these sums are not expressible as probabilities of a logical function of the A_i. There are, however, well-known algebraic identities expressing any $S_r^{(n)}$ as a linear combination of the $P_{[0]}^{(n)}, \ldots, P[n]^{(n)}$. (See, e.g., *Feller 1957*, 100.) In particular we have (dropping from here on the superscript '(n)' to improve readability)

$$S_1 = \sum_{r=0}^{n} r P_{[r]} \quad \text{and} \quad S_2 = \sum_{r=0}^{n} \frac{1}{2} r(r-1) P_{[r]}.$$

From the definitions it is clear that $P_{[r]} \geq 0$ and that $\sum_{r=0}^{n} P_{[r]} = 1$. This leads to the linear constraint system of equations and inequations in the $n+1$ variables $P_{[0]}, \ldots, P_{[n]}$ which, expressed in matrix form, is

$$\begin{bmatrix} 1 & 1 & 1 & \ldots & 1 & \ldots & 1 \\ 0 & 1 & 2 & \ldots & r & \ldots & n \\ 0 & 0 & 1 & \ldots & \frac{1}{2}r(r-1) & \ldots & \frac{1}{2}n(n-1) \end{bmatrix} \begin{bmatrix} P_{[0]} \\ \vdots \\ P_{[r]} \\ \vdots \\ P_{[n]} \end{bmatrix} = \begin{bmatrix} 1 \\ S_1 \\ S_2 \end{bmatrix}$$

$$P_{[r]} \geq 0, \ (r = 0, \ldots, n).$$

Asking for the maximum or minimum of a linear form in these variables, subject to these constraints, is a linear programming problem. Kwerel's paper treats only the case for which this linear form consists of the single term $P_{[r]}, r \in \{0, 1, \ldots, n\}$, and develops explicit expressions for the bounds. These include bounds for $P(A_1 \vee A_2 \vee \cdots \vee A_n)$ since this is equal to $1 - P_{[0]}$. These latter bounds are, of course, different from Fréchet's (see §3.5) since they are in terms of the sums S_1 and S_2 instead of the individual $P(A_i)$. As with Fréchet, Kwerel shows that his bounds are best possible ("most stringent") by constructing probability systems for which the bounds are attained.

The second of our two papers written by statisticians is *Kounias and Marin 1976*. The portion of this paper we shall be describing is its section 2.2 entitled *A method of deriving lower bounds of degree two*. Here 'degree two' refers to a restriction in which the given probabilities are of the conjunction (or intersection if events are represented by sets) of at most two events (i.e., of A_i and $A_i A_j$). The aim is to determine a lower bound for the probability of the union of n events, $\bigcup_{i=1}^{n} A_i$, in terms of the probabilities of A_i and $A_i A_j$, $i, j = 1, \ldots, n$ $(i < j)$. Note that here it is the individual probabilities, both of the A_i and $A_i A_j$, which are given and not their sums S_1 and S_2, as with Kwerel. The Kounias-Marin problem thus comes under what we are calling probability logic.

Their approach to the problem, carried out in the context of general probability spaces, is quite different from that originating with Boole. Indicator functions and their expected values are used: for ω an element of the probability space, the indicator function $\Psi_i(\omega)$ of A_i is defined by

$$\Psi_i(\omega) = \begin{cases} 1 & \text{if } \omega \in A_i \\ 0 & \text{if } \omega \notin A_i. \end{cases}$$

Thus the expected value of an indicator function of an event equals the probability of the event. The indicator function for the union $\bigcup_{i=1}^{n} A_i$ is $\max_{1 \le i \le n} \Psi_i(\omega)$.

Kounias and Marin remark that for the inequality

$$\max_{1 \le i \le n} \Psi_i(\omega) \ge b_0 + \sum_{i=1}^{n} b_i \Psi_i(\omega) - \sum_{1 \le i < j \le n} b_{ij} \Psi_i(\omega) \Psi_j(\omega) \qquad (2)$$

to hold for all ω in the space, a necessary and sufficient condition is that the following 2^n inequalities be satisfied:

$$0 \ge b_0, \quad 1 \ge b_0 + \sum_{i \in J_r} b_i - \sum_{\substack{i < j \\ i,j \in J_r}} b_{ij} \quad \text{all } J_r \qquad (3)$$

where J_r is a set of r integers contained in $\{1, \ldots, n\}$, there being $2^n - 1$ such nonempty sets. By taking expected values of the two sides of (2) the authors obtain

$$P(\bigcup_{i=1}^{n} A_i) \ge b_0 + \sum_{1 \le i \le n} b_i P_i - \sum_{1 \le i < j \le n} b_{ij} P_{ij} \qquad (4)$$

where $P_i = P(A_i)$ and $P_{ij} = P(A_i A_j)$. It is then stated (p. 309):

Thus among the lower bounds given by (4) the best is found by solving the linear programming problem. Maximize

$$b_0 + \sum_{1 \le i \le n} b_i P_i - \sum_{1 \le i < j \le n} b_{ij} P_{ij} \qquad (5)$$

with respect to b_0, b_i, b_{ij} $(1 \le i < j \le n)$ subject to (3).

We have a few remarks about Kounias-Marin's treatment of the probability bounds question.

(i) Since the procedure concerns only finitely many events and their Boolean compounds there is no need to invoke the strong condition of (Kolmogorov) probability spaces. By Theorem 4.71 probability logic validity suffices.

(ii) By the theorem of Rényi it is an immediate consequence that (3) is a necessary and sufficient condition for (2).

(iii) We note the advantageous use of this result (i.e., the one mentioned in (ii)), for by its use Kounias-Marin come out directly with their linear program in a form in which the parameters P_i, P_{ij} are present only in the objective function and not in the constraint system, thus circumventing a use of the duality theorem (as was needed, e.g., in our *1965* §4, and in §4.5 above.)

(iv) While it is true that the solution of Kounias-Marin's linear program would give the best *linear* bound, the question as to whether it is the best *possible* bound is not settled, for by starting with (2) they are of necessity restricting their search to linear functions.

Kounias and Marin conclude this section of the paper with their explicit results of computing, for the cases of $n = 2, 3, 4$, best linear bounds of degree two on $P(\bigcup_{i=1}^{n} A_i)$. The result for $n = 2$ is the elementary

$$P(A_1 \vee A_2) \geq P(A_1) + P(A_2) - P(A_1 A_2)$$

(which of course can be strengthened to an equality). The result for $n = 3$ is incorporated as part of Theorem 4.57 above. Their result for $n = 4$ involved finding the 22 vertices of the polytope of feasible points, and will not be reproduced here.

The value of $P(A_1 \vee \cdots \vee A_n)$ is a desideratum in many statistical applications. As an example (discussed by Kounias-Marin), suppose there are random variables $X_1(\omega), \ldots, X_n(\omega)$ and one is interested in the 'tail' probability

$$P(\max_i X_i(\omega) \geq c).$$

If A_i is the event corresponding to the set of outcomes ω for which $X_i(\omega) \geq c$, then $A_1 \vee \cdots \vee A_n$ is the event corresponding to

$$\max_i X_i(\omega) \geq c.$$

When full information is lacking approximations to, or bounds on, $P(A_1 \vee \cdots \vee A_n)$ are of interest. We have seen in §3.5 that Fréchet's upper and lower bounds are in terms of the $P(A_i)$. Since these bounds are the best possible under the circumstances they can be improved upon only with additional information. The results of Kounias-Marin incorporate, in addition to the $P(A_i)$, the joint probability values $P(A_i A_j)$.

A somewhat similar statistics application appears in *Rüger 1978*. (Unlike the papers of Kwerel and Kounias-Marin it was written with awareness of our *1965*.)

Rüger treats the following problem in statistics. Suppose there are n tests of a hypothesis H_0 with critical regions C_i with respective levels of significance α_i $(i = 1, \ldots, n)$. In terms of n, k, and the α_i, what is the best upper bound for the level of significance of the test *Reject H_0 if at least k of the n individual tests do.* Here the constraints are the one-sided inequalities $P(C_i) \leq \alpha_i$ (from the definition of level of significance) and the probability of the critical region for the test is

$$P(\bigcup_{I \in \mathcal{I}} \bigcap_{i \in I} C_i),$$

where $\mathcal{I}$ is the set of all k-membered subsets I of $\{1, \ldots, n\}$. Finding the best upper bound for the probability of this logically expressed event, subject to the given constraints, is a linear programming problem. Rüger finds the solution to be

$$\min\{1, \ \frac{1}{k}\sum_{i-1}^{n}\alpha_i, \ \frac{1}{k-1}\sum_{i=1}^{n-1}\alpha_i, \ \ldots, \frac{1}{1}\sum_{i=1}^{n-k+1}\alpha_i\}, \tag{2}$$

in which it is assumed that the α_i are arranged in non-decreasing order. For the case of $\alpha_1 = \alpha_2 = \cdots = \alpha_n = \alpha$ this reduces to

$$\min\{1, \ \frac{1}{k}n\alpha\}.$$

Morgenstern 1980, with equalities $P(C_i) = \alpha_i$ in place of $P(C_i) \leq \alpha_i$, reproduces these results of Rüger's by direct methods without linear programming. Each of the terms in (2) is obtained by a judicious application of the Markov inequality

$$P(X \geq k) \leq \frac{1}{k}E(X). \quad (X \text{ a positive random variable})$$

No doubt cued by Morgenstern's paper, Rüger in his *1981* now uses direct methods to derive best possible upper and lower bounds on

$$P(\text{at least } k \text{ of } n \text{ events } A_1, \ldots, A_n \text{ occur})$$

and also on

$$P(\text{exactly } k \text{ of } n \text{ events } A_1, \ldots, A_n \text{ occur})$$

in terms of n, k and $(\alpha_i,\ \beta_i)$, where

$$\alpha_i \leq P(A_i) \leq \beta_i. \quad (i = 1, \ldots, n)$$

Concerning his results (which will not be reproduced here) Rüger remarks that they "are relevant when treating combined tests or confidence procedures".

That, in particular cases, direct methods suited to the circumstances can advantageously replace more general uniform methods is not surprising. We recall that Fréchet's proof (§3.5) that his bounds on $P(A_1 \vee \cdots \vee A_n)$ were best possible was obtained without the use of linear programming.

§6.3. Circuit fault testing.   Fault trees

Probabilistic methods in connection with digital (switching, combinational) circuits are well-known and are encountered in many textbooks on the subject. What we shall be describing here is an unusual one which makes use of probability intervals. To describe it we shall need only a minimum amount of 'hardware' background material (i.e., circuit diagrams). Most of the discussion will be in terms of the logical representation. Our source for the material in this section is *Savir, Ditlow and Bardell 1984*.

Under consideration are circuits having n input lines and one output line. Operation of the circuit is representable by a Boolean (logical) function $\phi(A_1, \ldots, A_n)$, where A_i means 'input line i has a positive signal', and $\phi(A_1, \ldots, A_n)$ means 'the output line has a positive signal'.

Information about faults in a circuit represented by $\phi(A_1, \ldots, A_n)$ can be obtained by trying all 2^n possible sets of input signals (positive or non-positive) and seeing if the output line has the appropriate signal in accordance with the (truth-table) value of the function $\phi(A_1, \ldots, A_n)$. One can also do a probabilistic test, namely by randomly generating signals for the input lines so that each has a positive signal with given probability. In this case one needs to know, for given probabilities of the input lines having a positive signal, the probability of the output line having a positive signal. That is, one needs to know $P(\phi(A_1, \ldots, A_n))$ when given $P(A_i)$ for $i = 1, \ldots, n$. Theoretically this probability can be computed if—as is the case for the kind of testing here being used—the A_i are (stochastically) independent. For if they are mutually independent then the probability of any conjunction $B_1 B_2 \ldots B_n$, where B_i is either A_i or $\overline{A}_i$, is equal to the product of the probabilities of the conjuncts; and, supposing $\phi(A_1, \ldots, A_n)$ expanded into complete disjunctive normal form, its probability is the sum of the probabilities of its mutually exclusive disjuncts.

Both of these methods, the exhaustive and the probabilistic, involve exponential order calculations and, moreover, do not give much help in isolating the fault if there is one. To describe the method of Savir *et al.* by which they propose to ameliorate this situation (actually, we describe only part of it) it will be helpful to have a simple circuit diagram before us. In Figure 6.31 the symbol

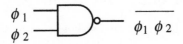

FIGURE 6.31

is schematic for a NAND gate G whose output (to express it in logical terms) is $\overline{\phi_1 \phi_2}$ if its inputs are ϕ_1 and ϕ_2. (As a truth-function NAND suffices to express all truth-functions.) If ϕ_1 and ϕ_2 are stochastically independent then the probability of the output is determined by

$$P(\overline{\phi_1 \phi_2}) = 1 - P(\phi_1 \phi_2)$$
$$= 1 - P(\phi_1)P(\phi_2). \tag{1}$$

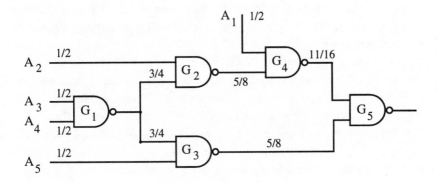

FIGURE 6.32

For the circuit diagram in Figure 6.32 suppose positive signals arrive, independently, at the input lines $A_1, \ldots, A_5$ with probability $\frac{1}{2}$. Formula (1) suffices to obtain the output probabilities (as shown) at all gates except that of G_5 where the input signals cannot be treated as independent. While Savir *et al.* take this for granted it can be formally justified by the following theorem which we take from our *1986*, p. 47:

(Theorem 0.93). *The following two conditions are equivalent:*

    (i) *For arbitrary events $A_1, \ldots, A_n$, if $A_1, \ldots, A_n$ are mutually independent, then so are $\phi(A_1, \ldots, A_n)$ and $\psi(A_1, \ldots, A_n)$.*

   (ii) *The sets of variables on which $\phi(A_1, \ldots, A_n)$ and $\psi(A_1, \ldots, A_n)$ depend essentially are disjoint.*

A Boolean function $\phi(x_1, \ldots, x_n)$ depends essentially on x_i if

$$\phi(x_1, \ldots, x_{i-1}, 0, x_{i+1}, \ldots, x_n) \neq \phi(x_1, \ldots, x_{i-1}, 1, x_{i+1}, \ldots, x_n).$$

Clearly if two logical functions have no variable in common then their sets of essential variables are disjoint. Referring to Figure 6.32, we see that the computation of the probabilities by the simple rule (1) stops at gate G_5 since its two inputs are not necessarily stochastically independent. This is apparent from the circuitry as well as from the above theorem—the logical functions corresponding to the two inputs to G_5 are (with some logico-algebraic simplification)

$$\overline{A_1} \vee A_2(\overline{A_3 A_4}) \quad \text{and} \quad \overline{A_5} \vee A_3 A_4. \tag{2}$$

Clearly A_3 and A_4 are essential variables for both logical functions.

To circumvent this dependency condition and reduce the computational complexity Savir *et al.* introduce what they call "the cutting algorithm" which alters the graph structure of the circuit so as to produce a "tree" (a connected acyclic graph). Note that in Figure 6.32 the output line of G_1 "fans out" to both G_2 and G_3 and that their influence "reconverges" at G_5. Only one cut is needed to convert this circuit graph into a tree, namely by introducing a new input A_C, as shown in Figure 6.33. Note that this amounts to replacing in the first formula of (2) the formula part '$A_3 A_4$' by 'A_C', converting it to a formula having no variable in common with '$\overline{A_5} \vee A_3 A_4$'. The new input line A_C is assigned a probability interval $[0, 1]$. This probability interval propagates (by simple evident rules) up the tree, combining with the output of G_3 to produce the output probability for G_5 as shown in Figure 6.33. By virtue of the independence of the input lines of a tree the rule for computing interval outputs is easily seen to be: if $[l_1, u_1]$ and $[l_2, u_2]$ are input intervals to a NAND gate, the output interval is $[1 - u_1 u_2, 1 - l_1 l_2]$. (Single value inputs, e.g., a, are treated as intervals $[a, a]$.) The authors show (Theorem 2, p. 81) that the interval obtained by use of their cutting algorithm necessarily includes the true value.

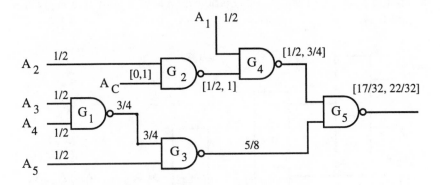

FIGURE 6.33

Savir *et al.* do not mention that there could be more than one way of cutting the circuit up into a tree, and that one could get different final output intervals. For example, if in Figure 6.32 the cut were made between G_1 and G_3 the resulting probability interval outputting from G_5 would be $[10/32, 21/32]$. Clearly a sharper interval can be obtained by using the maximum of the lower bounds and the minimum of the upper bounds. Another and more substantial improvement is obtained by using their "restricted range cutting algorithm". This takes advantage of the nature of the circuit to reduce the overly generous $[0, 1]$ assigned to the cut to either $[0, p]$ or $[p, 1]$; the value of p, and which intervals to be chosen, being determined by "the type of reconverging gate, the path inversion parity and the signal probability of the stem". (Not being directly related to our subject the details of this algorithm are omitted.)

In concluding this section we contrast Savir *et al.*'s problem with the basic type we are concerned with. We are given the probabilities $P(A_i)$, or probability intervals $a_i \leq P(A_i) \leq b_i$, and wish to find best possible upper and lower bounds on $P(\phi(A_1, \ldots, A_n))$. For the Savir *et al.* situation there is the additional information that the A_i are stochastically independent. For this case the linear programming model is not needed: by having the input signals independent, and by converting circuits into trees, the calculation of bounds is straight forward and theoretically simple. This ceases to be the case—even if the circuit were a tree—if the input signals were not independent. Fault trees are an example of this.

Such trees are the result of performing a risk analysis for the possible failure of a complex system (e.g., a chemical process, a nuclear power plant, a satellite launch). A simple schematic example of a fault tree is given in Figure 6.34.

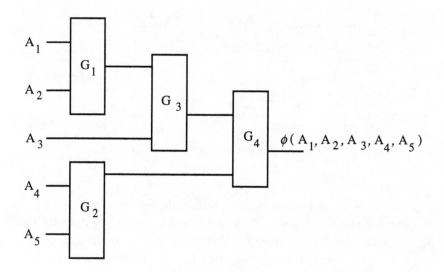

FIGURE 6.34

The rectangles could be either AND or OR gates. Chance input signals at $A_1, \ldots, A_5$ could result in a signal at $G_4 (=$ disaster). In many cases it is unrealistic to assume that the initiating events leading to system failure, $\phi(A_1, \ldots, A_5)$, are independent. Yet even without knowledge of the extent of this dependency the linear programming model can produce bounds for $P(\phi(A_1, \ldots A_5))$ if values, or intervals of values, for $P(A_i)$ are given. And if joint probabilities are known or can be estimated, then this data can also be included in the model. Optimal[1] bounds for $P(\phi(A_1, \ldots, A_5))$ could then be computed.

---

1. Perhaps 'pessimal' should be the word here since $\phi(A_1, \ldots, A_5)$ represents failure of the system.

§6.4. Network reliability

Material for this section comes from three papers which appeared in the journal NETWORKS. Although all three papers concern themselves in large measure with the matters of tractability and efficiency of the computations which result, we shall have little to say about this aspect of the papers. We begin with *Zemel 1982*, apparently the first paper to apply linear programming to network reliability problems.

Since telephone, computer, and organizational chain-of-command networks are well-known, we can start out immediately with an abstract model.

Under consideration is a system $\mathcal{G}$ (graph) which has n components $e_1, \ldots, e_n$ (edges) each of which can be, subject to chance, in an operative or inoperative (failure) state. The (chance) state of the system is specified by a vector $X = (x_1, \ldots, x_n)$ of random variables x_i, taking on the value 1 if e_i is operative and the value 0 if e_i is inoperative. It is supposed that for each state X of its components the system $\mathcal{G}$ is operative or inoperative. Then $\mathcal{G}$'s operation is represented by a *structure function* $\Phi(X) = \Phi(x_1, \ldots, x_n)$ which has the value 1, or 0, for each of the possible states of its components. Thus Φ is a Boolean function of a 2-element Boolean algebra, i.e., is a truth-function. For the application contemplated it is additionally assumed that Φ has the properties

$$\text{(i) } \Phi(0, 0, \ldots, 0) = 0$$
$$\text{(ii) } \Phi(1, 1, \ldots, 1) = 1 \tag{1}$$
$$\text{(iii) if } Y \geq X, \text{ then } \Phi(Y) \geq \Phi(X).$$

Here $Y \geq X$ means that each component of Y is (numerically) not less than the corresponding component of X. Item (i) implies that the system is inoperative if all of its components are inoperative; item (ii) that the system is operative if all of its components are, and (iii) that if the system is operative for a given state of its components, then it remains operative for a state having additional operative components. These three properties imply that Φ is a positive function of its arguments, i.e., is representable by a truth-function composed of only the truth-functions for *and* and *or* (Theorem 0.13 above).

For the systems $\mathcal{G}$ under considerations it is also assumed that there is a joint probability distribution function $F(x_1, \ldots, x_n)$, i.e., $F(X)$, defined on the states of $\mathcal{G}$. The *reliability* of the system is the probability r that Φ

takes on the value 1; that is, r is the sum of the values $F(X)$ for those X's for which $\Phi(X) = 1$.

Somewhat less abstractly described, a system $\mathcal{G}$ is a *probabilistic graph*: a collection of nodes with interconnecting edges each of which is operative, or inoperative, subject to chance. A *path* is a sequence of edges such that each member of the sequence has a common node with its successor in the sequence. A path is operative if each of its edges is. With each state X of the system one can associate a set $E(X)$ of edges, namely the set of those e_i for which $x_i = 1$, i.e., are in an operative state. If s and t are two (specified) nodes of interest a path connecting them is an *st-path*. One can define a 2-terminal *structure function* $\Phi(X)$ by

$$\Phi(X) = 1 \quad \text{iff} \quad E(X) \text{ contains an operative } st\text{-path.}$$

For a Φ so defined the reliability r is referred to as 2-*terminal reliability*.

Zemel notes that calculating r, even for simple situations such as that of the random variables x_i being independent, can be "tedious" and, in some cases, NP-hard, i.e., equivalent in computational complexity to the satisfiability problem of propositional logic. We quote (*Zemel 1982*, 441):

> In this article we analyze the problem of computing r when the function F is not completely specified. In particular, we assume that the only information available on this function is in the form of the individual bounds:
>
> $$a_i \leq p_i \equiv Prob[x_i = 1] \leq b_i, \quad i = 1, \dots, n, \qquad (2)$$
>
> for a given set of constants, $0 \leq a_i \leq b_i \leq 1$, $i = 1, \dots, n$. It is apparent that the relations (2) do not, in general, completely specify the function F. Consequently, the reliability r is not well defined. What we seek, then, is the best that can be hoped for under the circumstances, namely to calculate the best possible upper and lower bounds on r consistent with the relations (2). We denote these bounds by β and α, respectively. As shall be revealed shortly, the calculation of these bounds may in some cases be relatively easy and could be accomplished in polynomial time. In other cases, however, the task of calculating α or β may turn out to be NP-hard.

For comparison with our earlier discussion it will be convenient for us to convert Zemel's truth-functional formulation to a propositional one. State vectors $(x_1, \ldots, x_n)$ can be taken to be truth value assignments to sentences $A_1, \ldots, A_n$, where A_i stands for 'edge i is operative'. Instead of indexing the 2^n truth value assignments with numerals $1, \ldots, 2^n$, Zemel uses subsets of $N = \{1, 2, \ldots, n\}$; thus, for $S \subseteq N$, X_S is the state which has $x_i = 1$ if and only if $i \in S$. To each vector X_S we associate a constituent

$$X_S(A_1, \ldots, A_n) = A_1^{x_1} A_2^{x_2} \ldots A_n^{x_n},$$

where $A_i^{x_i}$ is A_i if $x_i = 1$ and is $\overline{A}_i$ if $x_i = 0$. Let Π be the set of those $S \subseteq N$ for which state X_S implies that $\mathcal{G}$ is operative. In place of Zemel's structure function Φ we have the sentence $\phi(A_1, \ldots, A_n)$ defined by

$$\phi(A_1, \ldots, A_n) = \bigvee_{S \in \Pi} X_S(A_1, \ldots, A_n).$$

The reliability r is then $P(\phi(A_1, \ldots, A_n))$.

Zemel goes on to state (citing *Hailperin 1965* for justification) that the best possible upper bound β for $P(\phi(A_1, \ldots, A_n))$ is obtained by solving the linear program find

$$[\beta =] \max \sum_{S \in \Pi} y_S$$

subject to ('$S : i \in S$' meaning 'all S such that $i \in S$')

$$\sum_{S : i \in S} y_S \le b_i, \quad i = 1, \ldots, n$$

$$\sum_{S : i \in S} y_S \ge a_i, \quad i = 1, \ldots, n$$

$$\sum_{S \subseteq N} y_S \le 1,$$

$$y_S \ge 0, \quad S \subseteq N.$$

The lower bound α is obtained by finding the minimum of the linear form subject to the same constraints. The dual form of this linear program is find

$$[\beta =] \min(\sum_{i=1}^{n} u_i b_i - \sum_{i=1}^{n} v_i a_i + w)$$

subject to

$$\sum_{i \in S} u_i - \sum_{i \in S} v_i + w \ge 1, \quad S \in \Pi, \ u_i, v_i, w \ge 0$$

$$\sum_{i \in S} u_i - \sum_{i \in S} v_i + w \ge 0, \quad S \notin \Pi, \ u_i, v_i, w \ge 0.$$

Zemel notes that, since the index S ranges over the 2^n subsets of N, the primal form involves exponentially many variables. Although its dual involves only $2n + 1$ variables, the number of constraint equations is now exponential. As they stand neither form, primal nor dual, can be solved in a computationally tractable manner, i.e., with the number of operations being a polynomial function of n. Taking advantage of the special nature of the Φ's under consideration (equivalently, of the $\phi(A_1, \ldots, A_n)$) Zemel shows that the dual can be reduced to a simpler form.

Briefly, let K be the subsets of N which are complements with respect to N of non-members of Π. (If $S \in \Pi$ then X_S implies $\mathcal{G}$ is operative.) Let Π^* and K^* be the subsets of Π and K consisting of their respective minimal elements (with respect to inclusion). Then (*Zemel 1982*, 442) for a Φ satisfying (1) (equivalently, for a $\phi(A_1, \ldots, A_n)$ which is a positive function of $A_1, \ldots, A_n$) the dual linear program takes the form:

$$\text{find } [\beta =] \ \min(\sum_{i=1}^{n} u_i b_i + w)$$

subject to

$$\sum_{i \in S} u_i + w \geq 1 \quad \text{all } S \in \Pi^*, \ u_i, w \geq 0,$$

for the upper bound, and

$$\text{find } [1 - \alpha =] \ \min(\sum_{i=1}^{n} u_i(1 - a_i) + w)$$

subject to

$$\sum_{i \in S} u_i + w \geq 1, \quad \text{all } S \in K^*, \ u_i, w \geq 0$$

for the lower bound α. In this form the number of variables is n and the number of constraint conditions is the number of S in Π^*, respectively, in K^*.

In a 2-terminal network system $\mathcal{G}$ (considered as a graph) the sets Π, Π^*, K, K^* have geometric significance. The elements of Π correspond to sets of edges which, when operative, contain an operative path from s to t (elements of Π are "pathsets"); the elements of K are "cutsets" and correspond to sets of edges such that if each edge of the set is inoperative then there can be no operative path between s and t—for if there were it

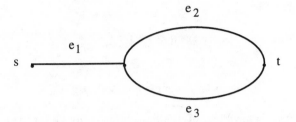

FIGURE 6.41

would have to lie in the N-complementary set of this element of K which, by definition, can have no such path. We illustrate these notions with a simple 3-node network depicted in Figure 6.41.

It is readily seen that (writing e_i instead of i)

$$\Pi = \{\{e_1, e_2\},\ \{e_1, e_3\},\ \{e_1, e_2, e_3\}\}$$
$$\Pi^* = \{\{e_1.e_2\},\ \{e_1, e_3\}\}$$
$$K = \{\{e_1\},\ \{e_1, e_2\},\ \{e_1, e_3\},\ \{e_2, e_3\},\ \{e_1, e_2, e_3\}\}$$
$$K^* = \{\{e_1\},\ \{e_2, e_3\}\}.$$

In terms of the propositional interpretation we have the associations

$$\Pi^* \quad \text{with} \quad \{A_1 A_2,\ A_1 A_3\}$$
$$K^* \quad \text{with} \quad \{A_1,\ A_2 A_3\}.$$

Moreover, the propositional form $\phi(A_1, A_2, A_3)$ is $A_1(A_2 \vee A_3)$.

We use this example to compare Zemel's simpler dual form with that of Adams-Levine's Example II discussed above in §3.6. Both treatments have (in general, as well as in this case) objective functions which are positive logical functions. Although the one is concerned with network reliability and the other with transmission of uncertainty in logical inferences, the results are closely related and (as we shall see) identical in some cases.

Consider the Adams-Levine uncertainty problem: given uncertainties ϵ_i for premises A_i ($i = 1, 2, 3$), what is the maximum uncertainty in the conclusion $A_1(A_2 \vee A_3)$? As a linear programming problem this is to find

$$[\beta =]\ \max P(\overline{A}_1 \vee \overline{A_2 A_3})$$

subject to

$$P(A_i) \le \epsilon_i. \quad (i = 1, 2, 3)$$

Equivalently, find

$$[\alpha =]\ 1 - \min P(A_1(A_2 \vee A_3))$$

subject to

$$1 - \epsilon_i \leq P(A_i) \leq 1. \quad (i = 1, 2, 3)$$

This is exactly the problem of finding the lower bound on the reliability of the network depicted in Figure 6.41 if the $a_i = 1 - \epsilon_i$ and $b_i = 1$, $(i = 1, 2, 3)$. In Zemel's form this is to find

$$[1 - \alpha =]\ \min(u_1\epsilon_1 + u_2\epsilon_2 + u_3\epsilon_3 + w)$$

subject to

$$u_1 \cdot 1 + u_2 \cdot 0 + u_3 \cdot 0 \geq 1 \quad (\{1\} \in K^*)$$
$$u_1 \cdot 0 + u_2 \cdot 1 + u_3 \cdot 1 \geq 1 \quad (\{2, 3\} \in K^*)$$

which, as may be seen from §3.6, is the same as the Adams-Levine version of the uncertainties problem. The sets $\{A_1\}$, $\{A_2 A_3\}$, corresponding to the elements of K^*, are precisely the minimal essential premise sets of Adams-Levine. It is not difficult to see that any network reliability question can be converted into a probabilistic inference question.

In general the network formulation is less general than that of the probabilistic inference formulation. For one thing, the premises of an inference can be more complicated than just a simple A_i (which corresponds to an edge in a network) and, in addition, the premises and conclusion could involve negation—to which there is nothing comparable in a network.

The second of our NETWORKS papers is *Assous 1986*. By conceiving of the problem as one of optimizing a flow through a network, and applying a central theorem of that subject (Maxflow/Mincut theorem of Ford and Fulkerson) Assous shows (p. 321) that Zemel's program has the solution

$$BUB = \min\{1,\ \min_{S \in K^*} \sum_{i \in S} b_i\} \tag{3}$$

$$BLB = \max\{0,\ 1 - \min_{S \in \Pi^*} \sum_{i \in S} (1 - a_i)\}. \tag{4}$$

He notes that although there are efficient algorithms for computing the minima in (3) and (4) the results aren't generally useful. He cites numerical examples of the bounds on the reliability being "extremely loose" when a_i and b_i are near 1. To improve the situation Assous considers additional information adjoined to $1 - b_i \leq P(\overline{A_i}) \leq 1 - a_i$, namely second-order bounds on edge probabilities,

$$1 - b_{ij} \leq P(\overline{A_i}\overline{A_j}) \leq 1 - a_{ij}.$$

Here the constraint conditions are now the same as that of Kounias-Marin (§6.2) with $\overline{A_i}$ replacing A_i. However, simple solutions of the form (3) and (4) are no longer forthcoming. Assous' heuristic methods which he introduces for obtaining reliability bounds in this case lie outside the scope of our study.

The third paper in our group is *Brecht and Colbourn 1986*.[2] Its concern is with (computer) networks whose links can fail, independently, with probability p. Because of this independence assumption exact computation of the reliability of such a network is theoretically possible but in practice intractable. Consequently efficient methods for obtaining useful bounds are sought. Many such methods, with varying rationales, have been proposed. Brecht and Colbourn introduce a procedure which, riding piggyback on any of these methods, produces a marked tightening of the bounds, as shown by a number of computed examples. Their method makes use of the linear programming approach discussed in connection with the two preceding papers. In outline the procedure has 3 steps:

STEP 1. For any two nodes x and y of the network obtain, by any available method, a pair of bounds $a(x, y)$ and $b(x, y)$ on the probability that there is an operational path from x to y.

STEP 2. From the graph of the network construct a complete graph, i.e., one in which any two nodes are directly connected by a link (edge). With each edge of this graph associate upper and lower bounds as obtained in Step 1.

Note that, unlike the edge probabilities of the original network, there is no reason to suppose that these represent independent probabilities. Nevertheless the first-order Zemel-Assous results, being independent of dependence, do apply, leading to

STEP 3. Obtain by (3) and (4) best possible upper and lower bounds on the reliability of an operational connection between any two given nodes of the graph.

---

2. The material is also included in chapter 7 of *Colbourn 1987*, a comprehensive treatise on the combinatorics of network reliability.

BIBLIOGRAPHY

Adams, Ernest W.

1965   The logic of conditionals, *Inquiry* vol. 8, 166–97.

1966   Probability and the logic of conditionals, *Aspects of inductive logic*, J. Hintikka and P. Suppes (eds.) (Amsterdam: North-Holland).

Adams, Ernest W. and Howard P. Levine

1975   On the uncertainties transmitted from premises to conclusions in deductive inferences, *Synthese* vol. 30, 429–60.

Assous, J. Yael

1986   First- and second-order bounds on terminal reliability, *NETWORKS* vol. 16, 319–29.

Bardell, P. H.

See Savir, J., G. S. Ditlow and P. H. Bardell.

Bayes, Thomas

1763   An essay towards solving a problem in the doctrine of chances, *Philosophical transactions of the Royal Society of London* vol. 53 (published 1764), 370–418.

Bernoulli, Jacobus (Jakob, Jacques, James)

1713   *Ars conjectandi* (Basel). Reprinted in Bernoulli *1975*, original page numbers included.

1899   *Wahrscheinlichkeitsrechnung.* Ostwalds Klassiker, number 108. Translation by R. Haussner of Part Four of Bernoulli *1713* (Leipzig: Akademische Verlagsgesellschaft).

1975   *Die Werke von Jakob Bernoulli* Band 3. edited by B. L. van der Waerden (Basel: Birkhäuser Verlag).

Bolzano, Bernard

    1837   *Wissenschaftslehre* 4 vols. (Sulzbach: von Seidelschen Buch-handlung).

    1972   *Theory of science.* Edited and translated by Rolf George (Berkeley and Los Angeles: University of California Press).

Boole, George

    1851a   On the theory of probabilities, and in particular Mitchell's problem of the distribution of the fixed stars, *The London, Edinburgh and Dublin philosophical magazine and journal of science,* series 4, vol. 1, 521–30. Reprinted as essay VIII in *Boole 1952.*

    1851b   Further observations on the theory of probabilities, *ibid.* vol. 2, 96–101. Reprinted as essay IX in *Boole 1952.*

    1851c   Proposed question in the theory of probabilities, *The Cambridge and Dublin mathematical journal* vol. 6, 186. Reprinted as essay X in *Boole 1952.*

    1854   *An investigation of the laws of thought, on which are founded the mathematical theories of logic and probabilities* (London: Walton and Maberly).

    1854a   On conditions by which solutions of questions in the theory of probabilities are limited, *The London, Edinburgh, and Dublin philosophical magazine and journal of science,* series 4, vol. 8, 91-8. Reprinted as essay XIII in *Boole 1952.*

    1857   On the application of the theory of probabilities to the question of the combination of testimonies or judgements, *Transactions of the Royal Society of Edinburgh* vol. 21, 597–652. Reprinted as essay XIV in Boole *1952.*

    1868.   Of propositions numerically definite, *Transactions of the Cambridge Philosophical Society* vol. 11, 396–411. Reprinted as essay IV in *Boole 1952.*

    1952   *Studies in logic and probability,* R. Rhees, ed. (London: Watts).

Brecht, Timothy B. and Charles J. Colbourn

    1986   Improving reliability bounds in computer networks, *NETWORKS* vol. 16, 369–80.

Calabrese, Philip G.

    1991   Deduction and inference using conditional logic and probability, *Conditional logic in expert systems,* I. R. Goodman, M. M.

Gupta, H. T. Nguyen, and G. S. Rogers, editors (Amsterdam: North-Holland).

Carnap, Rudolph
  1950   *Logical foundations of probability* (Chicago: University of Chicago Press).
  1962   Second edition with new preface.

Carnap, Rudolph and Richard C. Jeffreys, editors
  1971   *Studies in inductive logic and probability* Volume 1 (Berkeley, Los Angeles, London: University of California Press).

Charnes, A. and W. W. Cooper
  1962   Programming with linear fractional functions, *Naval research logistics quarterly* vol. 9, 181–6.

Čirkov, M. K.
  1971   On (the) method of probabilistic logic (in Russian) *Computer technology and questions of cybernetics* No. 8 (Publications of the University of Leningrad), 52–66.

Colbourn, Charles J.
  1987   *The combinatorics of network reliability* (New York, Oxford: Oxford University Press). *See also* Brecht, Timothy B. and Charles J. Colbourn.

Cooper, W. W. Cooper
  See Charnes, A. and W. W. Cooper

Couturat, Louis
  1901   *La logique de Leibniz d'après des documents inédits* (Paris). Reprinted 1961 (Hildesheim: Georg Olms Verlagsbuchhandlung).

Dale, Andrew I.
  1991   *A history of inverse probability from Thomas Bayes to Karl Pearson* (New York: Springer-Verlag).

Daston, Lorraine
   1988   *Classical probability in the enlightenment* (Princeton: Princeton University Press).

Dantzig, George B.
   1963   *Linear programming and extensions* (Princeton: Princeton University Press).

De Moivre, Abraham
   1756   *The doctrine of chances: or, a method of calculating the probabilities of events in play,* third edition (London: Printed for A. Millar). First edition 1718, second 1738.

De Morgan, Augustus
   1837   Theory of probabilities, *Encyclopedia metropolitana* vol. 2, 393–490. [Offprints dated '1837']
   1847   *Formal logic* (London: Taylor and Walton) Reprinted 1926, A. E. Taylor, editor (London: Open Court).
   1847a   Theory of probabilities, *Encyclopedia of Pure Mathematics* (London). Reprint of *De Morgan 1837.*
   1849   On the structure of the syllogism, and on the application of the theory of probabilities to questions of argument and authority, *Transactions of the Cambridge Philosophical Society* vol. 8, 379–408. Read 9 November 1846. The portion of this essay on probability was omitted from *De Morgan 1966.*
   1864   On the syllogism, No. IV, and on the logic of relations, *Transactions of the Cambridge Philosophical Society* vol. 10, 331–58. Read 23 April 1860.
   1966   *On the syllogism and other logical writings,* ed. Peter Heath (New Haven: Yale University Press).

Ditlow, G. S.
   See Savir, J., G. S. Ditlow and P. H. Bardell.

Dubois, Didier and Henri Prade
   1990   The logical view of conditioning and its application to possibility and evidence theories, *International journal of approximate reasoning* vol. 4, 23–46.

Feller, William
1957   *An introduction to probability theory and its applications* Volume I. Second edition (New York: Wiley).

Ferenczi, M.
1977   On valid assertions—in probability logic, *Studia scientiarum mathematicarum Hungarica* vol. 12, 101–16.

Forbes, James D.
1850   On the alleged evidence for a physical connection between stars forming binary or multiple groups, deduced from the doctrine of chances, *The London, Edinburgh, and Dublin philosophical magazine and journal of science* series 3, vol. 37, 401–27.

Fréchet, Maurice
1935   Généralizations du théorème des probabilités totales, *Fundamenta mathematicae* vol. 25, 379–87.

Freudenthal, Hans
1980   Huygens' foundations of probability, *Historia mathematica* vol. 7, 113–7.

Gaifman, Haim
1964   Concerning measures in first order calculi, *Israel journal of mathematics* vol. 2, 1–18.

Genesereth, M. R. and Nils J. Nilsson
1986   *Logical foundations of artificial intelligence* (Palo Alto: Morgan Kaufman).

Georgakopoulos, G., D. Kavvadias and C.H. Papadimitriou
1988   Probabilistic satisfiability, *Journal of complexity*, vol. 4, 1–11.

Goodman, I. R., H. T. Nguyen and E. A. Walker
1991   *Conditional inference and logic for intelligent systems* (Amsterdam: North-Holland).

Hacking, Ian
1974   Combined evidence, *Logical theory and semantic analysis*, ed. S. Stenlund, 113-24 (Dordrecht: D. Reidel).

1975   *The emergence of probability* (Cambridge: Cambridge University Press).

Hailperin, Theodore
1957   A theory of restricted quantification, I and II, *The journal of symbolic logic* vol. 22, 19–35, 113–29. Corrections, vol. 25, (1960), 54–6.

1965   Best possible inequalities for the probability of a logical function of events, *American mathematical monthly* vol. 72, 343–59.

1976   *Boole's logic and probability* (Amsterdam: North-Holland).

1984   Probability logic, *Notre Dame journal of formal logic* vol. 25, 198–212.

1986   *Boole's logic and probability.* Second edition, revised and enlarged. (Amsterdam: North-Holland).

1988a   Infinite truth-functional logic, *Notre Dame journal of formal logic* vol. 29, 28–33.

1988b   The development of probability logic from Leibniz to MacColl, *History and philosophy of logic* vol. 9, 131–91.

1991   Probability logic in the twentieth century, *ibid.* vol. 12, 71-110.

1992   Herbrand semantics, the potential infinite, and ontology-free logic, *ibid.* vol. 13, 69-90.

Hansen, Pierre, Brigitte Jaumard and Marcus Poggi de Aragão
1992   Boole's conditions of possible experience and reasoning under uncertainty. Les Cahiers du GERAD/Groupe d'études et de recherche en analyse des décisions/ G-92-33 (Montréal, Québec).

Hansen, P.
See Jaumard, B., P. Hansen and M. Poggi de Aragão

Hintikka, J., D. Gruender and E. Agazzi, eds.
1981   *Probabilistic thinking, thermodynamics and interaction of the history and philosophy of science.* Proceedings of the 1978 Pisa Conference on the History and Philosophy of Science, vol. II (Dordrecht: D. Reidel).

Hosiasson, J.
See Hosiasson(-Lindenbaum), Janina.

Hosiasson(-Lindenbaum), Janina
1940   On confirmation, *The journal of symbolic logic* vol. 5, 133–48.

Jackson, Frank, ed.
1991   *Conditionals* (Oxford: Oxford University Press)

Jaumard, Brigitte
See Hansen, Pierre, Brigitte Jaumard and Marcus Poggi de Aragão.

Jaumard, B., P. Hansen and M. Poggi de Aragão
1991.   Column generation methods for probabilistic logics, *ORSA journal on computing* vol. 3, 135–48.

Jeffreys, Richard C.
See Carnap, Rudolph and Richard C. Jeffreys.

Jevons, William Stanley
1892   *The principles of science* (London and New York: Macmillan). First printed in 1874. Second edition 1877.

Keynes, John Maynard
1921   *A treatise on probability* (London: Macmillan).

Kleene, Stephen C.
1952   *Introduction to metamathematics* (New York: Van Nostrand).
1967   *Mathematical logic* (New York: Wiley).

Kneale, William and Martha Kneale
1962   *The development of logic* (Oxford: Clarendon Press).

Kneale, Martha
See Kneale, William and Martha Kneale.

Kolmogoroff, A.
1933   *Grundbegriffe der Wahrscheinlichkeitsrechnung* (Berlin: Springer-Verlag). Reprinted 1946 (New York: Chelsea).

Kounias, Stratis and Jacqueline Marin
1976   Best linear Bonferroni bounds, *SIAM journal of applied mathematics* vol. 30, 307–23.

Krauss, P.
See Scott, Dana and Peter Krauss.

Kwerel, Seymour M.
1975   Most stringent bounds on aggregated probabilities of partially
        specified dependent probability systems, *Journal of the Amer-
        ican Statistical Association* vol. 70, 472–79.

Laplace, Pierre Simon
1820   *Théorie analytique des probabilités.* Third edition. (Paris: Ve.
        Courcier).

Lakatos, Imre, ed.
1968   *The problem of inductive logic.* Proceedings of the International
        Colloquium in the Philosophy of Science, London 1965, vol. 2.
        (Amsterdam: North-Holland).

Lambert, Johann Heinrich
1764   *Neues organon* (Leipzig: Wendler). Reprinted as first two vol-
        umes of *Philosophische Schriften* 1965 (Hildesheim:
        Georg Olms).

Leibniz, Gottfried Wilhelm von
1962   *Nouveaux essais sur l'entendement humain* A Robinet and H.
        Schepers, editors. (Berlin: Akademie-Verlag).
1971   *Sämtliche Schriften und Briefe* Sechste Reihe. Erster Band.
        (Berlin: Akademie-Verlag).
1981   *New essays on human understanding* Translated and edited by
        Peter Remnant and Jonathan Bennett. (Cambridge: Cam-
        bridge University Press).

Levine, Howard P.
See Adams, Ernest W. and Howard P. Levine.

Lewis, Clarence Irving
1918   *A survey of symbolic logic* (Berkeley: University of California
        Press).

Lewis, David

   1976   Probabilities of conditionals and conditional probabilities, *The philosophical review* vol. 85, 297–315.

MacColl, Hugh

   1877   The calculus of equivalent statements and integration limits, *Proceedings of the London Mathematical Society* vol. 9 (1877-8), 9–20.

   1878   The calculus of equivalent statements (second paper), *ibid.* 177–86.

   1880   The calculus of equivalent statements (fourth paper), *ibid.* vol. 11, 113–21.

   1897   The calculus of equivalent statements (sixth paper), *ibid.* vol. 28 (1896-7), 555–79.

Marin, Jacqueline

   See Kounias, Stratis and Jacqueline Marin.

McCluskey, Edward J.

   1986   *Logic design principles* (Englewood Cliffs: Prentice-Hall).

Maistrov, Leonid E.

   1974   *Probability theory. A historical sketch.* Translated and edited by Samuel Kotz. (New York and London: Academic Press).

Michell, John

   1767   An inquiry into the probable parallax and magnitude of the fixed stars, from the quantity of light which they afford us, and the particular circumstances of their situation, *Philosophical transactions of the Royal Society* vol. 57, 234–64.

Mill, John Stuart

   1843   *A system of logic, ratiocinative and deductive, being a connected view of the principles of evidence and the methods of scientific investigation* (London: John W. Parker).

   1879   Tenth edition of *1843* (London: Longmans, Green, and Co.).

Morgenstern, Dietrich
1980   Berechnung des maximalen Signifikanzniveau des Testes
        "Lehne H_0 ab, wenn k unter n gegebenen Tests zur Ablehnung
        führen", *Metrika* vol. 27, 285–6.

Nguyen, H. T.
See Goodman, I. R., H. T. Nguyen and E. A. Walker.

Nicod, Jean
1930   *The logical problem of induction* (New York: Harcourt,
        Brace).

Nilsson, Nils J.
1986   Probabilistic logic, *Artificial intelligence* vol. 28, 71–87.
See also Genesereth, M. R. and Nils J. Nilsson.

Ortel, Marvin and John Rossi
1981   An excercise involving conditional probability, *Mathematics
        magazine* vol. 54, 125–8.

Peirce, Charles S.
1867a   On an improvement in Boole's calculus of logic, *Proceedings of
        the American Academy of Arts and Sciences* vol. 7, 250–61.
        Reprinted in *Peirce 1984*, 12–23.
1867b   A review of *Venn 1881*, *North American Review* for July. Vol-
        ume CV, 317–21. Reprinted in *Peirce 1984*, 98-102.
1984   *Writings of Charles S. Peirce. A chronological edition* Volume
        2, (1867–1871). Editor-in-chief E. C. Moore. (Bloomington:
        Indiana University Press).

Poggi de Aragão, Marcus
See Hansen, Pierre, Brigitte Jaumard and Marcus Poggi de Aragão. *See
also* Jaumard, B., P. Hansen and M. Poggi de Aragão.

Prade, Henri
See Dubois, Didier and Henri Prade.

Prullage, Margaret M.
1976   A theory of restricted variables without existence assumptions,
        Notre Dame journal of formal logic vol. 17, 589–612.

Quine, Willard V.

1972   *Methods of logic*, third edition (New York: Holt, Rinehart and Winston).

Rényi, Alfred

1970   *Foundations of probability* (San Francisco: Holden-Day).

Rüger, Bernhard

1978   Das maximale Signifikanzniveau des tests: "Lehne H_0 ab, wenn k unter n gegebenen Tests zur Ablehnung führen", *Metrika* vol. 25, 171–8.

1987   Scharfe untere und obere Schranken für die Wahrscheinlichkeit der Realisation von k unter n Ereignissen, *Metrika* vol. 28, 71-7.

Savir, J., G. S. Ditlow and P. H. Bardell

1984   Random pattern testability, *IEEE Transactions on computers* vol. C-33, 79–90.

Schay, Geza

1968   An algebra of conditional events, *Journal of mathematical analysis and applications* vol. 24, 334–44.

Schechter, Murray

1989   An extension of the Charnes-Cooper method in linear fractional programming, *Journal of information and optimization sciences* vol. 10, 97–104.

Schneider, Ivo

1981a   Why do we find the origin of a calculus of probabilities in the seventeenth century? In *Hintikka, Gruender, Agazzi 1981*, 3-24.

1981b   Leibniz on the probable, *Mathematical perspectives*, J. W. Dauben, ed. (New York: Academic Press), pp. 201–19.

1988   *Die Entwicklung der Wahrscheinlichkeitstheorie von den Anfängen bis 1933* (Darmstadt: Wissenschaftliche Buchgesellschaft).

Scott, Dana S. and Peter Krauss

1966   Assigning probabilities to logical formulas, *Aspects of inductive logic*, J. Hintikka and P. Suppes, eds. (Amsterdam: North-Holland), pp. 219–64.

Shafer, Glenn

1976   *A mathematical theory of evidence* (Princeton University Press: Princeton and London).

1978   Non-additive probabilities in the work of Bernoulli and Lambert, *Archive for history of exact sciences* vol. 19, 309–70.

1982   Bayes' two arguments for the rule of conditioning, *The annals of statistics* vol. 10, 1075–89.

Smith, G. C.

1982   *The Boole-De Morgan correspondence 1842–1864* (Oxford: Clarenden Press).

Stigler, Stephen M.

1978   Laplace's early work: chronology and citations, *Isis* vol. 61, 234–54.

1986a   Laplace's 1774 memoir on inverse probability, *Statistical science* vol. 1, 359–78.

1986b   *The history of statistics* (Cambridge, Mass: Belknap Press of Harvard University Press).

Stoer, Josef and Christoph Witzgall

1970   *Convexity and optimization in finite dimensions* (New York Heidelberg Berlin: Springer-Verlag).

Suppes, Patrick

1966   Probabilistic inference and the concept of total evidence, *Aspects of inductive logic*, J. Hintikka and P. Suppes, eds. (Amsterdam: North-Holland), pp. 49–65.

Tarski, Alfred

1948   *A decision method for elementary algebra and geometry.* Prepared for publication by J. C. C. McKinsey. (Santa Monica: The Rand Corporation). Second edition (Berkley and Los Angeles: University of California Press).

Terrot, Charles Hughes
    1857   On the possibility of combining two or more probabilities of the same event, so as to form one definite probability, *Transactions of the Royal Society of Edinburgh* vol. 21, 369–76. Reprinted in *Boole 1952*, 487–96.

Todhunter, Isaac
    1865   *A history of the mathematical theory of probability from the time of Pascal to that of Laplace* (London and Cambridge: Macmillan and Co.). Reprinted 1949, 1961 (New York: Chelsea).

van Heijenoort, Jean
    1967   *From Frege to Gödel. A source book in mathematical logic, 1879–1931* (Cambridge, Mass: Harvard University Press).

Venn, John
    1881   *Symbolic logic* (London: Macmillan).
    1894   Second edition. Revised and rewritten.
    1971   Reprint of second edition. (New York: Chelsea).

Walker, E. A.
    See Goodman, I. R., H. T. Nguyen and E. A. Walker.

Whately, Richard
    1844   *Elements of logic*, 8th edition (London: B. Fellowes).

Wilbraham, Henry
    1854   On the theory of chances developed in Professor Boole's "Laws of thought", *The London, Edinburgh, and Dublin philosophical magazine and journal of science* series 4, vol. 7, 465–76.

Witzgall, Christoph
    See Stoer, Josef and Christoph Witzgall.

Zemel, Eitan
    1982   Polynomial algorithms for estimating network reliability, *NETWORKS* vol. 12, 439–52.

INDEX

301